THE RUSH THAT NEVER ENDED

BY THE SAME AUTHOR

The Peaks of Lyell, 1954, 1959, 1967

The University of Melbourne: A Centenary Portrait, 1956

A Centenary History of the University of Melbourne, 1957

Gold and Paper, 1958

Mines in the Spinifex, 1960

A History of Camberwell, 1964

The Tyranny of Distance, 1966

Across a Red World, 1968

The Rise of Broken Hill, 1968

THE RUSH THAT NEVER ENDED

A History of Australian Mining

GEOFFREY BLAINEY

MELBOURNE UNIVERSITY PRESS

First published 1963
Reprinted, with corrections, 1964
Second edition 1969
Reprinted 1974
Third edition 1978
Printed in Hong Kong by
Dai Nippon Printing Co. (Hong Kong) Ltd for
Melbourne University Press, Carlton, Victoria 3053
U.S.A. and Canada: International Scholarly Book Services, Inc.,
Box 555, Forest Grove, Oregon 97116
Great Britain, Europe, the Middle East, Africa and the Caribbean:
International Book Distributors Ltd (Prentice-Hall International),
66 Wood Lane End, Hemel Hempstead, Hertfordshire HP2 4RG, England

National Library of Australia Cataloguing in Publication data

Blainey, Geoffrey Norman, 1930–
The rush that never ended.
Index.
Bibliography.
ISBN 0 522 84145 7.
1. Mines and mineral resources—Australia—History. 2. Mineral industry—Australia—History. I. Title.
622'.0994

PREFACE TO THE THIRD EDITION

The book was first published in 1963 and reprinted with minor corrections a year later; but within a few years the boom in Australian mining made the final chapters seem antiquated. Accordingly, for the second edition in 1969, a long chapter was written on the new mining fields. At the same time some dates and statistics in earlier chapters were corrected, the story of the famous Egerton lawsuit of the 1870s was amended, and new snippets of information were grafted onto parts of the narrative. Whereas the original book was almost entirely concerned with metal mining, the discoveries of the 1960s—including oil and natural gas and phosphate—widened the scope of the second edition.

In this third edition a postscript discusses some of the main changes and events of the years 1969-77: the wild Poseidon boom, the finds of uranium and natural gas, the rise of the conservation crusade, and the increasing importance of Australia in the mining world.

November 1977

PREFACE

For more than a century Australia has been one of the world's important mining countries, but no history of the mines has been written. This book is about the precious and base metals that were mined vigorously in Australia. It also touches on iron because Australia's steel industry was nursed and nourished by a silver-lead mine. It omits coal, for that industry is and was radically different from metal mining. Coal was less a speculative industry, was not an outback industry, had no metallurgical problems; its industrial tensions differed, its markets differed, and it was not such a dynamo of Australia's growth. This then is a history of Australian metal mining.

It has not been easy to spin into a narrative events on so many fields mining so many metals scattered in so many places of a wide continent. Mostly I tell the story in terms of the main fields, for each field had its own distinct obstacles of terrain, climate, distance and isolation, its own peculiar problems in the wealth of its lodes or the perversity of its ores, its own heroes and villains, and its own influence on the wider region or colony of which it was part.

One hundred books could be written on this theme, and the author's own interests and bias will inevitably intrude. I have perhaps emphasized too much the discoverers, but in courage and strength they often were giants and their feats have been neglected. If not neglected, they have been cloaked with such mystery that the pattern and logic of mineral discovery have themselves been overlooked. Above all the prospectors were workingmen, and while the life of the wages miner is not well recorded, the life of the prospector is.

I may not have written enough on trade unionism, and if so it is partly because I think the class struggle and the industrial friction on the metal fields have been exaggerated. I may have written too much on the gambling and investing side of mining, partly through curiosity about it and partly through a belief that it illuminates uncannily everything from rich strikes to bitter strikes.

In quoting statistics I have tried to be frugal by academic standards. Official figures for the population on early mining fields or the treasure they won are deceptively detailed and specific. It is hard to compute accurately for any year of the last century the wealth Australia earned from metals or the export of metals. The statistics I do cite are misleading unless we remember that the Australian £1 of 1863 was in purchasing power as irrelevant as the rouble to our £1 of 1963. Sharp inflation harms the historian particularly, because it breaks one of his most useful measuring rods, money.

Finally, the expense of travelling and of gathering material for this book was generously met by Mount Isa Mines Ltd, the Chamber of Mines of Western Australia, Broken Hill Mine Managers' Association, Broken Hill Proprietary Co. Ltd, and the University of Melbourne. A list of personal acknowledgments, a bibliography and a glossary of common mining words appear at the back of the book.

GEOFFREY BLAINEY

1963

CONTENTS

ILLUSTRATIONS

MAPS

PLATES

Plates 2, 3 and 4 are reproduced from the originals in the Holtermann Collection in the Mitchell Library, Sydney, with the Trustees' permission.

INTRODUCTION

Across Australia are the ruins and scars of a lost age. Amidst anthills a Cornish boiler lies in the sun, in low red ranges an iron chimney spreads its long evening shadow. Long tropical grass conceals a barrow wheel and the rubble of a miner's fireplace. On old goldfields small holes and mounds of clay cover river flats like a vast graveyard.

Across the wide continent are living mines with nostalgic names, Great Boulder, Hercules, Rum Jungle, and Nobles Nob. There are metal towns that glow on a dark plain at night, railways that stretch straight toward a distant coast, cages that rush men down deep shafts, gardens watered by pipelines that coil across desert, noisy mills and red furnaces and chimneys like spires. There are acres of metal refineries and smelters by the seaboard, mines in the forest below snowed mountains and mines by the warm ocean, but most mines are out in the arid wastelands. They have long yielded much of the world's precious and base metals, and they have moulded much of Australia's history.

Australia's map is criss-crossed with the spidery tracks of the metal seekers. One track ran around the rim of the continent from the Palmer and Charters Towers in north Queensland to Kalgoorlie and beyond. It ran west from Pacific Ocean to Indian Ocean and south to the Southern Ocean, and it was followed by hundreds of gold-seekers on foot or horse in one of the greatest treks of any land in the nineteenth century. A track that should be famous, it has no name.

The isolation of new mining fields carved lines of transport. Cobb and Co.'s mail coaches started on the goldfields and the early grid of inland railways was shaped much by mining fields, while from an Australian leadfield the world's first flying doctor made his first flight. Gold finders created or spurred nearly every tropical port from Rockhampton to Port Hedland, and south of the tropic every big port was enriched by the flow of metals.

All Australia's large inland cities of the nineteenth century were mining cities, and gold made Melbourne for half a century the largest coastal city in the land. Though Melbourne today is the heart of the poorest metal region its skyscape is still etched by metals. Its tallest buildings belong to companies that won their first Australian fortunes from mines. The spire of its highest cathedral honours a copper king, its international centre of medical research honours a gold magnate and its most celebrated hotel, Young and Jackson's, was built with New Zealand gold.

This was the last continent found in Europe's long search for treasure and perhaps in no other continent has European colonization been so affected by the winning of metals. In two periods metals were Australia's most valuable export. New mining regions virtually rescued every Australian colony at least once from depression. They influenced racial policies, unionism, religious life, equalitarian laws, and politics. One small fact illustrates this pervasion; half of Australia's prime ministers were either the sons of men who were attracted to Australia through gold or were themselves once residents or representatives of the mining fields.

Out of the ground came wealth that created new industries. Humble silver and copper companies so prospered that they founded Australia's heavy chemical industry and one of the world's large iron and steel industries; they made paper and fertilizers and paints and built aircraft and ships. Wealth from Australian rocks created financial dynasties in London and Melbourne, and a fortune from Queensland gold was indispensable in finding Britain's rich oilfields in the Middle East.

Australian prospectors found or pioneered new mining fields from the Rand and Rhodesia to New Zealand and the Klondike. Australian mining investors opened Malayan tinlands and New Guinea and Fiji goldfields, and there is hardly a mining field that has not used Australian innovations in metallurgy.

Above all, the mining fields were the stage and backdrop for hundreds of thousands of lives, and this is the story of how the discoverers, diggers, miners, promoters, engineers, and gamblers faced their world of mercurial wealth.

PART I

The Gold-Seekers

1

PICKPOCKETS' GOLD

MAPMAKERS OF EUROPE and navigators of the Indies once thought Australian seas washed the isles of gold. Even after navigators had seen the north-west coast of Australia it was named on one map the coast of gold. Unknown coasts were treasure-lands; imagination shaped and gilded them. Then slowly Dutch and British voyagers tarnished the gilt, and Australia turned from a land of reward to a land of punishment when Great Britain dumped convicts and guards at Sydney in 1788.

The imagination of the ancients had more truth than the knowledge of the moderns, but for two generations the settlers did not know that their prison had bars of gold. Convicted London thieves idled away the afternoons with their masters' sheep within sight of reefs of gold, and exiled Dublin pickpockets heard water rushing down gullies that held more gold than all Dublin's pockets.

Australia, nevertheless, was quicker than North America to reveal its mineral wealth. It was quicker because many of its richest mineral lands were closer to the coast, grew good pastures, and were occupied early by sheep and shepherds. Within thirty years of the first settlement Sydney sheep owners had crossed the Blue Mountains to the inland plains and their sheep nibbled grass near gold nuggets and reefs of gold in milky quartz. Within fifty years shepherds behind the small town of Melbourne were watching sheep by golden creeks, and on the Adelaide plains the sheepfolds stood near outcrops of copper that were stained with the brilliant blue and green of the peacock's tail. The more men walk over ground that is rich in minerals the higher the chance that they will find and recognize the minerals. The nearer they are to settlements and towns the higher the chance that someone will explore and mine those deposits that come to the surface of the earth. All this is obvious and therefore rarely perceived.

A public official, James McBrien, trailing his surveyor's chain in February 1823 in the granite country east of Bathurst, thought he saw 'numerous particles of gold' in sandy hills by the Fish River and recorded his find in a field book. The gold was sparse, barely worth a search. It was sluiced from the sand a generation later, when

McBrien's field book and its inky smudges and erasures lay forgotten in a government office: the first authentic record of gold in the continent.

Shepherds were the main inhabitants of the gold lands and they had the leisure to examine the earth on which they walked. Perhaps a hundred pastoral labourers found gold before 1850, and the shepherd McGregor near Wellington found so much gold that he was known as 'the gold-finder'. For several years he chipped away with hammer at a quartz reef on his master's sheep run, and perhaps as early as 1845 he carried his treasure to Sydney, where street strollers saw it in the window of Cohen the jeweller. His mine—for mine it was on primitive scale—yielded him perhaps £200 sterling of gold in the opinion of Simpson Davison, an experienced miner who later examined the reef. Ironically, the gold-finder was in prison for debt when the gold rushes erupted in 1851.

The lonely shepherds in cabbage-tree hats found gold in solid quartz rock rather than in the alluvial gravels and shingles; the quartz reefs jutted above the surface of the ground and were easily seen, whereas the gold in the alluvium was usually buried in earth and clay. They chipped the rock with mallet or chisel and won little gold for their effort. The mining of hard rock requires machinery and equipment, and the shepherds had none. Moreover, they had no desire to interest men who knew more about mining than themselves. Many shepherds were eccentric, silent men, and even those who were garrulous with drink were probably silent about gold, for they usually found it on land owned by the Crown and occupied by their employer.

Thousands vaguely knew by the mid-1840s that gold was being found in Australia but few knew exactly where. And few seriously bothered to enquire, for in English law all deposits of gold and silver belonged to the Crown. Though the precious metals found in mines of copper or lead or iron were exempted from this law after England's bloodless revolution of 1689, gold and silver mines remained 'royal mines'. The Crown demanded its gold as late as 1796 when soldiers expelled hundreds of Irish peasants who washed in wooden bowls the golden sands of a stream in County Wicklow. The archaic law retarded the mining of gold in Australia. Why should we mine gold, complained two Australian mining entrepreneurs in 1848, when any gold we find will 'necessarily lapse to the Crown'?

Perhaps geologists could have convinced governments that they should mine the gold or allow private citizens to mine it. But geology was a new and erratic science. Not until the end of the previous century did a British engineer, William Smith, observe while surveying canals that principle that virtually made geology a science, the principle that each stratum of rocks could be identified by the fossils it held. The new geology bowed to botany and zoology in the hierarchy of the natural sciences, and colonial governments employed botanists but not geologists or mineralogists, and their explorers exulted in trees and grasses and plains but not in the minerals they concealed. Caravans of explorers passed over rich mineral regions in south-east Australia and saw nothing to suggest that their dray tracks would soon be followed by the galloping gold escorts.

A few geologists visited Australia. A self-styled count of Poland named Paul Edmund de Strzelecki called at Sydney in 1839 and quenched his curiosity about the mysterious continent by walking seven thousand miles. In the Wellington district where McGregor was soon to hammer gold he saw native silver and specks of gold, and in his enthusiasm he hoped that mines would some day be worked and the obstacles of the royal metals overcome. The further he walked the more his enthusiasm sank, and when in London in 1845 he wrote his learned *Physical Description of New South Wales and Van Diemen's Land* he lamented that 'the scarcity of simple minerals was such as might have discouraged the most ardent and persevering mineralogist who ever devoted himself to the science'.

Sir Roderick Murchison, an English aristocrat who had turned from hunting foxes to hunting fossils, saw the crates of rocks which Strzelecki brought to London and glimpsed strong likenesses between Australia's coastal spine of mountains and Siberia's Ural Mountains. As Siberia had the largest goldfield in the world, Australia's eastern mountains might also contain gold. Receiving gold specimens from Australian admirers in 1846 he urged Cornish miners emigrating to Australia to pan for gold in the rivers. Colonel Helmersen of St Petersburg repeated the advice to Australians. These famous geologists were ignored by Australian shepherds, Cornish emigrants, and the governments in London and the colonies.

William Branwhite Clarke, a gold-fearing Anglican clergyman, taught in the King's School near Sydney and preached to the bush sawyers. He longed to visit the mountains that stood smoky blue in the west, for he had studied geology under Professor Sedgwick at Cambridge and had studied rocks in France where he first caught the rheumatic fever that finally exiled him to the warmth of Sydney. In 1841, two years after reaching Sydney, he crossed the mountains with field book and geological hammer, and near the western road at Hartley he was amazed to see specks of gold in granite rubble. In the same area he later found gold embedded in quartz, the largest piece weighing perhaps a pennyweight. He did not visit these places again, and they have not yielded payable gold to this day, and to Clarke they were more a revelation of God's wonders than riches for Man. He admired Wordsworth and loved Nature, and the trilobite named Trinucleus Clarkeii fascinated him more than gold. He was in holy orders, and gold was 'the cry of Mammon' in a country where he saw no honesty in servants and no morality in masters. He believed gold mining was morally and economically undesirable. 'There is no instance,' he wrote dogmatically in 1849, 'of a man making his fortune by opening a gold mine.'

At his parsonage near Sydney Harbour Clarke collected many specimens of gold given him by country parishioners and wellwishers, and he sent specimens to Professor Sedgwick for the museum at Cambridge. On 9 April 1844 he even showed his best specimen of gold to Governor Gipps. The Governor was apathetic, as well he might be, at the sight of such a tiny fragment of gold. While he has gone down in history as the author of that pithy sentence, 'Put it away, Mr Clarke, or we shall all have our throats cut', Clarke himself was the author. He saw gold as a moral menace. Theology seasoned his geology.

'Put it away, Mr Clarke.' It has been one of the top tunes in the hit parade of Australian history, and historians have long insisted that Gipps and the New South Wales governors prevented the rise of gold mining during the convict era, and that there could be no gold rushes until the governors relented. However, the evidence for an official ban on gold mining is weak. The likelihood of such a ban is also weak, for gold mining was not unfit for a convict society; and in fact much of the world's gold was being mined by slaves and serfs in Brazil and Siberia. It was the ancient English law, not the governors in Sydney, that more likely choked the mining industry by discouraging the capitalist and mining engineer and the humble discoverers of gold.

The first metal mines in the continent were opened in the new colony of South Australia, a thousand miles by sea from Sydney. They were found and mined so soon because they were not royal mines and because they lay nearer the coast than any other metal deposits in the settled districts of Australia. Lead was found in boulders on the Glen Osmond hills in 1839, only four miles from the town of Adelaide, and the landowner vowed he would not even sell for £30,000; he would have been wiser to sell. On park-like plains forty miles north of Adelaide a pastoralist's son found copper ore while picking wildflowers, and a pastoralist found copper nearby while riding in drenching rain in search of straying sheep. 'My first impression,' he wrote, 'was that the rock was covered with a beautiful green moss.' Dismounting, he found carbonate of copper. The pastoralists were suffering from low wool prices and, eager for a new income, they quietly bought the Crown land at Kapunda that contained the copper. Learning that the copper was rich, they hired three miners and set out in 1844 in a covered dray to open the mine with formal ceremony and 'an interesting address on mining'. Their mine inspired shepherds and shearers to look elsewhere for the beautiful green and blue stains that signified carbonate of copper. Copper is easier to see than gold, and so many copper deposits were found by 1850 that South Australia was exporting more copper than wool and wheat.

The fame of South Australia's copper kings flew east. Men who had ignored minerals became alert. At the busy port of Geelong lead was found near the town reserve and a copper lode was traced into a hill. Cornish miners went to lead and copper lodes near Yass and Bathurst, across the mountains from Sydney. One copper mine lay at Cornish Settlement, only a few hours' canter from the creek that became Australia's first goldfield, but the Cornish miners there did not take Murchison's advice to wash the gravel for gold. Nevertheless, the little mines amongst the gum trees were symbols of a new interest in metals.

Travellers who rode overland from Sydney to Melbourne in 1848 saw by the road at Mittagong the proudest symbol of the new mining industry. There in the quietness of the bush, a hundred miles from Sydney, were smoking chimneys and the flames of a tiny furnace smelting iron ore. One of the promoters of these Fitzroy Ironworks was William Tipple Smith, a mineralogist. He had sent gold ore to Sir Roderick Murchison in 1848, and on the table of his Sydney office he displayed a specimen of gold ore that was perhaps a subtle persuader to those who came to buy shares in the iron mine. Smith knew where gold had been found in the Bathurst district and he would have tried to open a gold mine but for the archaic law. As he could not own a gold mine perhaps he could be rewarded if he found one for the government. So he showed the Colonial Secretary in the summer of 1848-9 a handsome piece of gold in quartz. The secretary rightly wanted to see the gold deposit before he paid the reward, and strangely Smith backed out. He probably thought that the shepherd who had actually found the gold would get the reward, but whatever his motive he told others what he would not tell the government. He gave those secrets recorded in a guide to gold-finders dated 19 April 1850: 'Lewis Ponds Creek and Yorky's Corner, these are very deep in which is the Quartz Rock which contains the G . . d.' Smith died of paralysis at the close of 1852, just one year after hundreds of diggers had plundered the gold in the rock pools of Yorky's Corner.

All these scattered incidents, so close and yet so far. It was not enough to find scattered nuggets of gold or to break gold from the jutting reefs. The treasure hunt had to become an industry, and that required the discovery of large and rich deposits of gold. Now the deposits of gold that were most easily mined were those alluvial layers that had been eroded over the centuries from the hard rock and had been buried in the gravels a few feet beneath the earth's surface. A chance excavation in the gold regions could have found such a deposit. Dead shepherds no doubt were buried in shallow graves only a few feet above such layers of gold. Holes were dug for posts and trenches for the walls of houses near the gold.

In the green valleys of north-eastern Victoria David Reid grazed sheep and grew grain, and in the summer of 1845-6 he built a flour mill and large water wheel. To race the water under the wheel his men dug a narrow channel from the creek 150 yards away, and in the channel a labourer saw yellow grains. 'Master,' he said, 'this looks like gold.' Reid examined the grains and pronounced them mica, arguing that no gold had been found in that part of the world. A brickmaker digging for clay on the creek flats found more yellow grains but Reid did not change his mind. He preferred to harvest the golden grains of wheat.

Exactly two years later, half-way across the globe, a Californian settler built a sawmill and water wheel on the wooded slopes of the cold Sierra Nevadas. His wheelwright, James Marshall, finding that the tailrace was not deep enough to carry away the flood of water, made his men dig it deeper and then at nights let a flood of water scour the ditch. On a cold January morning he inspected it and thought he saw a flake of gold. He took it in his hand. It was no larger than a melon seed but was certainly gold. Walking along the channel he saw more flakes glittering on the bedrock and shouted to the millhands: 'Boys, by God, I believe I have found a gold mine.' He had found more than a gold mine, and within a year Americans were going overland, sailing around Cape Horn, or hurrying across the isthmus at Panama to reach the richest goldfields the world had known. On street corners by the Atlantic coast they sang 'Oh, California, that's the land for me'. From London and Sydney and Hamburg tens of thousands of gold-seekers went to Sacramento with the washbowl on their knee.

In Australia the water wheel turned slowly in summer at the flour mill of David Reid, and who could say what trifling cause preserved that water race from the spades of gold diggers. An enterprising shepherd or a curious visitor could have mined the gold, just as a visitor who had once won gold in Georgia taught the millhands to wash gold and so closed forever the sawmill by the Sacramento. But no visitor disturbed that quiet Australian valley near Beechworth even when news of Californian gold reached Australia in the first months of 1849.

The first word of the fabulous gold of California seemed a hoax in the quiet Australian ports. A keg of gold ore from San Francisco silenced the doubters in Sydney. Newspapers from California with stories of the first finds reminded Melbourne's *Argus* of Aladdin's treasures and stirred hope in men who were gloomy at the low price of Australia's wool. Ships waited briefly by Australian wharves, advertising fast trips to the gates of gold.

Then on to this quivering stage walked a shepherd boy named Chapman with some 38 ounces of gold that he had found on Glenmona station near the present town of Amherst, a hundred miles north-west of Melbourne. Not the first man to find gold in those ranges, he walked into the Melbourne shop of Brentani the jeweller in the first month of the California excitement. Brentani examined with quick pulse the gold, and bribed the boy with gifts of clothes and money to reveal where he had found it. From Melbourne the little party set out with horse and dray, but near the sheep run the lad became uneasy, maybe fearing his employers' anger if he revealed the site of the gold. He disappeared and the dray returned to Melbourne empty.

News of the gold spread, rumour and fact ingrown. 'Gold there is, at all events, and of such quality and quantity as the world, save in the case of California, never before witnessed,' wrote the Melbourne *Argus* on 6 February 1849. That same day the superintendent of the Port Phillip district, C. J. La Trobe, sent black troopers to disperse those men who were breaking the law by trespassing on the pastoralists' lands. At the shepherd's hut the troopers found thirty or forty men had already gathered with pick-axes, shovels and hoes to dig the gold, and baskets and saddle bags to carry it away, but they had so far dug only one shallow barren hole in search of the gold. Officialdom was too efficient. It dispersed the intruders before they could dig down to fantastically rich deposits of alluvial gold only a few feet below the grassroots, and troopers guarded the parched grasslands until all excitement had gone. Then officialdom regretted its efficiency. Search was made for Chapman in Sydney, where he was said to be hiding, but there was no trace of the lad who alone knew where the gold existed. But three years later, when the gold rushes had really begun, La Trobe himself learned the site of the gold, for he reported to his superiors in London that a man and a lad had just dug sixteen pounds' weight of gold close to Chapman's hut.

The gold rush to Chapman's hut was not entirely thwarted. The Governor of New South Wales, who ruled Australia's eastern coastline and far inland, decided a month later that a competent geologist should survey the mineral resources of his territory. He made that decision because of the shepherd's gold, the mysterious gold of Smith the iron promoter, and above all because of the success of the copper mines in neighbouring South Australia and the promise of base-metal mines near the Blue Mountains. His decision was endorsed in London, and Samuel Stutchbury sailed for Sydney as geological surveyor. He was to report on mineral prospects, and so enable the government to stimulate and control the mining of metals.

The arrival of that fussy, elegant, impractical geologist from England could have been a decisive event, but was not. He rode from Sydney to examine mines in January 1851, and because there were no known gold mines he first examined the copper mines near Bathurst. Eighteen days after he rode west a fat man on a borrowed horse rode up the same road. He was Edward Hammond Hargraves, one of thousands of Australians who had followed the gold to California. As a prospector he was less successful than a score of shepherds who had been before him, but he shrewdly accepted the state of the law and undermined it by a remarkable foray in publicity.

2

'A LITTLE EXCITEMENT'

IRISES FLOWER and grass shoots on a forgotten road north-west of Bathurst in New South Wales. Once the Wellington Inn stood by the road, and there on the evening of 12 February 1851 a giant of a man and a few specks of metal created excitement and amusement. The man was Edward Hammond Hargraves, and he held a slip of paper enfolding three or four grains of gold.

The lodgers and drinkers were eager to see the gold. One examined it through a magnifying glass, another through a glass tumbler, and could barely see the yellow grains. Told that they were living in a gold region as rich as California they scoffed. But the stout visitor was so assured and affable that he was not derided openly, even by those who had no fear of a man weighing about eighteen stone.

At the hotel at Guyong that morning Hargraves had borrowed a tin dish, small pick, and bricklayer's trowel before riding in search of gold. Guided by the publican's son, John Hardman Australia Lister, he had ridden north twelve or fourteen miles down the valley of the Lewis Ponds Creek. The river was dry that summer, but about midday they found a waterhole and made a pot of tea and hobbled their horses. Hargraves then announced that he would wash gold. Eagerly watched by his guide he walked down to a bar of rock in the dry creek and scraped sand and gravel into the tin dish. Back at the waterhole he scooped water and swirled the dish with rhythmic motion of the arms. He was practising the age-old art of panning, and if there was gold in the dish it would work its way down through the lighter gravel to the bottom. Anxiously he palmed with his hand the top layer of gravel from the dish. At last only a film of sediment remained in the dish, and peering in he saw a grain of gold. With his penknife he lifted the grain and put it in a slip of creased paper. Again he took a dish of gravel and laboriously washed it by the waterhole, and again he saw a grain of gold. Six dishes he washed, and all but one yielded a grain of gold. 'This,' he said, 'is a memorable day in the history of New South Wales. I shall be a baron, you will be knighted, and my old horse will be stuffed, put into a glass case and sent to the British Museum.' Hargraves did not become a baron, but he did become a justice of the peace.

Hargraves was aged thirty-four, an Englishman by birth and an Orangeman by politics, and had lived nearly half his life in Australia. He had been a sailor, station overseer, farmer, publican, shipping agent, cattle owner, succeeding at none of these trades. In 1849 he went to California to dig gold, and boasted of his success. 'The greater our success was,' he wrote later in his ghosted autobiography, 'the more anxious did I become to put my own persuasion to the test, the existence of gold in New South Wales.' In fact he was not a lucky digger; he was not even energetic. A man of his weight found it exhausting to shovel gravel or work the cradle all day, and of the year he spent in California he actually dug for gold only a few months. He liked instead to potter about the camp or to yarn around the fire, and he would discuss with his friend Davison whether Australia might be rich in gold. They discussed places where gold had been found, and McGregor the gold-finder was often in their talk. When in March 1850 Hargraves was lodging in San Francisco to avoid winter on the goldfields, he wrote to a Sydney merchant: 'I am very forcibly impressed that I have been in a gold region in New South Wales, within three hundred miles of Sydney.' Always restless, the idea of finding payable gold in his own land obsessed him. With the approach of his second winter in California he thought of snow and iced winds and hastened his plans to return to Sydney. Concern for his wife and five children who were struggling at the timber port of East Gosford (N.S.W.) added to his restlessness. And so, having failed again in a new occupation, he walked the wooden sidewalks of San Francisco for the last time and joined disappointed Australians who were sailing home.

Hargraves had unruffled confidence. Back in Sydney in the hot January of 1851 he boasted openly that he would find a goldfield. He told the jeweller to whom he sold his gold dust. He told a Sydney solicitor, James Norton, and even persuaded him to write to the Colonial Secretary requesting a grant to equip him for his journey. When this grant was refused he persuaded a Sydney businessman named Northwood to lend him money to buy a horse and provisions. Northwood lent £105 on condition that Hargraves shared his gains. He was to find that Hargraves was as shrewd with his sums as his tongue.

Travellers on the road that led west from Sydney saw a stout man with florid face and black moustaches jogging towards the Blue Mountains on 5 February 1851. Hargraves was riding west to find gold in places where gold had already been found. He had heard from the Colonial Secretary that gold had been found in quartz at the Carcoar sheep run of the wealthy squatter, Thomas Icely, and he knew himself that a shepherd had found gold at Wellington. Learning near Bathurst that Icely was absent from his estate, he went towards Wellington. And it was on the road to Wellington that he halted at the inn of the widow Lister and changed his plans. The mantelpieces were crowded with specimens of copper and pyrites and quartz, for a copper mine was working nearby and men had sought for gold and many minerals. And down the creek which flowed past the hotel was Yorky's Corner where the shepherd had found the gold which W. T. Smith showed the Governor in 1849.

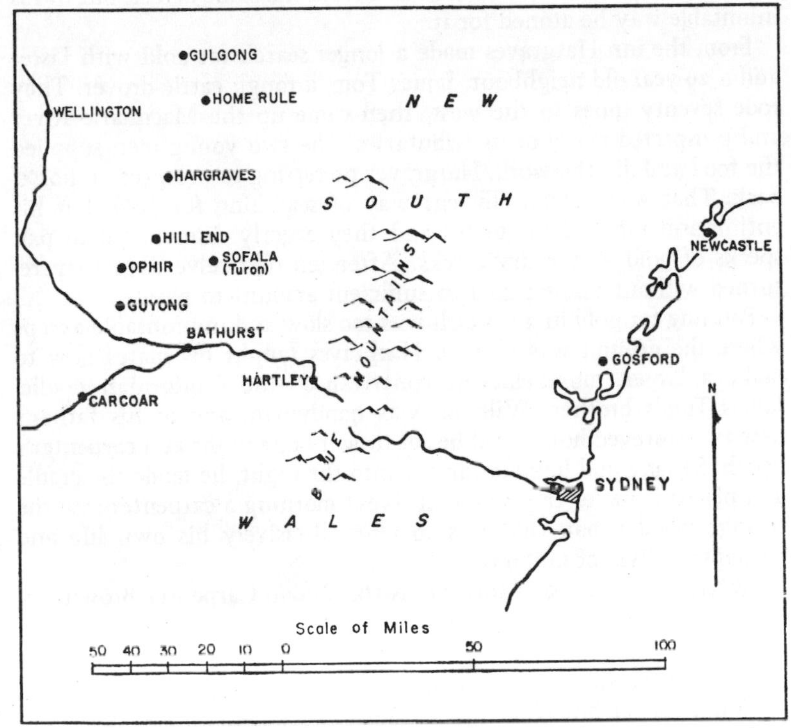

1 Early gold towns in New South Wales

Hargraves certainly made his famous trip of 12 February in order to examine an area where he knew gold had already been found. Hence his boast to his young guide at the waterhole before he had even washed a dish of dirt that the gold lay under his very feet. Hence his journey in the direction of Yorky's Corner, which he would have reached but for the scarcity of water with which to wash the gravel. Hence, too, his decision to return home early that day for fear that the shepherd in the neighbourhood, who had already found gold, would disturb him. There seems no reason to believe Hargraves' story that he had visited the Lewis Ponds Creek in search of lost bullocks in the 1830s, that he had noted when he reached California that its gold regions had similar rocks, and had therefore returned to the Australian creek to prove his wisdom as a geologist and skill as a gold-finder. Hargraves in fact made a less spectacular but none the less sensible observation, that Australia's gold was not being mined because few who sought it knew where to search or how to search. Indeed, Hargraves suffered from the same defect, but in his inimitable way he atoned for it.

From the inn Hargraves made a longer search for gold with Lister and a 29-year-old neighbour, James Tom, a tough cattle drover. They rode seventy miles to the west, then came up the Macquarie River and prospected many of its tributaries. The two young men supplied the food and did the work, Hargraves preferring to prospect on horseback. That was not an efficient way of searching for gold, but his enthusiasm infected his mates and they eagerly dismounted to pan specks of gold in the dry creeks. After ten or twelve days they returned without finding gold in sufficient amount to pay.

Panning for gold in a tin dish was too slow and unprofitable except where the ground was rich, so Hargraves taught his mates how to make a larger, more efficient contrivance—the Californian cradle. James Tom's brother, William, was handyman, and at his father's new two-storeyed homestead he set to work one night at a carpenter's bench. Sawing and hammering far into the night, he made the cradle and placed it under the verandah. Next morning a carpenter saw the strange wooden box that was to affect decisively his own life and perhaps the lives of millions.

'What, in the name of fortune, is that?' said Carpenter Brown.

'That is a bird-cage,' said Tom.

'What is it for?'

'To catch birds.'

'What sort of birds?'

'Very valuable birds.'

'Where do you catch them?'

'On the Canobla mountain; every one caught will be worth a pound,' replied the cheeky lad. The carpenter was puzzled at the answers, but recalled on oath forty years later that he believed them.

They placed the cradle on the bank of the creek below 'Parson' Tom's house, shovelled in gravel, and Hargraves showed them how to rock the cradle so that the heavy gold was retained while the lighter gravel and earth escaped. As they found no gold in the valley the three Tom brothers loaded the cradle weighing a hundredweight on a horse and went to the auriferous country ten or more miles away. They reached a long sheet of water on Lewis Ponds Creek about midnight. Next day one brother waded into the water and filled buckets with gravel from the creekbed; another brother carried the buckets to the cradle and a third rocked the cradle. Every bucket yielded a few fragments of fine gold, and at home that night they weighed it on the family medicine scales and tallied sixteen grains.

Hargraves meantime had gone over the mountains to Sydney to ask the government for a reward. He knew that gold mining was a lottery and that even if he found a rich field he himself might dig little gold. He knew too that all gold belonged to the Crown and that it was wise to depend on the Crown for a reward. In Sydney he waited three hours in a wet overcoat to interview Deas Thomson, Colonial Secretary, and optimistically said he had found a goldfield. Asked for evidence he flourished the creased sheet of paper holding the few grains of gold. Thomson could barely see them and was sceptical. Hargraves, unruffled, asked the government for £500 to compensate him for his costly search—a search on which he could hardly have spent one-fifth of that sum. As for the rest of the reward he would await the generosity of the government. Twelve days later the government told him to say where he had found his gold. Hargraves ingeniously replied with two letters, one naming the golden streams and the other requesting a loan of £30 to buy a new horse. Hargraves got the horse and rode west to meet the government's mineral surveyor and show him the goldfield.

While Hargraves was negotiating in Sydney, John Lister and William Tom found what he had failed to find. On the morning of 7 April they had gone north to Yorky's Corner to search the area where the shepherd had found his nugget. About two miles downstream from Hargraves' first find they hobbled their horses and camped. 'I was buttoning the brace of my trousers,' recalled Tom, when suddenly gold glistened from a crevice of rock. He picked it up and Lister cut it with his knife to see if it was gold. Elated, they fetched the cradle hidden in the scrub and washed fine gold near the site of the nugget. Lister rode his horse further down the creek and saw a two-ounce nugget on the ground. Returning to the homestead when their food was exhausted they balanced the gold on the scales with sixteen new sovereigns borrowed from the widow Lister. They had found four ounces of gold, had found a payable goldfield, had found their Ophir.

Hargraves learned by post of their discovery before he left Sydney to meet the government geologist, and instead of meeting him he visited Guyong and purchased the nuggets of gold on 6 May. He sent some pieces by mail to the Colonial Secretary and held the remaining gold for a more important mission. In company with the Toms he visited the new gold diggings for the first time, quietly breaking the news that he had told the government of the discoveries. The Toms and Lister protested that their new diggings would be invaded by intruders, and argued that they could win a fortune if they could work alone in the creek. 'Oh nonsense,' said Hargraves, 'if you work you will dirty the water, and when shepherds come to water their sheep you are sure to be discovered.' If all they wanted was secure tenure of the richest ground, Hargraves could easily arrange it. Impudently he wrote on a sheet of paper the words 'On Her Majesty's Service', named the richest point of the creek Fitz Roy Bar, named the area Ophir, and gave the bearers exclusive right to the gold.

To name the diggings after the biblical city of gold and to name the richest point after the Governor of the colony suggests that Hargraves was a master of publicity. Indeed his behaviour that day was the prelude to a shrewd campaign. As he had pegged his own claim on the government's files rather than on the banks of the creek, he was in an unusual position. Instead of being secretive he was talkative. The more men he could attract to Ophir the higher the chance of finding a widespread goldfield. And the richer the goldfield the higher the prospect of a favourable report from the government geologist and a generous reward from the government. Whispered conversation in the inn at Guyong warned Hargraves to be discreet but he rode to Bathurst with the news on his sleeve. In that busy country town he lectured one evening on his discovery, displayed pieces of gold, and told his excited audience where to find it and how to mine it.

A senior public servant, the local Commissioner of Crown Lands, heard the news and hurried in alarm to Ophir to find seven or eight men digging for gold in the creek. He told them they were trespassing on Crown land. The law-breakers defiantly produced the unauthorized authority from Hargraves. The official, Charles Green, turned back to Bathurst and wrote excitedly on 8 May to the Colonial Secretary suggesting ways of preventing the labouring classes from deserting their jobs. He had no sooner returned to Bathurst than Hargraves left the town at the head of a troop of horsemen, and at Ophir he showed them likely places to dig and how to wash for gold. Soon the thirty miles of track to Ophir were marked with the hooves of horses and the wide tracks of drays. The *Bathurst Free*

Press hailed this 'second California', and copies of the paper were read disbelievingly in front of hundreds of stoves and hearths. A man reached Bathurst with a piece of gold worth £30. It was smaller than a matchbox but it was magnified in a thousand minds.

Charles Green, getting no reply to his desperate letter to Sydney, wrote again five days later. He said that the 'excitement in Bathurst among *all* classes is intense'. Hundreds of men had gone or were about to go to the diggings, and so many were taking firearms that he feared that all the violence of California would be seen in the steep valley of Ophir. Thomas Icely, the rich pastoralist, decided that he should see this magic valley and rode out to find hundreds of men pitching camp or digging gold in pieces the size of a sovereign. Frightened that men would loot his property on their way to the diggings he went home and removed all plate and treasure and refused to stay in the house while the excitement was raging. As he went to Sydney he counted 138 men in one day hurrying west to the diggings.

At nervous Icely's homestead near Carcoar, Hargraves at last met Samuel Stutchbury, the government's mineralogical surveyor, and escorted him to Ophir. On 14 May he reached the goldfield and wrote a hasty note to the Colonial Secretary that he had just had time to prove that gold definitely existed. Hargraves galloped back with the letter to Guyong and gave it to the driver of the Sydney mail coach. Then he settled in at the hotel to be lionized by successful diggers and plagued by intending diggers. On 18 May he wrote to Deas Thomson with cunning restraint: 'The effect of my appearance in the district has caused a little excitement amongst the people', and he estimated that 500 were digging gold.

It was well for Hargraves to conceal his part in fanning the excitement. Solely through his publicity the diggings had been rushed weeks before a rush could have been expected. Through his campaign the government faced a crisis, and what is more it did not even know that there was a crisis. Ophir was 170 miles from Sydney, and the only chain of contact was the mail coach and a mountainous road that was slippery and boggy in winter rain. So the letter written anxiously by Commissioner Green at Bathurst on 8 May did not reach Deas Thomson in Sydney for another five days, and even then caused no alarm. Thomson still doubted that gold existed west of the ranges. He had shown Hargraves' latest specimens of gold to the American Consul, who learnedly declaimed that the mountains of New South Wales were not high enough to yield gold. He showed the gold to a miner fresh from California and he too scoffingly said that the gold had come from California; he even suggested that two of the specimens had come from the Feather River and the other from the Yuba. Such testimony was so convincing that the Executive Council met in Sydney on 13 May and did not even discuss the topic that was on every tongue on the other side of the mountains.

Two days later—two precious days—the government could no longer discount the news. The *Sydney Morning Herald* carried word of gold into every house in Sydney and every ship at anchor. The second letter arrived from Commissioner Green predicting Californian violence on the new diggings. Even then the government did not act. The crisis was hard to believe, harder to tackle. Next day Thomson enquired of his legal advisers whether the existing laws would enable him to control the diggings. And before he had received a reply he sent letters to Bathurst ordering the Inspector of Police and Commissioner Green to go out and halt the digging of the Crown's gold.

Deas Thomson and Governor FitzRoy feared desperadoes and petty thieves and released convicts would quickly gather at the goldfields. Shepherds and drovers would leave their sheep, farmers would abandon their plough, seamen the sea and schoolmasters the schools, policemen would desert their beats and jailers would desert their prisoners. Law and industry might be submerged in the rush for gold.

Where could the government turn for guidance? Neither Great Britain nor any British colony had had to face such a crisis. California had gold rushes, but the gold was found just after the United States had taken the territory from Mexico and the temporary military rulers did nothing to regulate the goldfields. California therefore gave no precedent and no guidance, except a reputation of lawlessness.

It was too late for the government to ban the digging of gold. Hundreds of men were digging and could not be stopped, as Commissioner Green sadly reported on 25 May: 'I have served many of the numerous parties now digging for Gold with Notices to desist, to which notices they pay but little attention.' Thomson knew that the colony did not have enough policemen and soldiers to drive away the diggers, and even if it did have the soldiers they might drop their arms and themselves dig for gold. When pastoralist James Macarthur proposed martial law and when Major-General Stewart suggested that the government mine the gold itself, they were told by Thomson that it 'would be madness to attempt to stop that which we have not physical force to put down'. The Governor's answer was more vivid, and he told London that it was as hazardous to stop the waves of the sea as the rush to the diggings.

All they could do was to sanction and regulate the rush. Their one baton against the mob was the licensing system. By insisting that every digger buy a licence they could raise money with which to police the goldfields, and they could issue licences on such terms that thousands would be discouraged from seeking gold.

To drug the excitement and deter gold diggers was a vital aim of official policy. It was so crucial that on Saturday, 17 May the Governor and Colonial Secretary and commander of the forces decided to issue no edict on gold for fear that it would aggravate the excitement. They still doubted if the goldfields were rich and for five days they hesitated to act. On 22 May the Governor wrote his first despatch to England on the new discoveries and suggested that Hargraves was exaggerating the gold for his own ends and that his gold came from California. But before his despatch was put on board a ship bound for Valparaiso, another report arrived from Stutchbury the geologist. With surprise the Governor read that many men were winning an ounce or two a day with only a tin dish and that gold would be found over a 'vast extent of country'. So wrote Stutchbury in pencil 'as there is no ink yet in the City of Ophir'.

With that pencilled note on his table the Governor could no longer dither. Invoking a sixteenth-century lawsuit, Queen Elizabeth v. the Earl of Northumberland, he proclaimed the Crown's right to all gold found in New South Wales. No man could dig that gold without buying a licence, and the government charged 30 shillings a month for the licence in the hope that it would be high enough to drive unlucky diggers back to their normal jobs and yet not so high that it would provoke defiance.

At first the fee seemed likely to be defied. Even before the first licences were sold the police magistrate at Bathurst heard rumour that diggers intended to resist the fee. He believed the rumour was true and wrote hastily to Sydney. On 30 May the government ordered the Inspector of Police to prepare to send reinforcements to the goldfields in horse-drawn carriages. A detachment of soldiers stood by in Sydney, 'ready to march to the Gold fields at short notice'.

The success of the licensing system and perhaps the stability of the colony depended on John R. Hardy, new Commissioner of Crown Lands for the gold district. With ten henchmen, armed and mounted, he had to issue the licences, administer the diggings, settle quarrels, and serve justice in court of petty sessions. He reached Ophir on 3 June and saw over a thousand men camped along the creek. He had no tent, no table, not even a candle by whose light he could write at night the reports that the government nervously awaited. On his first morning he went down the creek and collected the tax with so little trouble that he complained only of the tedium of repeatedly writing out licences, weighing the gold the diggers offered in payment, and marking out the area to which he decreed each man was entitled. His sole excitement was restraining a muscular butcher from invading other men's ground in search of gold. The butcher poached on other claims whenever they seemed to be rich, and when he was warned a second time he snatched a spade and threatened to hit the commissioner. 'I instantly collared him,' wrote Hardy, 'put him in handcuffs, and marched him off the ground.' Even that scuffle did not disturb the commissioner and he reported that Ophir was as quiet as the quietest English town.

Hardy's tact made the licensing system effective. He let newcomers work unlicensed for several days in the hope that they might dig enough gold to buy their licence, and over two thousand came and went without buying a licence. Hardy was too tolerant in the eyes of the government. The Colonial Secretary insisted that all men buy a licence before they lifted a spade of earth, and Hardy protestingly complied. He was a democrat and he thought the system was unfair, but the government had intended it to be unfair. The licence was a tax on the unsuccessful digger, forcing him back to his farm or forge. It was a passport to be issued to no man who had left his previous employer without leave. It was an identity card that was not transferable—on the butt of the licence was a description of the digger, the colour of his beard, the tattoo on his arm, described with a leavening of tact because 'much offence might be given by attempting a perfect exactness'. As a crisis measure the digger's licence was masterly in design and initial execution. The tragedy was that like many taxes it remained long after the crisis had passed. In 1854 it was to incite an even graver crisis—armed rebellion at Ballarat.

The skies blessed the government and cursed the gold-seekers in the first cry of the rush. It was winter and nights on the diggings were frosty and ice stiffened the morning puddles. Slush on the roads, and the high prices of provisions, deterred hundreds of men from travelling west. Hundreds were deterred by the licence, which after Hardy's initial generosity was firmly enforced. The policeman who walked his beat examining the licences and smashing the cradles and tools of the unlicensed men might complain that diggers escaped behind rocks or into the bush, but compared to later diggings the narrow-walled valley was tailored for officialdom.

Fortunes were won in a fortnight at Ophir. Some men earned as much in a fortnight of gold-digging as they had earned in a year digging drains. However, there was little gold for the weak or the lazy. Men had to lever away heavy boulders that covered the river flats, and then dig through several feet of coarse gravel to the layer of washdirt. After rain the creek became a torrent and flooded holes laboriously dug to within inches of the golden bottom. Many men dug hole after hole and found only traces of gold. Some got £300 from the clayslate bottom of their first hole and found no more in months of digging. Commissioner Hardy reckoned on arriving at the diggings that half of the 1,500 men had averaged £1 a day; 'You may depend upon this as a fact,' he said. A pound a day was high money to men who had earned four shillings a day in flour mills or on wharves.

In the dark, lights glowed in tents or through chinks of wooden huts occupied by men from every occupation. Decayed gentlemen boiled mutton in a black pot and proud farmers lay on soiled blankets under drays. Seamen and men who had never seen the sea sat on logs with their quart-pot of tea. Thieves who knew the ways of thieves slept with their firearms by claim and cradle. Jewish merchants called at tents to buy gold from bearded shepherds who denounced in Gaelic the accuracy of their scales. Clamour and singing betrayed the sly-grog shop, sharpening of knives the butchery, the sound of the anvil that blacksmith's humpy where diggers' horses were shod by the light of the forge. Men of every trade were now manual labourers. 'I could get anything done even if I wanted my watch mended, or my hair cut, or clothes mended, there were men whose trade it was,' wrote Gideon Lang of his quartz mine. Most of these grimy hands would have been better washed and plying their old trade. The strong men, the common labourers, found most gold.

Ophir's gold lay in shallow ground but many of the parties came with elaborate equipment. Four men walking beside a jolting dray would halt at unclaimed ground, and from the dray would come heavy crowbars and picks, shovels and buckets, axe and tomahawk, tin milk dishes and a wooden cradle for the gold washing and maybe a water-lifter for baling a flooded hole, not to mention the paraphernalia of the camp. They would cut saplings and rig a tarpaulin and by nightfall the bread was in the camp oven, a leg of lamb in the black iron pot, and the kettle lid rattling in steam. A party of four from Sydney spent about £37 on equipment and licences before they put a pick in the ground. Others who came with only a swag on the shoulder, bought pick and shovel and pan, and borrowing and contriving managed to live and dig.

Down the steep hill called Blacksmith's Pinch in July 1851 rode a captain of the Royal Navy to see the famous Ophir before his ship sailed from Sydney. Rising from blankets and opossum skin rug on a frosty morning he walked down the ravine with a gold buyer to see the desolation of waterlogged holes and mounds of gravel. He was surprised to see only one drunken man and the widespread respect for his own naval rank, and delighted to witness near sunset a sailor hand the gold buyer a flat nugget of gold weighing over 4 pounds and acclaimed as the best nugget yet found on Ophir. Back to Sydney went Captain Erskine with this lump of gold he had bought and a bundle of notes that became one of the shelf of new gold books selling in London: *A Short Account of the Late Discoveries of Gold in Australia; with notes of a visit to the gold district.*

Erskine visited Ophir in the third month of the rush and even then it was half looted of its gold. Only two hundred men dug where perhaps two thousand once had dug, and Ophir was in its long decline. Today, by the weeping willows and the gravel drifts, a lonely obelisk spells the names of the gold-finders, for there is neither ink nor pencil in the city of Ophir.

The government fears even in May 1851 that Ophir will soon decay. Police-Inspector Scott warns that the 'vast herd' may soon have nowhere to dig. In fear of disorder the government calls on Hargraves to find new fields. At a meeting of the Executive Council on 3 June he heartens the Governor with his irresponsible prediction: 'I imagine the southern parts of this Colony, Goulburn, Gundagai, and Murrumbidgee, from what I have heard, to be more rich than the western.' He is made Commissioner of Crown Lands with fee and expense account of £2. 10s. a day and leaves with servant and horses to find new goldfields. Leading the packhorse through rough country he covers nine hundred miles in two months, the ashes of his campfires close to auriferous ground that he doesn't see. Always preaching that his corner of the continent was studded with gold, he finds gold hard to find.

As Hargraves' tour provides no new fields, the Governor invites the Reverend W. B. Clarke also to search for gold. He curses the gold that has lured masters from his parish schools and made dissolute the lucky diggers from his flock, but taking his bishop's blessing he leaves his comfortable parsonage with its library and catalogued museum and rides inland to find goldfields and lost souls. On Sundays he preaches under wattle trees or in shepherds' huts and on week days he moves through the bush observing the rocks and testing his strange theory that gold-bearing regions have definite relation to the quadrature of the circle.

Clarke's tracks cross those of Hargraves and so do their reports, one claiming a spot is rich in gold and the other claiming it worthless. Clarke is often right, and some of the hundred promising regions he reveals to the government become famous. But while the success of his earthly predictions pleases him, the success of his spiritual predictions does not. He has long preached that gold is a symbol of lust and now he sees the poison in homes and lives. He sees many drunken men in Sydney's streets that were known for their quietness, and often the employers of tradesmen tell him: 'We can do nothing for the men are all drinking.'

The government need not have worried about finding goldfields for the restless diggers. Its own licensing system forced many to search for new deposits far from Commissioner Hardy's vigilant troopers. By June 1851 men were leaving Ophir and pushing north-east across the ranges to the Turon River to shovel lush turf where sheep and cattle had safely grazed. The gold was sprinkled at least twenty miles along the Turon 'as regular as wheat in a sown field', reported Hardy when he overtook the wandering diggers. He thought any hard worker could win ten shillings a day—twice a labourer's pay—and it was common talk that three men with five hired labourers had won £1,800 in a month. It was clear to Hardy that the gold within forty miles of Bathurst had no limit, and on 10 July he found a spot where sixty unlicensed diggers had each won about £3 for the day. 'I was beset by a crowd,' he added, 'all thrusting their pound notes into my face, and begging me to mark their boundaries.' Though the gold was scattered and hundreds of men therefore evaded the tax, seven of every ten licences issued in the colony by October had been issued on the Turon. It was premier field for several years with perhaps a peak population of six or seven thousand in 1852.

Travellers descending into the green valley of the Turon saw wispy white smoke rising up the hills from tents and open fires and heard the hollow sound of axes and the murmurous noise of the cradles. The main town of Sofala was new and unpainted with tents and drays and wooden stores in crooked rows. Its streets were wheel ruts and hoof pads in the mud. Hotels and lodging houses had calico walls that swayed in the wind. It was a camp of effervescent wealth, and though thousands of pounds in gold went out each Tuesday morning in the government's armed escort to parents and families in Sydney, there was still enough soiled paper money for the forty public houses, enough gold dust to fill jars in butchers' and drapers', and sovereigns and silver to bulge purses of men who crowded in on Saturday night. One former Californian vowed that seven of every eight diggers squandered their money in hotels on Saturday night. About 20 per cent of the men were teetotallers, another miner noted tactfully. Most men wore the long curved bowie-knife to prise gold from cracks and wedges in the bedrock and they carried arms to defend their gold, and with so many knives and guns and grog shops the visiting clergyman wondered at the infrequency of murder and assault.

Bank clerks in Bathurst weighed gold from the Turon and marvelled at its abundance. They ceased to marvel after a day in July 1851. On a sheep station fifty miles north of Bathurst, near the present ghost town of Hargraves, an aboriginal was casually watching his master's sheep. Passing an outcrop of white stone he cracked it with his tomahawk and saw so much gold that he rushed to the homestead of his master, Dr Kerr. They hurried back and broke the gold-studded quartz into several lumps and took two hundredweight in all to the homestead and then on a long journey to Bathurst. The doctor arrived in a cart pulled by two greys in tandem, and bystanders lifted the tin trunk of the gold and staggered under the weight. One lump was a foot in diameter and too heavy for the scales, so was taken to a second bank and finally weighed. Over half the stone was quartz, but 1,272 ounces was gold. It was larger than any mass of gold reported from California and larger than the celebrated 70-pound lump in the St Petersburg museum. 'Bathurst is mad again! The delirium of golden fever . . .' diagnosed the local *Free Press* on 16 July. The *Sydney Morning Herald* predicted a sensation such as 'was never produced before' when the news reached London. It already knew the sensation in Sydney, for that day it sold 4,150 newspapers—perhaps an Australian record. And one copy went that night to a ship bound for Java with the Governor's excited despatch for England in its mails.

Kerr and the black gold-finder had even more excitement. Just as the doctor had apparently confiscated the gold from its finder, so the Crown temporarily confiscated the gold from the doctor because he had no licence to mine £4,160 of the Crown's treasure. Meanwhile he had rewarded the aborigines, Jemmy, Daniel and Tommy, with two flocks of sheep, a team of bullocks, a dray, and utensils and trinkets, and they 'soon appeared dressed in the first style, and riding about like other gentlemen'. News of their elevation to the gentry reached the Apsley Aboriginal Mission at Wellington where they had once lived, and most natives deserted the mission for the goldfields, and the missionary prayed for their safe return: 'Return they did, all whom death had spared,' he wrote finally with pride and pathos.

This golden reef by the Meroo Creek became the new mecca. A thousand licences were issued by the end of October to men chipping at the white reef and digging for gold that had eroded from the reef. Twenty-four yards from the reef Brenan dug through clay and found a £1,156 nugget, and his neighbours found nuggets close to the grassroots—one of them a deep yellow piece of 157 ounces that took the majestic name of 'King of the Waterworn Nuggets'.

Bathurst might be mad but the colony was not. The mails got

through, the clerks sat at high stools in the counting houses, and there were men to tend newborn lambs in frosts. Thousands of men went digging one month and were at their old jobs the next. Of the colony's 190,000 people possibly no more than 6,000 were on the diggings on any one day in the winter of 1851; of the colony's work-force perhaps not more than one in twenty was on the diggings. Nevertheless, that wise civil servant Deas Thomson could not be complacent. When the days lengthened and the sun dried the roads, thousands of men might leave their firesides. And with harvest and shearing coming, could the colony spare them?

Thomson went west in August to find an answer. Driven in a gig over slippery roads to Turon he gloried in the beauty of the valley at sundown as he watched the gold-washing. Questioning diggers he decided the average man was winning so much gold that it 'would be very undesirable on every ground' to publicize it. He still feared the goldfields might explode. He feared famine if the harvest should rot and he feared wide insolvencies if the sheep went unshorn. Should he ban gold mining in the summer, or double the licence fee? These schemes tossed in his brain as he sheltered from days of rain at Carcoar and wrote confidentially to the Governor. After twenty-three pages he reluctantly decided his plans were too dangerous to enforce.

The sky cleared and Thomson rode forty miles to Ophir. It tired him, a staid man of fifty-one, to climb around boulders on the creek bed and pleased him to see the diggers so quiet and respectful. When he called again at Sofala he was cheered to see only one drunken man (not even riotously drunk) and to see little groups gathered in prayer around Methodist preachers. His faith in the digger was high as he returned to Sydney. His fears for the harvest were also high, for in the three weeks that divided his two visits to Sofala another 1,500 men had arrived.

Summer ripens the grain. The harvest is reaped and the sheep are shorn. Delighted, if a trifle bewildered, Thomson questions a respectable citizen from the goldfields: 'Is it the practice of many diggers to go away from the diggings to shearing and harvest?' The citizen answers, 'Yes; they look at that as a sort of holiday, as many of them can get together and spin unlimited yarns at such seasons; it is a favourite occupation with them.'

The year 1851 was one of crisis for New South Wales, and Deas Thomson had guided his colony with gentle hands.

3

VICTORIA'S RUSHES

LESS THAN ten miles east of the hut where the Victorian shepherd Chapman had found his mysterious gold, a creek cut deep into green plains. Snug in the hollow of this Deep Creek, Donald Cameron had the homestead of his Clunes pastoral run, and from the long verandahs earthen paths curved to garden and orchard. Cameron thought a load of gravel would grace those paths, and he mentioned his idea to another squatter, William Campbell, who suggested that the reefs of white quartz a quarter-mile upstream would provide white pebbles. Cameron dallied in collecting the pebbles. Some time later Campbell was thinking of Californian gold and the shepherd Chapman's gold when he suddenly thought of the reefs on the banks of Deep Creek. Inspecting them in March 1850 he saw specks of gold in crevices. Cameron the squatter was not impressed: 'I never considered it to be of any value,' he wrote.

Along the ungravelled paths to Cameron's homestead came a German physician and geologist, Dr Georg Hermann Bruhn, who had ridden west from Melbourne in search of gold in the month that Hargraves had ridden from Sydney. In the dark ranges near Daylesford he had found a quartz vein, and in loose lumps of rock he had seen two specks of gold. Blazing a tree he rode hopefully on, asking all he met the persistent question: had they found gold? Near Cameron's station shepherds led him to the golden reef on the banks of the creek. Proudly, if impersonally, he wrote, 'Should gold not exist in the locality mentioned by Dr Bruhn, he not only perils his character but his professional abilities.' His character was not imperilled. Ten years later the reefs sustained Australia's largest gold mine.

Bruhn was eccentric and impractical, travelling in search of gold without mining tools. His search was for knowledge, not gain, and so he spread the news of gold. His most eager listener by the firesides of huts and homesteads was James Esmond, a bush sawyer near Burn Bank village who had returned from San Francisco on the same ship as Hargraves and knew how to wash gold and the delight of finding it. 'Happy Jim' they called this alert bushman with the strange mongolian face, and happily he went in a spring cart with his mate Pugh to find the gold. Their cart slid down the valley to Cameron's high-roofed homestead, and they saw gold in rock and earth. Their first reaction, unlike Cameron's or Bruhn's, was silence. They rode one hundred miles to Melbourne and said nothing of gold in a city that was greedy for news. The ironwork bought for their cradle, they rode to Geelong to buy provisions, and in the port they guardedly showed specimens of quartz to a gold broker who in turn introduced them to a newspaper reporter. They did not reveal the site of their gold until he promised not to publish the news for seven days.

Within seven days the news was stale. The pastoralist Campbell wrote about the gold to a friend in Melbourne who announced the news on 8 July. That same day Donald Cameron sat in his Clunes homestead and wrote to tell the Lieutenant-Governor, La Trobe, that he had just seen half a dozen men digging the banks of the creek. At the village of Burn Bank, ten miles away, the storekeeper was about to spread the news of gold to promote the trade of his iron-roofed emporium. And in Melbourne a committee which had offered 200 guineas to incite the finding of a payable goldfield found many claimants bellowing for the reward.

There was no sudden rush to Clunes. At the end of the month only fifty men and two women were camped in tents and covered drays and under carts, yarning at night of gold by open fires fanned by chill winds. A washerwoman came to preside over a line of flapping clothes and red-coated troopers came, walking in pairs through Victoria's first gold camp. By day the sides of the creek were busy with the cradles or swirling dishes of golden earth men had carried down the hill balanced on their heads. And at the foot of the reef men scraped with knives the red earth from clefts in the quartz, and even fractured the hard rock itself with picks and crowbars. It was gouging more than mining, so primitive that the gold barely paid, but by the start of August the men were more adept, and reliable observers computed the average earnings of nearly a hundred men at the ducal wage of ten shillings a day.

Meanwhile gold was being found within a day's walk of Melbourne in a small creek that flowed into the Yarra River. The finder, Louis Michel, ran the Rainbow Hotel at the corner of Swanston and Little Collins Streets in Melbourne, and had been so perturbed by the exodus of customers and cronies to Ophir and Turon that he decided only gold would lure them back. With a party he went thrice into the ranges east of Melbourne, and though most of the men gave up Michel persisted. He and William Habberlin made a last search, creeping from Melbourne before dawn on a winter morning to avoid being followed by other prospectors. In thick scrub about eighteen miles from Melbourne the young publican dug a small hole and gave his mate a dish of earth which he panned in the manner described in Sydney newspapers.

> 'Do you see anything, Bill?' said I. After once or twice saying no, he suddenly exclaimed: 'Your Worship, here's the clickerty'. Eagerly examining the residue, we found it contained ten small grains of gold.

Next morning, 14 July 1851, they had hot coffee and damper at a nearby hut occupied by a bushman named Ginger, quietly washed more gold from their creek, and walked to Melbourne with light swags in time to wash, change, and take the gold to the scales of a city chemist. Their goldfield's proximity to Melbourne excited men who had resisted the call of Ophir. Michel left the bar of the Rainbow and chased his own rainbow on Anderson's Creek at Warrandyte, and within a month eighty to a hundred men were scraping the slate bedrock a few feet below the surface for small nuggets. A few won 15s. to £1 a day, but many barely earned rations and many earned nothing. The evening fog that settled on the Yarra Valley became a pall of gloom to men who returned to camp each night with neither gold nor hope of gold.

Clunes for all its isolation still attracted most men, and on the road gold-seekers would pass through a town at the foot of timbered Mount Buninyong. It was perhaps the biggest inland town in the colony with a boarding school and Presbyterian pastor and physician, and of course a blacksmith. Hiscock was the blacksmith, and he got the gold mania. He first looked for gold after the excitement of Chapman's gold in 1849 and the Californian news intensified his unskilled searchings. 'I did not know how to go to work to find it,' he recalled. 'Many a time I have gone about on my hands and knees searching the ground for gold.' Nor did he recognize the appearance of gold, and he gave the mailman so many samples of what he thought were auriferous rocks to take to the experts in Geelong that the mailman refused to take more. He said the jewellers laughed at his samples of fool's gold. But the blacksmith was no fool. On Friday, 8 August, he carried a crowbar and a milk dish along the main road and began to break rocks of quartz. To his delight he saw gold gleaming in the marble stone, and when he washed the nearby earth in his milk dish he saw flakes of alluvial gold that were doubly precious. On the Sunday he took his gold to the coach driver and was reassured that he had at last found real gold. Proudly he told his townsmen, but the day being the Sabbath and Buninyong being ruled by the minister of the stone kirk, he refused to disclose the exact spot that day.

The fifty miles of road from the port of Geelong to Buninyong soon

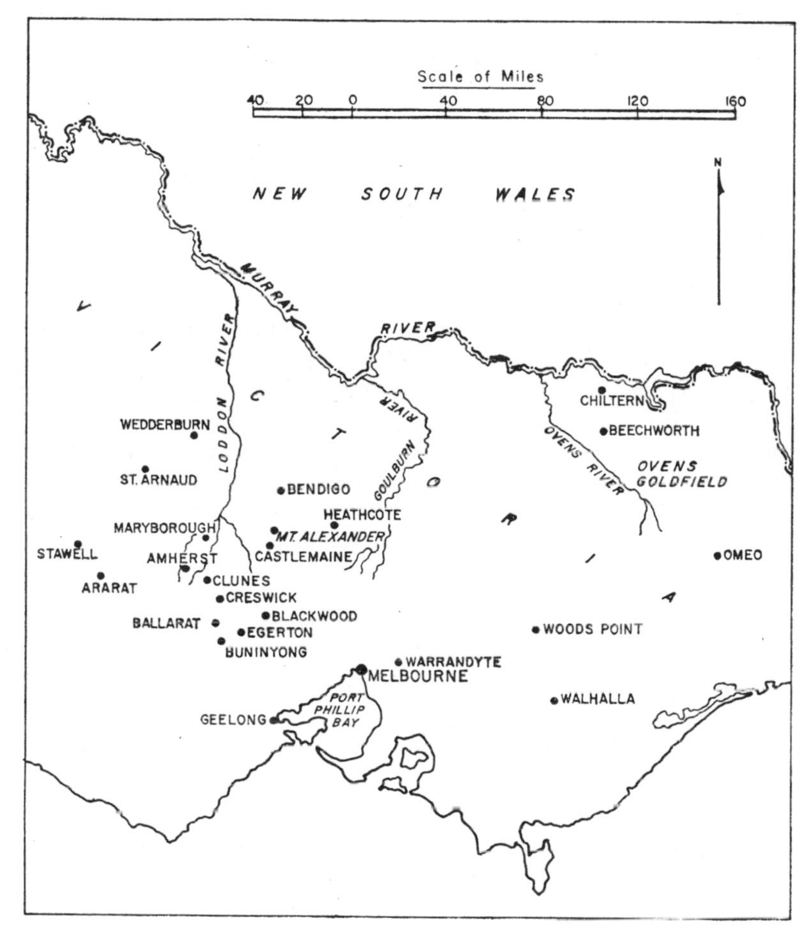

2 Victorian gold towns

had many bullock waggons hauling food and swags. Walking beside one dray was an old man of seventy-five, John Dunlop, who as a child had worked for his bread in England before Australia was first colonized. He went with five Geelong men who, perhaps out of kindness for his old age, took four or five days to walk to the diggings. A fortnight in the rain and sleet won them ten shillings-worth of gold, and three of the men abandoned the search. Dunlop persisted, sending his mate Regan to see whether Clunes was richer and to learn the best way of digging gold. Regan returned through rain and snow and was fortunate that the road crossed one of the richest zones of gold the world has known, the gold of Ballarat. Under the wattle trees he found gold. With soaked feet he hurried seven miles to the camp at Buninyong and led Dunlop next day to Ballarat. By the end of September the colony at last had a field that might rival the Turon. 'The whole town of Geelong is in hysterics', reported the Melbourne *Argus*. 'Gentlemen foaming at the mouth, ladies fainting, children throwing somersets, and all on account of the extraordinary news from Buninyong.' Eight hundred to a thousand men were digging near Ballarat when commissioners and troopers issued the first licences on 21 September 1851.

The Governor of the new colony of Victoria had decided to copy the licensing system of New South Wales, persuaded more by the need for uniformity than by the innate merits of the licence. And the officials who sold the licences were as tactful as Hardy had been at Ophir. Mr Fenwick rode to Warrandyte with four mounted constables and a Crown bailiff and, finding the diggings poor, issued free 'Permits to Search' for the month of September, and sent the bailiff and a black trooper into the sodden ranges in a futile quest for richer gold. Even Ballarat with all its riches had to be handled softly. When the commissioners and troopers arrived, 'stump orators' climbed on tree stumps and incited the diggers to refuse to pay for licences. A few men came to the gold commissioners' tent, bought their licences, and on leaving the tent 'were struck and pelted by the mob'. As they were protected by the black troopers others were emboldened to pay, and by 2 p.m. on the first day the commissioners had no more licence forms. Nevertheless, Captain Dana of the police peered from his tent and was alarmed at the ruffians swaggering about. Writing to La Trobe he confessed that if a disturbance should occur he must withdraw his men, for they were far outnumbered. In Melbourne La Trobe read these nervous letters and wrote curtly on the back of one: 'A party of Police must be mustered at all hazards to proceed immediately.' But when the police rode into the green valley, Ballarat was quiet. Only a fool would bother to fight when so much gold sparkled a few feet below the grass.

Eight feet square of ground was allowed each man, and he dug through soft earth and gravel, red and yellow clay, and hard clay studded with pebbles: neat layers like blankets on a bed. The richest gold lay in a thin seam of blue clay rarely more than five inches thick. Sometimes this blue clay was ten feet down, but some holes descended thirty feet and found neither blue clay nor gold. But each man had many chances in the lottery and some men dug three holes in barren ground and were rich at last in the fourth.

La Trobe saw Ballarat when it was barely one month old. Standing on the piles of slippery clay at the top of the shafts he watched a team of five men who dug 136 ounces of gold one day and 120 ounces the next, worth possibly £4,000 measured in our money. He stooped to see one digger wash eight pounds' weight of gold from two milk dishes of clay, heard of another party getting thirty-one pounds troy of gold in one day. When he rode from the field he estimated 2,500 men were busy. A fortnight later Commissioner Doveton estimated the numbers had doubled and was 'happy to add that the greatest order prevails on Sunday throughout the Diggings'. The Superintendent of Police was more pessimistic, lamenting on 22 October that the scum of Melbourne and the convict scum of Tasmania was hurrying to a camp where men carried or concealed some £80,000 of gold, an unconscious invitation to crime. Ballarat, however, had little crime, for its thousands of pits were soon deserted and only three hundred people stayed to celebrate its first Christmas.

Ballarat could not compete with Mount Alexander, and it is doubtful if any goldfield could have equalled Mount Alexander within six feet of the surface. A name meaningless to Victorians today but magical to millions of Englishmen a century ago, the granite Mount sat like a lion above the hills of Castlemaine, less than forty miles north of Ballarat. In the mountain's morning shadow Christopher Thomas Peters was a hut-keeper on Barker's sheep station and on 20 July he cracked a piece of quartz from a reef that crossed the ground where the sheep were penned at night. The white stone glittered with gold. Sensibly he told only his closest friends, a hut-keeper and shepherd and bullock driver, and three weeks after giving notice to their employer they quietly disappeared into the park-like forest. They were ignorant of alluvial mining, they had probably not seen crude gold before, but they worked the rock with hammer and chisel so strenuously that each man earned a year's wages in a month. They were undisturbed at their hideout a mile from the Melbourne road until they were seen by passing squatters, one of whom was that itinerant busybody William Campbell of Strathlodden, who told the Governor. As the news was out the most literate of the four miners informed the Melbourne *Argus* of their find, but the reporter who edited the item was weak in his geography and announced that the gold was at Western Port, which was east instead of north-west of Melbourne. Peters and his mates used the precious reprieve to gouge more gold, and in mid-October they sold 98 ounces to Dalgety's merchant store in Melbourne for the poor price of 45s. an ounce. When the gold commissioner finally arrived he found six pounds of gold in their hut, guessed they had found much more, and demanded 10 per cent of the estimated gold as penalty for having no licence. So the finders of Mount Alexander were rewarded.

Diggers who had energy to break the Sabbath by climbing Mount Alexander late in 1851 saw at a glance the sweep of this huge goldfield. To south and west they saw white tents, mounds of yellow clay on yellow grass, blue smoke of diggers' camp fires, tracks winding through the scrub, the diggings of Fryer's Creek, Forest Creek (Chewton), Campbell's Creek, Sailor's Gully, Castlemaine, Ranter's Gully and Cobbler's Gully. The diggings spread over fifteen square miles at the foot of Mount Alexander, and gold was even found close to the 'old road made by Sir Thomas Mitchell's drays' when he came south from New South Wales on exploration fifteen years before. It was not, however, the extent of these fields that enticed gold-seekers, nor their closeness to Melbourne; they were seventy-five miles away whereas Ballarat was one hundred by the popular route through Geelong. The richness and shallowness of the gold was the bait. Men found nuggets camouflaged by dust on the surface of the soil. They scraped away eight to twelve inches of black soil and found gold studding the clay. They dug three or four feet into the clay and found nuggets of gold wedged into cracks in the slate. 'A pound weight of gold a day is small remuneration for a party,' La Trobe wrote home, and some parties found five and six pounds a day. In seven months government escorts alone carried to Melbourne and Adelaide £2·4 million of gold from the fabulous Mount Alexander. What men carried privately in saddle bags and hidden purses, what they concealed in wool drays or the harness of horses in order to evade thieves and bushrangers, is not known. And how much gold the luckiest teams dug is not known, but one parcel sent to Melbourne by Eddy and Gill and three others weighed 3,008 ounces. In seven weeks these humble men with pick and shovel had won treasure worth nearly £50,000 at our price of gold, and worth even more to them in terms of what that gold could buy.

For those who found no gold at Mount Alexander there was a consolation prize to the north. On the sheep run of Ravenswood was an out-station with a large shepherd's hut and stockyard on a grassy flat. A shepherd known as Sailor Bill once occupied that hut and he was such a fighter or so boasted of his prowess that he was nicknamed after the notorious English pugilist of the day, William Thompson, alias Bendigo the boxer, famous victor of Deaf Burke and Tom Paddock. The hut and yard were known as Bendigo's and so was the chain of waterholes along the flat, and in the Bendigo creek gold was visible to the naked and sober eye. Who saw it first a select committee of parliament could not decide after examining twenty-three witnesses in 1890, and its decision was not made easier by the memories that had faded or by the curious fact that, even in 1890, thirteen people sought rewards for their real or imagined parts in finding that wonderful field.

Bendigo was named after a boxer and possibly found by a woman. The overseer at Ravenswood station had ridden across country to the village of Buninyong at the end of the winter to hire shearers and when he returned with a piece of gold the curiosity of his wife Margaret Kennedy was aroused. This Gaelic woman rode out on the cart that supplied food to the outlying huts of the shepherds, and at Bendigo Creek she either saw gold in the gravel or was shown it by the shepherds. She returned with the wife of a barrel-maker and some food and tools, and each morning the women emerged from their mia mia made of a linen sheet stretched on a pole, and washed gravel on the banks of the creek. The women were joined by the two shepherds, an old stocky fellow with red face and no whiskers, and a young fellow in a leather apron, and when an owner of the sheep station rode up he was astonished to see the sheep untended and the two shepherds crouching over the waterhole and washing gravel in a pan in which they usually kneaded flour for bread-making. And on the bank of the creek he saw two or three ounces of gold gleaming on a torn shred of blue shirt.

A pastoralist went down the Melbourne road to the Porcupine Inn and with liquor in him boasted that his shepherds were winning gold by the quart-pot. A man at the station with one arm and no name heard the news and spread it as he passed through Mount Alexander. William Sandbach had just driven sheep into the station for shearing when he heard boasts of gold, and next morning with others he left at sunrise and carried his swag over the Big Hill and down through the shaded veil of ironbark trees to the chain of ponds. There he saw two men and two women washing gold and instantly there ran through his mind the hymn of Bishop Heber, 'Where Afric's sunny fountains roll down their golden sand'. Henry Frencham, a journalist in red shirt and knee boots, broke the news in the *Argus* on 13 December as a Christmas box for the nation, and the name of Bendigo a Christmas later was known from Greenland's mountains to India's strand.

La Trobe thought there was no limit to Victoria's gold. 'Meantime,' he lamented, 'the whole structure of society and the whole machinery of Government is dislocated.' He had only forty-four soldiers and could hardly expect reinforcements from Tasmania where convicts had to be guarded, or from New South Wales where gold had to be guarded. His police force was at half strength, and on New Year's Day 1852 only two of Melbourne's forty municipal constables remained on duty in a city full of revelling diggers. Although he had ordered the gold commissioners to sell no gold licences to seamen and public servants and men who had absconded from hired service, they still absconded and quietly got a licence. Both the Postmaster and Surveyor-General predicted chaos. The Superintendent of Police said he had offered higher pay to his fifty-five constables, but fifty of them were determined to dig.

In Port Phillip Bay ships rode at anchor without crews and without hope of hiring them. Seamen earning £2 a month could earn that in a day at Bendigo and they swam or smuggled themselves ashore. On 6 January 1852 a total of thirty-five foreign ships lay off Melbourne, and only three had full crews. The ship *Rattler* blew into port to land a passenger and lost eleven sailors. The *City of Manchester* of 1,200 tons waited in vain in the bay for thirty of its crew of forty to return from the diggings and sail her to South America. The ship *Steboneath* was almost ready to sail to London with wool when thirty-five of its forty-seven men deserted her. La Trobe anxiously informed London that there was nothing to stop a warship or a privateer from anchoring in the bay and holding the city and its gold-laden safes to ransom.

In the ports of Melbourne and Geelong shopkeepers and employers found the relationships of society reversed. Calling at the blacksmith to shoe their horse they found his door locked and knew where he had gone. Their children returned from school to report the master had gone, their housemaids left without notice to marry rich diggers. Preachers looked down from pulpits and denounced avarice to congregations empty of men. La Trobe reported that in some suburbs of this city of 25,000 people 'not a man is left' and women were forgetting their feuds and living in neighbours' houses for protection. And up country squatters paid fanciful prices for diggers to shear the flocks and then waited hopefully for a flood of men to come from Britain. 'Until then God help us,' wrote one.

Sir William Denison, Governor of Tasmania, watching the people of his island 'getting mad and flocking away in the steamers to Melbourne in search of gold', decided to lock away a letter from the Reverend Mr Clarke, the geologist, pointing out places on his island where gold might be found. As Governor of a colony that still took convicts from Britain he was entitled to quake at the tumult. 'I sincerely hope that we may not find enough to gild a sixpence,' he confided to Deas Thomson in Sydney, but the hopes and predictions of this bluff soldier were not fulfilled. He thought the alluvial gold would soon be finished, and he thought Australian gold would attract few free migrants from England except 'the loose Fish of society'. In fact the fish swam out in shoals.

London, 1851. London first heard of the southern hemisphere's Ophir three months after the rush began. Migrants who boarded ships for Australia, their passage paid by a government, could not believe the first stories that men were digging gold like small potatoes. The wave of people continued to flow from Ireland to the United States of America, not even the most fanciful tales of gold turning their thoughts to Australia. London newspapers reported the gold of New South Wales during the last month of the Great Exhibition in the Crystal Palace, and some of the thousands who filed through the glass pavilions paused to see the copper and lead and the minute specimens of gold that were displayed in Australia's name, and could not have been impressed. As Australian mail took three or four months to reach London the news of the run of great finds from Ballarat northwards had not reached Europe at the end of 1851.

In Paris on Sunday, 16 November, Parisians crowded in the Champs Elysées to see the drawing of the lottery for two golden ingots, but workingmen who envied the French vine-dresser his first prize of £16,000 did not know that in Victoria alone more gold than that shining lump was being dug by workingmen each day. In Oxford at the close of 1851 the dons lamented that their university awarded no first class honour that year, but perhaps that didn't matter because some of the gifted graduates a year later would be caked with clay in narrow holes in the Australian bush. In London's shivering Christmas thousands saw the pantomimes, *King of the Golden Seas* at the Surrey playhouse, *The Orange Tree and the King of the Golden Mines* at Sadler's Wells, and some of them would spend the next Christmas at their own golden mines. And there were two pick-pockets, female and incorrigible, who had stolen a purse with £4. 10s. 4½d. from a woman in a Paddington omnibus and whose penalty was to be transported for ten years to Australia; this was almost transportation for life, wrote the London *Times* on 12 December in criticism of a harsh penalty. Transportation to a land of gold soon ceased to be punishment, and the last British convicts exiled to eastern Australia landed at Norfolk Island in 1855.

Europe heard of the riches of Ballarat and Bendigo early in 1852, and the news made thousands restless. It affected poets as much as fishmongers. Alfred Tennyson said he would have gone to Australia but for Mrs Tennyson. Henrik Ibsen wrote in his play *Love's Comedy* of a 'Ballarat beyond the desert sands', and some men of letters did more than imagine the riches of Ballarat. From Gravesend in 1852

William Howitt, famous author of *Visits to Remarkable Places*, sailed in his sixtieth year to visit another remarkable place. On the same ship a bouncy poet with long corkscrew curls, Richard Henry Horne, carried copies of his famous poem 'Orion' along with a portable blacksmith's forge and cart and cradle that he believed would enable him to make his fortune in Australia. A few weeks later Thomas Woolner the sculptor sailed, farewelled by the Rossettis and Holman Hunt and Ford Madox Brown, who painted the departing ship in his famous canvas *The Last of England*.

The emigrations of poets and novelists was a symbol of the wanderlust that gold implanted in romantic minds. For the first time Australia rivalled the United States as a field of migration from the British Isles. For several years more people left the United Kingdom for Australia than for the United States, though Irishmen still preferred North America. From the British Isles in the ten years after 1851 half a million people sailed to Australia; about one person in every fifty in the British Isles sailed south. Such was the wealth of the gold colonies that they could subsidize the fares of nearly half the people who sailed to Australia in that decade.

Gold had a magnetism which the welfare state has dulled. To win gold was the only honest chance millions of people had of bettering themselves, of gaining independence, of storing money for old age or sickness, of teaching their children to read or write. The 1840s had been a decade of revolution and misery and famine in Europe, and now across the globe was a gigantic lottery in which all had a chance and the strong-armed labourer the highest chance. Gold was the magic formula in an age without football pools or state lotteries or social services. Moreover, gold had an intrinsic attraction to a generation that handled gold as currency, knew its touch and beauty, and had that love of sovereigns that made the miser of fiction counting his bag of sovereigns a convincing reality.

Thousands who had resisted the call of California in 1849 did not resist the call of Port Phillip in Britain's year of gold fever, 1852. They boarded sailing ships at Gravesend and Liverpool and Plymouth with their own provisions for three or five months at sea, mostly young men who would never see parents or England again, and drifted with the wind into the oceans in one of the great migrations of the age.

For the thousands who boarded ship at English ports the long sea trip to Australia was more dangerous than life on the goldfields. Hundreds of young men were buried at sea, sewn in their hammocks, in each year of the gold trek. Heavily laden sailing ships ran aground on the Australian coast and bodies were tossed in the surf for days after. Men went mad with the boredom of five months at sea, with no sight of land to be expected from Europe to Cape Otway in Victoria, and took their own lives within days of sighting land. Robert Young, a clergyman, sailed from Plymouth in the new mail steamer *Adelaide* in 1853 with crowds of gold-seekers, burying some at sea on stormy evenings, and penning in his book *The Southern World* their excitement when they first saw the land of gold: mothers holding up babies to look at the shore, men shaking the hands of men to whom they had not spoken in months, men in tears as they thought of the wide waste of water severing them from loves and homes, young men shouting and leaping, grey-haired men throwing up hats, all envisaging a fortune and a quick return to Britain. On the last Sunday of the voyage his sermon in the saloon vainly denounced the folly of laying up treasure on earth.

Gold-seekers remembered to old age the excitement as their ship came up Port Phillip Bay. Across the water a smooth line of hill etched the western horizon with the long spur of Mount Macedon like a hump and the timbered peak of Mount Blackwood, and in gaps of the hills rolled roads to the goldfields. At the end of the bay was Melbourne, caged by the masts of ships. Few ports in the new world were busier. In May 1853 twenty ships arrived from London, twenty-three from other British ports, seventeen from the United States, seven from India and Mauritius, two from Cape of Good Hope, and they joined a fleet of ketches and schooners arriving daily from Australasian ports. Most ships transhipped cargoes to lighters and barges that went eight miles up the Yarra River, and the narrow river was sometimes as jammed with small ships as the Melbourne wharves and warehouses were jammed with cargoes. In the town the hotels and lodging houses were so crowded that migrants spent their first night trying to sleep on their luggage or in discarded packing cases on the wharves. Melbourne grew so fast that thousands of migrants never got beyond the city, became blacksmiths and coachmen and piemen and joiners, and died in old age without seeing the goldfields they crossed the world to dig.

Melbourne was a second San Francisco, more boisterous for it was nearer the diggings. In the first years of the diggings it became a fun parlour. On Sunday wedding parties raced down the streets in open carriages. On Monday saloons and drinking dens were as busy as if there had been no week-end, serving French champagne or Jamaican rum to lucky diggers and luckier prostitutes. After dark, theatres opened and diggers' ladies with flushed faces and low dresses and gold collars sat in the tobacco smoke of the dress circle with a Bendigo lord in a tartan shirt. William Kelly described in *Life in Victoria* how the diggers oathed and jested during the acting of *Hamlet*, one lowering a bottle of brandy on the thong of his stockwhip to the King of Denmark, another jesting the grave-digger on the depth of the sinking, and others pelting small gold nuggets at Hamlet and the ghost as they bowed nervously from the footlights in the final uproar.

Such scenes were spared most young Englishmen fresh from the sea. After gazing at nuggets and gold dust arranged seductively in gold buyers' windows they bought their colonial outfit and packed their swags and marched up Elizabeth Street in red or blue shirts to the latest rush, their swags of blankets chafing their flesh, tomahawk and drinking pot conspicuous on their belts, and polished leather leggings branding them as new chums. James Armour, straight from Scotland in the spring of 1852, was one of tens of thousands who hurried along the rutted road to Bendigo, anxious lest the richest gold be gone, anxious for his safety in forests where bushrangers and rogues watched from the foliage. Forty shipmates went with the Scot, and some lagged behind with their unaccustomed load. It rained and they were clumsy at lighting fires with wet wood, and they slept uneasily, steam rising from wet blankets and their feet cold and swollen in boots worn all night. Heavy in the feet after eight days, they tramped into Bendigo for the luxury of a bed of gum-leaves and a tent of calico and rags and the hope of fortune.

Wrapped in their hopes and fears and often sick for home, many gold-seekers saw little on the journey to the diggings. The leisured visitors saw more and wrote more about what they saw. Lord Robert Cecil visited Bendigo as a lad of twenty-two, long before he was to become Prime Minister of Great Britain and Marquis of Salisbury. With a carpet bag at his feet he rode in a spring cart with a shrieking woman and a Californian digger who carried pistols and whose chief phrase was 'put a bullet through his brain'. Parts of the road were so rough that they had to get out and push, but they almost preferred walking because they could escape the cloud of fine dust raised by the wheels and were not in danger of being capsized by the drunken driver who gulped brandy from a large bottle. At last, on the second day, the Californian snatched the reins and with repeated whipping of the horses and threats to put a bullet through their brains they reached Mount Alexander. In bright moonlight Cecil walked to the gold commissioners' camp through rows of tents and campfires 'without approaching pitfalls or being molested by brigands', he wrote with surprise. He was even more surprised as he went about the fields in black coat and white top hat to be spoken to uncivilly only once by drunken diggers, and their offence was merely to shout 'Who stole the donkey?'.

Armed bushrangers molested the main roads to goldfields and cut-throats murdered their mates in tents and golden holes, but these signs of violence were only one side of the page, the side the newspapers liked to print. The goldfields' chief commissioner noted on 1 October 1852 that the crime rate was no higher on the diggings than in the whole colony. The Reverend Thomas Raston, missionary at Bendigo in 1853, knew that violent crimes and horse stealing and burglary were rife but insisted that morality was higher on the diggings than elsewhere.

Visitors were surprised at the way the goldfields respected the Sabbath. Perhaps they should not have been surprised. The police forbade the mining of gold on Sunday and public opinion agreed; for most diggers wanted one day in which to rest or wash their clothes or yarn with friends or write home to Leith or Londonderry. Moreover diggers eagerly enforced the ban on Sabbath mining, for if a few men sank holes while their neighbours rested they could easily steal gold from the richer claims.

On Sundays ordained clergymen or Cornish diggers preached from carts or stumps or pulpits in vast tents to some of the largest congregations that had ever assembled in the land. Revivalist crusades converted hundreds at the height of new rushes. Wesleyan lay preachers such as Jimmy Jeffrey were famous for their wit and imagination:

'You diggers mark out a claim,' he would say, 'and put down your pegs near to a mount, say that it is Mount Alexander, or Mount Tarrengower, or the Wombat Hill, and you go to work in the hope of finding the gold, and some of you come on a rich patch, and others sink "shicer" holes; 'tis terribly uncertain about finding the gold; but I'll lay you on to the best place. Here, you diggers, come mark out a claim by Mount Calvary.'

For two years Victoria forbade the selling of alcohol on the diggings, and the traffic in sly grog and drugged grog was immense. Gold Commissioner Gilbert saw shops at Bendigo and Chewton sell grog that had been spiced with laudanum, acetate of lead, and a poisonous herb plucked from creek beds, and he reported that some diggers who drank drugged nobblers lost their consciousness and their gold. Grog cabins were rough gambling houses in which two-up was played with halfpence, they were hideouts for gangs who robbed travellers and for the detested night fossickers who stole gold from rich claims in the darkness. 'In coming down from the Diggings I passed from 100 to 150 drays on the road from Melbourne, and I conjecture that fully one-half of their loading was composed of spirits and fermented liquors,' said Commissioner Gilbert at the end of 1852, but he was powerless to interfere when teamsters invariably produced dockets claiming that the liquor was going to innkeepers beyond the goldfields. As the grog trade was illegal and the grog itself was often doped and as the grog tents were havens of criminals and violent crimes, the evil was widely condemned, and it was easy to deduce from such publicity that drunkenness was almost universal. But it is likely that the Victorian goldfields, even in their first years, were more given to church-going and Sabbath quietness and alcoholic temperance than modern Australian outback towns with a similar preponderance of men.

Australian goldfields were probably more orderly than California's, for the central governments in Sydney and Melbourne were strong when the first rushes began and moreover were not distant from new diggings. But police and soldiers were troubled to keep pace with the ceaseless tides of men rushing to new fields. The size of each claim on Australian goldfields was small in comparison with California, and so diggers quickly exhausted their gold in their small area of ground and had to move elsewhere. When a new goldfield was found huge numbers hurried there, knowing that there was room for thousands of small claims. Thus Australian diggers were unusually mobile, and their ceaseless rushing meant that police protection was often slow in reaching a new rush. Diggers thus often united in order to protect themselves, and vigilante committees that captured thieves or murderers and even punished them were not uncommon on new Australian goldfields from Castlemaine to Coolgardie. 'I have seen upwards of 700 diggers at once chasing a thief, or thieves,' wrote Reverend W. C. Currey at Mount Alexander in 1853.

Many Victorians feared riots on the goldfields once the flood of British seekers arrived. They argued that alluvial gold would soon be exhausted and that poverty would nettle thousands of men gathered in the narrow flats and gullies. Officials on the goldfields estimated 30,000 adult men were there in June 1852, and 100,000 in 1855, but the gold was far from exhausted. Prospectors had fanned out from existing diggings, shepherds and carters had been alert on the sheep runs, and so new goldfields arose to ease the strain on the old.

Gold was found at Omeo in the cold mountains of Gippsland and in the winter of 1852 there was a small rush to what then was the most inaccessible field in Australia. Dogged in the valleys by swollen streams and blinded on the heights by sun on the snow, Omeo men then crossed the mountains to the Ovens River, a hundred miles north-west. There two Californians fancied the valleys near Reid's sheep run or heard from a shepherd of the gold found when the flour mill was built in 1846; and they borrowed mining equipment and set out with a shepherd named Howell to find gold. They found it on the third day in a creek in front of a bark-roofed hut, the birth of the mining town of Beechworth. It was 160 miles north-east of Melbourne, close to the main road to Sydney, and thousands came from New South Wales late in 1852 and pitched tents in the flats.

The three Beechworth discoverers, their claims surrounded by newcomers, decided to seek new ground. The Reids, curious to know how they fared, rode out one day and saw a man hastily hide a pint pannikin under the bank of Reid's creek. David Reid rode up and called, 'What luck have you had, Howell?' 'Oh, just middling,' said the shepherd. 'Oh,' replied the pastoralist, 'that is all moonshine. What is that you've got under the bank?' The shepherd produced the pannikin, filled with 14 pounds' weight of gold.

Reid's Creek was rushed by thousands, and David Reid sold goods in a hut by the creek and bought gold at 56s. an ounce and earned huge profits. Many early diggers carried to the store a prospecting dish covered with a cloth and balanced on their heads, and placed on the scales fine gold that weighed 700 or 800 ounces. They extracted much of that gold from a depth of ten and twelve feet, and they had to bale water and fight floods before they reached that depth. Most claims were only twelve feet square but many yielded £2,000 and more for one week's work. And many more yielded nothing.

One valley was no sooner looted of gold than another was found. A Sydney man found gold at Yackandandah, Canadians found it at the Buckland, and Johnson persisted with his idea that the Woolshed Creek would be rich in gold. He paid six men to sink a wide hole or paddock in the valley but as the hole deepened the water poured in and the men could not master it. Johnson spent all his money, paid his men their last wages on the Saturday night, and prepared to abandon his search. The men thought well of him and promised to work another week on the chance of finding washdirt. On the Friday they struck thousands of flat grains of gold on the granite bedrock, perhaps the richest strike in the north-east. In a week they got £15,000 of gold and Johnson reputedly shouted his men twenty dozen bottles of champagne, and could well afford them, for he is said to have made a clear profit of £70,000.

The year 1853 was memorable at the Ovens and sensational at Bendigo. 'One vast diggings,' William Westgarth the merchant wrote of Bendigo in 1853, and some statistics—suspect like all goldfields statistics—suggest that the Mount Alexander district produced 662,000 ounces of gold that year. Eagle Hawk Gully was so rich that Commissioner Gilbert saw gold glittering on the surface soil, so rich that the barrow road or no-man's strip between each claim was coveted by all, and when one man left his claim in order to wheel away a load of auriferous gravel to be washed, his neighbour would quickly shovel the gravel of the dividing wall into his claim. Not far away at the White Hills men crawled and crouched along drives twenty-five feet underground and gouged out a greyish white seam of hard grit that often carried three ounces of gold to the bucket—a seam so rich that they would drive unseen into their neighbour's ground and mine his treasure while he was hopefully sinking his shaft. At Golden Square at nightfall men would climb up home-made ladders and wash the orange earth from face and hands and hair and sometimes see a grain or two of gold in the dirty water.

From Bendigo and Ballarat radiated new goldfields that inflamed the hopes of the thousands who landed monthly from Hamburg and Liverpool and San Francisco. To the east of Bendigo thousands rushed to the lagoons of Rushworth and the ancient hills of Heathcote—17,000 men on the McIvor field in the winter of 1853. South of Bendigo, down towards Ballarat, rich gold was found on the flanks of the Great Dividing Range at Daylesford and Creswick. Then, moving clockwise came Smythesdale, Beaufort, Avoca, Maryborough and other rich diggings that became permanent towns. At the foot of Mount Tarrangower, Captain John Mechosk, a German gentleman who had opened German Gully diggings near Fryerstown and the riches of Kingower, employed nineteen men to prospect for him and they found the famous goldfield of Maldon in 1853 after he had spent £1,900 on rations and wages. And on the red soil of the spurs of Wedderburn and Mount Korong fragments of white quartz as thick as wildflowers tempted the pastoral workers and passing Californian miners to dig shallow shafts. When a pastoralist rode out to see the men at work one asked him, 'Would you like a drink of tea, master?' and the master stooped to take the pannikin of black tea from the ground and almost dropped it when he felt its weight. Concealed in the tea were some of the first nuggets from one of Australia's richest surface goldfields, a field which almost a century later could still produce nuggets including a rugged lump worth £1,300 and known as the Wedderburn Dog.

The discovery of magnificent nuggets of gold fanned the excitement. Most of the alluvial gold in Victoria was in small pieces or specks, like pollen or fine dust, like melon seeds or gunshot, or in ragged coarse grains. In some ground, however, great lumps of alluvial gold lay buried, their sides often smoothed by the flow of water, their shapes familiar or grotesque, often with pieces of quartz or ironstone attached as relics of the reef or hard rock from which the gold had long ago eroded. A belt of country stretching from Steiglitz through Ballarat to Wedderburn—known in later times as the Indicator Belt—yielded fantastic nuggets. At Mount Moliagul in 1869 a Cornishman named John Deason found an inch below the surface of the ground a huge mass of gold which, melted of impurities, weighed 2,284 ounces and was sold to a Dunolly bank for the princely sum of £9,436. Named the Welcome Stranger, it was probably the largest recorded nugget in the world.

The Welcome Stranger had been preceded by the Welcome Nugget, a dazzling lump of 2,217 ounces found by twenty-two Cornishmen not far from the site of Ballarat East railway station in 1858, and it in turn had snatched the record from a long series of beautiful nuggets. At the White Horse Gully in Bendigo in 1852 diggers found one foot under the ground a quartz-encrusted nugget which they sold for £2,100. Nearby another beautiful specimen was bought by the Victorian parliament at the fabulous price of £4. 17s. an ounce and presented to the Queen. On Fryers Creek near Mount Alexander in 1855 two young men capped their short stay of three months in Australia by finding a lump weighing over a thousand ounces and shaped like a heron which they sold in England for £4,080. Few early nuggets were publicized, and many diggers concealed them to avoid attracting thieves or broke them in order to carry the heavy pieces more easily. We can only guess from the numerous nuggets catalogued after 1855 that far more were found but never recorded in the first years when the diggers swarmed over virgin ground. News of nuggets spread by tongue or newspaper were magic to the diggers, and the finding of so many lumps of gold at the depth of sixty feet at Canadian Gully in January 1853 did much to give Ballarat its second lease of life.

4

BALLARAT'S DITCH OF PERDITION

DIGGERS WHO VISITED Ballarat late in 1852 saw it reviving. The exodus from Ballarat to new rushes had burnt itself out, and diggers who knew the charming green valley with the lambs bleating and the wattles in gold came back and saw miners descending unusually deep shafts. Some diggers had not been content with the layer of golden gravel a few feet below the surface. Suspecting there might be a deeper layer they sank deeper and found it. Others followed a shallow layer of golden gravel and found it getting deeper. Parties sank fifty and sixty feet and found rich gold; neighbours sank deeper and found gold; and Ballarat diggers became deep miners. They had found the amazing systems of deep leads, the buried rivers of gold.

Thousands of years ago rivers and creeks had collected in their bends and rapids the grains and nuggets of gold eroded from hard rock on the hills. Then the volcanoes spurted out their lava, burying many of the old rivers. On the western side of Ballarat successive flows of lava had hardened into blankets of basalt, and no gold was seen by the first prospectors. On the eastern side of Ballarat the basalt rock dammed old streams and the valleys and slowly they filled with gravel and sand and clay. On shallow layers of clay rested the gold that sustained the first rush to Ballarat, but at first there was no sign of the far richer treasure that lay in the buried old streams below.

The modern creeks and gullies at Ballarat gave no indication of the course which the buried rivers had taken. The search for the old rivers, or the deep leads as the miners named them, was therefore more of a gamble than any other branch of gold mining. Dangerous to life and purse, it was still tantalizing. It demanded new methods of mining and organization and is, I believe, the key to understanding the armed rebellion of Ballarat miners in 1854.

Those buried watercourses had drained the country for a longer span of time than the existing watercourses and held much more gold. Miners who penetrated them with deep shafts found them heavy with water, like a vast herringbone of agricultural drains. Waterworn boulders lay in the beds of these streams. Trunks and branches of trees were preserved in the silt. The miner's pick found the fossilized leaf of the eucalypt, the fossilized fruits of cinnamon and flame tree and plants that had flourished when the climate of the land was warmer. Where the old river flowed gently, dropping thirty or forty feet to the mile, the gold was spread evenly through the gravel that rested on the hard rock. Where the river had narrowed or curved, or crossed a bar of resisting rock, huge nuggets collected, glittering in the light of the candle. They were the erratic prizes for which miners risked life and fortune.

On the Gravel Pits Lead and the Canadian Lead in 1853 small parties of miners led the descent to the ghostly rivers. Usually, four men held each claim, and 24 feet square was the maximum area allowed to any party. They selected their block of ground below which they believed the buried river would pass, and noting carefully if gold was hauled from nearby claims they either sank their shaft or waited for promising signs before sinking. The early sinking was easy. The shallow hole passed through clay, usually soft and not too wet. One man broke it with a pick and shovelled the clay into a bucket. On top another turned the windlass that hauled up the bucket. Another two men were perhaps in the nearby scrub splitting timber to line the shaft. At a depth of thirty or forty feet the shaft usually ran into water, and the miners took turns to bale with a heavy bucket day and night. Often they were too exhausted and needed more hands to man a night shift. More partners were admitted to provide funds and timber and to bale out the increasing flow of water. Down went the shaft, a battle with water and time. Often five months, sometimes nine months, passed before that shaft reached the golden bed of the river which miners called the gutter.

The small syndicates faced grave obstacles. Most shafts passed through zones of drifting sand from which water poured. Four men often hauled up the buckets of water day after day in an effort to lower the water level enough for them to descend and continue sinking. Often drift sand spilled into the shaft as fast as they could shovel it out, and in defence they had to lay slabs of stringybark, four to six feet long and two to three inches wide, on the walls of the shaft to hold the loose ground. Sometimes they spent weeks in ramming tough clay behind the timber slabs to hold back the water. After passing through one sand drift the shafts on many leads went down in the soft clay only to meet another drift, another inrush of water and silt. Some men were lucky to escape, some did not escape. Even when the hole had at last reached the gutter and the gold glistened in the gravel the water could still burst from behind those wooden slabs rammed with clay. Sometimes the windlass at the top could not haul up the water in bullock-hide buckets fast enough, and the syndicate bought horse and chaff at enormous prices, and the horse walked all day around the whim hauling up buckets.

A few bold Cornishmen and Californians, the innovators in Victorian mining, sent for steam engines and pumped out the water. Sand sometimes clogged their pumps. Even the steam engine did not always defeat the water. Some leads could only be worked when many shafts were close together with many men and horses or engines baling and pumping water; only through co-operation could the teams master the water. If a shaft reached the bottom and the water became too heavy, the gold commissioner compelled the men in the twelve adjacent shafts to bale out water day and night—on pain of forfeiting their claim.

Water was only one enemy in those buried streams. When the north wind rushed hot from inland, men working 100 or 150 feet underground could barely stand the heat and humidity. As very few holes or shafts were connected to one another and the air did not flow down them, miners had to erect calico windsails above their shafts to divert the wind downwards. When a storm was gathering and the barometer fell, the carbon dioxide gas drifted from the old watercourses like a poisonous mist. The shafts in fact were giant barometers, and when the air pressure declined gases escaped from decaying organic matter in the wet river gravels.

When the shaft reached the old river gravels the miners put out drives to each part of their small claim. In effect they excavated an underground chamber, installing heavy logs of timber to support the treacherous roof. In the humid air with the dank smell in their nostrils they excavated with pick and shovel the gravels of the old watercourse. They carried it in American buckets or Cornish wheelbarrows along the drive to the shaft where it was loaded into a bucket and hauled to the surface. Mining the paydirt, and extracting the gold in puddling machine or cradle, might occupy only a week or two. It was the sinking of the hole to reach the paydirt that took months.

Sinking shafts that were smaller in surface area than a grave and often as deep as an 18-storey lift-well, the lucky got fantastic rewards. In the Canadian Lead they plucked nuggets like pebbles on the beach. In the deep Blacksmith's Hole eight men got over £24,000 of gold from a layer of greenish boulders and gravel that rose to a height of about four feet above the bedrock. One tub yielded about 50 pounds of gold and the gravel was so rich that owners of adjacent shafts drove in tunnels secretly and stole it. One tunnel drove right into the Smithy's shaft. As the underground roof of the drives tended to collapse, the miners had to leave large pillars or blocks of washdirt to strengthen the ground. These pillars were so rich that two successive parties reopened the Blacksmith's Hole and at a depth of about 110 feet got handsome gold. And they too left enough gold in the slender pillars for Chinese miners to reopen the honeycombed workings. The Ballarat mining registrar vowed that this tiny claim, as small as a boxing ring, held about one ton weight of gold. It was only one of the jeweller's shops in Ballarat's buried rivers.

Most shafts, however, missed the hidden target. They either found no gold or gravels poor in gold. It was the riskiest form of mining in Australia. 'The deep sinking is nothing more nor less than a species of lottery,' said H. W. Silvester, a Ballarat digger, to a parliamentary committee in 1853. It was a lottery not only because so many holes were barren but because in 1854 the average shaft was 120 to 160 feet deep and took five to eight months to sink. Most miners spent half a year and much money in getting nothing. 'Ballarat had pretty well cleaned me,' wrote James Stewart to his folks in Scotland in 1856. 'We were eight months on the hole and there were £502 expenses and £1. 10s. per week for boarding and we did not get above £40 out of it. I must try my luck a little longer.' He spoke for most who sank deeply at Ballarat.

Official policy increased the cost of probing for deep leads. The gold commissioners gave to parties who were sinking shafts down a hundred feet no more ground than to parties who were sinking ten feet. That allowance after April 1853 was 144 square feet a man, and the maximum area one party could hold in the one claim was 576 square feet or four men's ground. As they had to expand their working party to eight, ten, or twelve men to cope with the water, they received no additional ground. Thus each man in effect received much less ground than a party sinking a cheap and shallow hole. As each party sank its own shaft on its own small claim, more shafts were sunk than the wealth of the leads justified. Huge sums of money and sweat were wasted in sinking the rows of expensive shafts, side by side, along the lead.

On moonlit nights at Ballarat one traced the subterranean course of the main leads by the sight of men turning the windlasses or horses walking the circle and drawing up the heavy wooden baling buckets; one could hear the splash of water as the buckets were emptied, hear the creak of a windlass or the sound of the axe as the neat slabs of wood were split to line the shaft. This energy required capital. 'I am now going down with a hole that will be at least 120 feet deep, and we are quite uncertain as to what we shall find at the bottom; and yet we are a party of eight, and during the time we are sinking we shall have to expend a large sum in licenses and keep.' So spoke Silvester in September 1853, and he spoke for thousands. As the shafts went deeper the working partners often ran out of funds, bought their last pie from the travelling pieman and their last twist of tobacco, and then approached a storekeeper or wealthy miner and sold part of their share for a lump sum or a weekly allowance. Imperceptibly company mining was becoming normal at Ballarat: small companies of ten or twelve working shareholders and perhaps an equal number of sleeping shareholders. A royal commission visiting Ballarat at the end of 1854 saw the remarkable rise of syndicates and companies, saw that 'share dealing has become a multifarious business', and sensed that what was peculiar to Ballarat would soon become common on most goldfields. Democrats might lament the eclipse of the humble digger but the complexities of mining demanded a new organization.

It is forgotten that Ballarat was a unique field on the eve of its rebellion of 1854. Its miners resented the licence fee because individually they got a smaller area of ground than any other diggers received. They resented a tax which hit the unlucky miner because by the nature of the gold deposits a high proportion of Ballarat men were unlucky. 'This Balaarat, a Nugety Eldorado for the few, a ruinous field of hard labour for many, a profound ditch of Perdition for Body and Soul to all,' wrote the Italian miner Raffaello. The licence was gravely unjust to Ballarat miners who earned no gold for six months and who sometimes had to borrow to buy their licence. The Ballarat deep leads nourished the most permanent mining camps in the colony, and as it is always easy to collect taxes from a stable population, Ballarat flinched at the licence raids that became frequent. The nature of mining at Ballarat aggravated these raids. The water in the shafts damaged or destroyed the licences, or miners mislaid their licences when they changed from wet clothes. These were offences in the eyes of the law, for the miner had to carry his licence or pay the penalty. Above all the miner down the deep shaft could be bullied unintentionally by the dutiful policemen on their rounds. Called on to produce his licence he had to climb up a ladder a hundred feet or more, with only footholds in the clay or the timber lining, merely to spend ten seconds in showing his document to the policeman, and so the operations in those deeper shafts were often delayed for half an hour or more. The hardships of the licensing system were vexing on the shallow diggings but they were infuriating on the deep leads of Ballarat.

Buried rivers created other tensions. As teams of miners could not predict whether their claim was above the rich gutter they sensibly waited to see if nearby shafts reached the gutter before they sank a deep shaft. Thus four men would hold a claim and employ one man to turn over a few shovelfuls of clay daily, waiting shrewdly to see results from the neighbouring shafts. This practice was known as shepherding, and along the supposed course of Ballarat's many leads the shepherders idled away the days, sanctioned by law but cursed by opinion. Men sinking nearby shafts cursed the shepherders who did not share in baling the water from the buried river. And hundreds of poor diggers cursed the shepherders because they occupied promising ground without using it, often selling it for high sums if the adjacent claims proved rich.

The buried rivers attracted steam engines, and the steam engines created dissent. Talbot and Emery sank a shaft to tap the rich Gravel Pits Lead near busy Bridge Street in Ballarat, and on the flat ground the golden gutter was deep and the water troublesome. Sending for a steam engine they pumped their own shaft so efficiently that they made it easier for nearby shafts to beat the water. Nevertheless, some diggers were incensed by the throb of the pumps. They argued that the machine would enable the owners to win too much gold, and would dispense with the need for so many working partners in each claim. They would have smashed the engine if Talbot had not guarded it with firearms.

The tension subtly increased hostility to the distant government and its local servants. Food was dear on the goldfield, but all around was rich volcanic soil that could grow grain and vegetables and milk if only the government sold it. And in the words of the historian Geoffrey Serle, oppression made many miners conscious that they had no political rights, no vote, no way of removing their grievances.

Even the men who administered Ballarat were troubled by the new mode of mining. The commissioners spent most of the day visiting claims and settling mining disputes. At Ballarat the mining disputes were harder to settle amicably, for when one party encroached on the gold claim of another party far underground it was often impossible to say where the exact boundary of each claim ran. Moreover, the ground in dispute was often intensely valuable and only entered after the most arduous work, and the commissioners frequently settled disputes in a way that angered the miners. Their verdicts, no matter how unjust, were irrevocable. In September 1854 miners formed a Gold Diggers' Association to plead for a fairer settling of disputes. A month later 250 men on the Gravel Pits Lead ceased work as an ineffectual protest at a commissioner who permitted one team of miners to abstain from baling water from their shaft. The Ballarat mines had advanced beyond the capacity of the law and its representatives to control them.

Inside the stores decked with gay flags men spoke with indignation late in 1854. With its relatively stable population Ballarat could sustain indignation much longer than shallow diggings such as Castlemaine and Beechworth with their transitory, flitting people. Ballarat had a longer memory of injustice and had more injustices to remember. Over ten thousand men lived in Ballarat's straggling canvas camps towards the close of 1854, and the atmosphere was explosive.

Four hundred miles north of Ballarat the Turon valley in New South Wales threatened riot. Many men were incensed that the licence fee of 30s. a month had been extended to goldfields tradesmen and servants, and doubled for foreigners. The Turon men protested to Deas Thomson and got a vague, if courteous, reply. In February 1853 hundreds resolved to dig unlicensed. The publicans who had to buy licences for their barmen and grooms stirred the diggers. The foreigners were justifiably angry. Men bought powder and cast shot from lead, and orators uttered sedition. The leader of the rebels was Mr E. Marjoribanks, an Englishman of good blood, groggy breath and violent tongue. The military leader was a German, Captain Müller, who organized his forces in the balmy air of February 1853, and marched them down the valley, four deep, to the music of drum and fife. Charles Green, the gold commissioner, thought four hundred of the marchers might be willing to fight his thirty-two armed policemen and he waited nervously. As a trial of strength the meeting of diggers surrendered four delegates who refused to pay their tax. Green tried them and boldly fined them each £1. The meeting paid the fines. From Sydney the government despatched men of the 11th Regiment in brilliant uniforms and their arrival seemed to hush the rebels.

A select committee of the Legislative Council studied the grievances and issued a masterly report which parliament adopted in October. The diggers' complaints were fairly remedied, and the licence fee fell to 10s. a month. New South Wales hoped to restrain men from trekking to the richer Victorian diggings, where the licence fee had been recently reduced.

Along the Great Dividing Range in Victoria there were scuffles and agitations. Given the rawness of the police, their arbitrary powers, their incentive to catch unlicensed publicans and diggers by the fact they got half the fines imposed, one wonders that troubles were so few in rough camps where thousands of men carried guns. Commissioner Armstrong went around Mount Alexander with 'a firestick in his hand, burning down tents' without proof that they sold sly grog—and there was indignation but nothing more. At

Fryers Creek a mounted policeman called at a tent for a glass of water and was told to drink in the creek. As he went away cursing, a drunken digger shouted 'There goes a broomstick,' and the constable turned and fired point blank, killing one man and wounding another. Vicious incidents were recounted and embellished in diggers' tents at night.

While the drums played on the Turon there was indignation at Reid's Creek on the Ovens. Assistant Commissioner Meyer went with six armed troopers to adjudicate in a mining dispute. He gave his verdict that one party of men should cease mining the disputed ground, and he sent two troopers with cocked guns into the hole to evict the intruders. One trooper slipped and his musket fired, killing a spectator. From hundreds of shallow holes men rushed to the spot and disarmed the police and pelted them, then moved to the official camp and smashed more firearms. Their protests at the inept administrators might have moved the government in distant Melbourne, but the protests fell to a whimper when the diggers rushed away in thousands to new diggings.

The Bendigo diggings, largest in Australia in 1853, were renamed Sandhurst in time for a military display such as Turon had seen. In the winter of 1853 thousands of men at Bendigo agitated for a lower licence fee. In the main street three miles long many canvas shops and stores carried inflammatory placards calling on the diggers to pay only 10s. a month. Red ribbons, the sign of the rebels, were hoisted outside hundreds of tents. Five thousand men went to one meeting of protest and a few diggers fired volleys into the sky to show their strength. Wesleyans from many diggings around Bendigo held their quarterly meeting on 27 September 1853 and protested that the vast majority of diggers could not afford to buy a licence and had no other jobs to fall back upon. Other goldfields swelled the protest and La Trobe temporarily reduced the licence fee.

The agitation against the licence was strong but peaceful, and Lieutenant-Colonel Valiant, who personally commanded the soldiers despatched to Bendigo, reported on 10 October that the diggers were the 'most orderly and well disposed body' that he had ever seen in all the lands in which he had served Her Majesty. He thought the diggers' complaints were genuine and just. Meddling in politics, he suggested that sooner or later the licence fee would have to go. The Victorian government did not agree, and in December 1853 it fixed the fee at £1 a month.

In August 1854 Sir Charles Hotham, a sailor, arrived to govern Victoria. He was autocratic, vainglorious, unimaginative, but not incompetent. Touring the goldfields he was perturbed, even personally insulted, by the common evasion of the licence. His colony was drifting into a commercial crisis and the Crown's revenue from customs house and land sales was falling, and his government was therefore all the more eager to gather revenue from licences. At the same time the gold output was falling and diggers were therefore more determined to evade or fight the licence fee. He then instructed his police to inspect licences twice a week in the hope of catching more lawbreakers. On most diggings the raids merely incensed people.

Ballarat, with its unique form of mining, suffered more than any other goldfield from the government's policies and was aroused more by the police's incessant licence raids. Any small incident could ignite the friction. On 8 October 1854, a miner named Scobie was murdered near the Eureka Hotel. The publican Bentley was suspected of murdering him but was discharged by the corrupt local magistrate. Angry men burned the flimsy hotel. Three of the men were arrested, charged with riot, and convicted on 25 November. That day the newspaper, *Ballarat Times*, encouraged revolt. Already the government had sent more police and soldiers to Ballarat, for protest meetings were becoming larger and more bitter, but the massing of armed men in the government camp only incited miners. On Tuesday, 28 November, Bishop Goold in Melbourne was so alarmed about the mounting tension in Ballarat and the militant role of his Catholic parishioners that he rode all night in haste to reach the canvas city. That same evening more soldiers in horse-vans reached Ballarat and were hooted and pelted, one van being overturned and the ammunition captured. On the following day, Wednesday the 29th, thousands of miners met at the Bakery Hill to hear wild oratory and pass resolutions refusing to buy more licences. A few men burned their licences at the meeting.

Thursday was hot, the last day before the summer. The hot wind blustered from the north, skimming yellow dust from the clayheaps and flapping the calico sails above hundreds of deep shafts. The breeze diverted by the sails down the stifling shafts did not stir much air, stirring instead the tempers of the miners at work. If the barometer was falling, and weather conditions suggest it was falling that day, then the gases would probably have been escaping from the washdirt in the drives below, making miners faint or anxious to end work early. The day was made to try tempers and send men up the shafts. And the government officials chose that day to make another raid for licences. On the flats hundreds of men defied the police. The soldiers fired shots over the heads of the crowd and took prisoners.

1 Some of Ophir's last gold diggers, 1898
(N.S.W. Government Printer)

2 The shaft sinkers, Trunkey goldfield, N.S.W., in the 1870s
(Holtermann Collection, Mitchell Library)

3 Proud gold miners at Hill End, N.S.W., in the 1870s
(Holtermann Collection, Mitchell Library)

4 N.S.W. gold towns resembled wild-west film sets
(Holtermann Collection, Mitchell Library)

News of the incident was shouted down a thousand shafts, the noise of the firing ricocheted along the lines of tents, and men gathered at Bakery Hill for another meeting of protest and decision.

Their course was now clear. If they refused to buy licences they would have to arm and organize in their own defence. Peter Lalor, 27-year-old Irish engineer who was mining down a 140-foot shaft on the Eureka Lead, emerged as a leader at the tense meeting:

> I called for volunteers to come forward and enrol themselves in companies. Hundreds responded to the call . . . I then called on the volunteers to kneel down. They did so, and with heads uncovered, and hands raised to Heaven, they solemnly swore at all hazards to defend their rights and liberties.

Twelve miles across the timbered ranges the goldfield of Creswick received delegates from Ballarat and heard their oratory and jingles,

> Moral persuasion is all a humbug,
> Nothing convinces like a lick in the lug.

Hundreds of Creswick men worked deep leads, suffered the same grievances as Ballarat, and resolved to give the government 'a lick in the lug'. On the hot windy morning when Ballarat had its licence hunt, a procession formed outside a grog shanty at Creswick and marched behind a Hanoverian band playing the Marseillaise. With speeches from tree stumps their generals gathered recruits and fire-arms and crowbars, and 300 or 400 men marched four deep up the road towards Ballarat. The road was dust and the air sweaty, and a thunderstorm drenched the marchers and turned many back to Creswick.

On the Friday, 1 December, blue smoke piped from thousands of breakfast fires in the mild morning air, a day for action on Ballarat. One of the busiest places that morning was the Eureka Lead, a buried watercourse that ran south for a mile and then turned west. The Irishmen had most of the shafts and tent-stores and restaurants at the place where the lead changed course, and their ground was chosen as a defensive post by Lalor's rebels. During the day hundreds of armed men gathered there, organizing patrols to tour the other leads and collect arms and protect the miners. They surrounded their parade ground and tents with a simple fence, made from the slabs of timber 4 to 6 feet long with which they lined the sides of their deep shafts. In the yard the miners drilled and planned, awaiting the next military raid on working miners or on themselves.

On Saturday evening, 2 December, more than a thousand men were in the stockade at Eureka, and more than four hundred soldiers and police were in their camps on the hills two miles away. The miners' stockade lacked sufficient tents to house its armed men, and late in the evening many men went to sleep in their own tents along the leads. Lalor went to bed in the stockade at midnight. At 1.50 a.m. on the Sunday morning whisky was freely given to the rebels, and some were intoxicated on their last night. A few captains were suspicious that the issue of whisky was a government plot, so they led their armed miners from the stockade. By 3 a.m. only 120 men armed with guns and pistols and pikes remained behind the slab fence. Ballarat lay in darkness, hardly a light from the tents and huts where thousands slept.

A few stirred in their sleep, awakened by the noise of a company of men moving past. In the pale light before the sunrise soldiers and police attacked the stockade. A hail of shots fell on the defenders. Captain Wise of the 40th Regiment was mortally wounded and 'as he lay on his back he cheered them on to the attack'. Lalor was wounded with a musket ball in the left shoulder and was concealed under a pile of mining slabs while the slaughter went on. The fighting lasted less than half an hour but killed an uncounted number of miners, perhaps thirty, and five soldiers. The troops wrecked the stockade and set fire to tents and fired on unarmed miners nearby, and by breakfast that Sunday morning the rebellion was bleeding.

On Monday the miners' funerals were followed for miles by thousands to the burial ground by the swamp. That day there were isolated clashes and indiscriminate firing by the soldiers. Martial law was proclaimed in the district for a few days, and captured rebels were taken to Melbourne to face trial for high treason.

The spirit of rebellion lived on, and few miners or diggers throughout Victoria bought new licences. In Melbourne juries refused to convict the Eureka prisoners. Finally in the winter of 1855 the government reformed its goldfields laws. Instead of the digger's licence it issued for £1 a year a 'miner's right' that entitled the miner to dig gold and vote at parliamentary elections and make his own mining laws. To raise revenue which the licences had once provided, the government collected 2s. 6d. on each ounce of gold exported, a duty equivalent to about 3 per cent. And so the rigid system of controlling the diggings that had been born in the crisis across the Blue Mountains in 1851 was abolished in Victoria, and soon wiped from the lawbooks in the other auriferous colonies.

In Australia's quiet history Eureka became a legend, a battlecry for nationalists, republicans, liberals, radicals, or communists, each creed finding in the rebellion the lessons they liked to see. Collectively their fascination magnified the effects of the episode. The effects were easy to magnify, because from 1856 four of the five existing Australian colonies gained the essentials of the British parliamentary system and virtual control of their domestic affairs, and to many this seemed one of Eureka's achievements. In fact the colonies' political constitutions were not affected by Eureka, but the first parliament that met under Victoria's new constitution was alert to the democratic spirit of the goldfields, and passed laws enabling each adult man in Victoria to vote at elections, to vote by secret ballot, and to stand for the Legislative Assembly.

The Ballarat riot probably quickened political democracy in Victoria but its mark was clearest on the goldfields. In 1855 the miners began to make their own mining laws and settle disputes in their own courts. Such in essence was the Californian system, but whereas it had arisen in California because there was at first no mining law and no authority to make or administer laws, it arose in Victoria at the command of the central government. The government created local courts in the main mining districts, and all men who held a miner's right could elect the nine members of the court. The local court made all mining laws for its district, decided how much ground each man could hold, on what conditions he could retain it, and even interpreted its own laws. This system was probably the high tide of Australian democracy. The miner in moleskins, for long hunted and herded, was lord of his own goldfields.

In 1857 the Victorian government wisely handed the judicial powers of the local court to a new court of mines, with a judge presiding and miners or storekeepers acting as assessors to assess the damages or compensation in disputes; and the local courts themselves became known as 'mining boards', with the same strong powers of law making over a wide area. In southern New South Wales the new goldfields of Kiandra, Adelong, Young, Forbes, and Araluen elected local courts, and Gympie in Queensland elected its own court, and they tended to adopt the liberal Victorian rules during their brief life. Their influence, however, was faint compared to that of the Victorian mining boards which together with the judicial decisions of the Victorian mining judge, Robert Molesworth, shaped mining law for most of Australia. And when Arthur C. Veatch came to study Australia's mining code half a century later as the personal representative of Theodore Roosevelt, President of the United States, he praised a code which seemed to him in advance of America's code.

The miners' moots that first sat at Ballarat and Bendigo and the Victorian gold towns in the winter of 1855 rapidly granted larger areas of ground to men who had to combat water or great depth. Ballarat's court encouraged the mining of its deep leads by allowing miners to unite their claims and by permitting one man to hold shares in many claims. Its court and mining board were so liberal in difficult areas that by 1863 a man could hold 120 times the maximum area of a decade previously.

The new mining code was flexible, imposing duties and privileges on the miner in relation to the nature of the ground he worked. It usually aided the humble digger in areas where he could win gold with his primitive methods, and it helped large parties of working miners in deeper ground where an alliance of capital and labour alone could succeed. The Eureka rebellion thus paved the way for the rapid and orderly growth of capitalist mining and the accumulation of large fortunes in few hands. This was most marked on Ballarat, the scene of the rebellion, and significantly one of the driving capitalists of the new era was the chief rebel, Peter Lalor.

5

GOLD'S CONQUEST

FOUR YEARS after the Ophir rush Australia's working goldfields were erratically spaced along a thousand miles of Australia's Great Dividing Range. They stretched like a crescent from Ararat in western Victoria to Rocky River and Puddledock on New England's chilled tableland. They had mostly been found in 1851 and 1852. Thereafter, gold prospectors who vanished with their pack-horses into timbered mountains invariably saw old campfires and shallow holes and axe-wounds on trees to prove that someone had tried the gullies for gold before them. The pessimists said the richest goldfields had been found.

The pessimists were wrong. Even in Victoria, the scarred heart of the gold country, they were wrong. Gold that nestled near the grass-roots of green sheep runs had mostly been found, but there were still untouched goldfields in mountain forests or deep below the plains. Those fields were harder to find. Edward Hill and his men gave up struggling through gorges before they finally found the gold of Blackwood in 1855, and so isolated was their valley a mere forty miles from Melbourne that their white tent stood alone for a month before their monopoly of gold was disturbed by the first of maybe 20,000 diggers. James Law's party of four sank eleven barren holes to an average depth of 50 feet before they found the rich lead at Navarre in 1859. In the same year the Thompson brothers and two mates worked hard before they found the rich gold of Psalm-Singers' Gully that opened Inglewood field and spurred a wild rush of perhaps 30,000 men.

In the second half of the fifties there were huge rushes beyond the western perimeter of Victoria's old goldfields to St Arnaud, Dunolly, Amherst, Beaufort, Ararat, Stawell, and Landsborough, and in the north-east to Chiltern and Rutherglen's deep leads near the Sydney road. Ararat in its busiest week of 1857 may have held as many as 50,000 people in a straggling metropolis of canvas and bags and weatherboard. Other fields were said to hold about 30,000 men in thirty square miles for a fleeting week or two. Many river flats attracted 7,000 or 10,000 in their first month of fame. The peak population of these rushes was so large, not because these fields were more attractive than Mount Alexander and Ballarat and those of 1852, but because Victoria had more men eager to join in a rush and because in the new fields they generally had to spend longer sinking shafts and so stayed longer.

Victoria's great rushes really ceased in the early 1860s, though even in 1906 and 1907 the finding of a string of rich nuggets at Poseidon and Rheola near Tarnagulla led to rushes of several thousand men. The discovery of distinct goldfields also virtually ceased in the early 1860s, the packhorse bells of the fine Irish prospectors in the mountains from Woods Point to Walhalla sounding the knell. However, important discoveries in old fields did not cease; the richness of gold which was trapped in solid rock was barely appreciated in 1860; and on Ballarat and many western fields were vast rivers of alluvial gold that were covered not by clay and gravel but by basalt from extinct volcanoes. In the first decade of Victorian gold the discoveries had been mainly horizontal, on the surface of the ground. Now the new discoveries were vertical, far below the surface. Discovery had not ceased, it had merely entered a new and costlier dimension.

The Victorian goldfields by 1860 had probably yielded a quarter of the gold they have so far yielded, and they were shedding their surface glamour. Although the Victorian gold statistics are imperfect—for much gold evaded the customs houses and the exported gold was of varying purity and value—the high tide of gold probably ran from 1852 to 1856 with an annual output of about 3 million crude ounces. Thereafter the gold output fell, and from the late 1850s to the late 1870s it was quartered. Likewise the number of Victorian goldminers probably reached a peak of about 140,000 in 1858 and thereafter slumped just as rapidly. But as an export Victoria's gold was more valuable than wool in each year until 1874.

The exodus of tens of thousands of Victorian miners seemed a national tragedy. Politicians ascribed the decline to poor mining laws, lack of capital for company mining, and scarcity of water for alluvial mining. Historians say the accessible gold was being exhausted and some point to the low average income of miners in the late 1850s—imperfect statistics that proved nothing. But the goldfields did not decline solely because gold was becoming harder to win. The causes of decline were complex and subtle. In every year of the 1850s thousands of diggers had gone home to Liverpool or Boston or Canton or become farmers and tradesmen and labourers in Victoria, and their departure from the goldfields was compensated by the flow of new diggers from abroad. But at the end of the 1850s Victoria had fewer immigrants, and so fewer new men went to the goldfields.

Moreover the discovery of new fields in New South Wales and New Zealand not only diverted British migrants from Victoria but attracted Victoria's more skilful and wealthier miners—the very men Victoria needed.

The rise of the goldfields had thrown cities and farms and sheep runs out of balance, depriving them of men, but ultimately these pursuits were so stimulated by gold's wealth and immigration that at the end of the 1850s they were depriving the goldmines of labour and capital and accelerating their decline. It is likely that the growth of Australian trade and manufactures dislocated the goldfields rather than the decline of the goldfields created problems in the cities. Recent scholars have had a favourite question: where did the Victorian miners go after the goldfields declined? They have ignored the more important question: why did the goldfields decline? It is fair to answer that the goldfields declined partly because miners went to rushes in other colonies or more permanent jobs in Victoria, and took their capital with them. Victoria's goldfields could have supported in 1860 a far larger population than they were called on to support, and indeed they revived dramatically in the depression of the 1890s when they could again attract idle capital and labour.

Gold had stimulated such an expansion of Australia's economic fabric that the economy became strong enough to cushion mining's fall. Possibly no other country in the world had been so quickly transformed by metals; the normal growth and achievement of several decades were crammed into one. Australia ceased to be a land of exile in British eyes and became a respectable field of migration and investment. Australia's population almost trebled in the first twelve years of gold. For once Australia almost competed with North America in luring European migrants, predominantly young men, in full vigour.

The swift growth of population widened the market for Australian manufactures and foodstuffs. It stimulated farms and factories and workshops and cities. Gold drew population into the interior and attracted railways from the ports; Bendigo and Ballarat in 1862 got the continent's first upcountry railways, and cheap transport stimulated farming. Goldmines were a vast market for timber, candles, boilers and engines and pumps, and even in the 1880s Victoria's mines used more horsepower than all its factories, and were the mainstay of the engineering industry. Australia ceased to depend on the sheep and in both the 1850s and the 1860s Australia exported more gold than wool. Above all, whenever the price of wool was low, and colonies were depressed, gold always revived to stimulate the economy.

As port for the main goldfields Melbourne had 130,000 people in 1861, a university and railways and telegraphs and gasworks and the amenities of the age. Gold made Melbourne the largest city in Australia and for half a century it led Sydney and entrenched itself as the nation's financial capital. Gold confirmed the dominance of the south-east corner of the continent; Victoria itself had half Australia's population in the late 1850s. Later in the century gold dispersed population to the outback and the remote coastlines of the north and west. Each colony in turn got from gold and metal discoveries the same stimulus that Victoria received in the fifties.

Gold checked and for a time reversed Australia's tendency to become a land that favoured the big man. Whereas Australia's first natural asset, the sheeplands, was grasped by a few thousand men, its second rich natural asset, the goldlands, was divided amongst hundreds of thousands of men. The hope of gold was shared by every man, and Australian society became more optimistic, individualists, and fluid than it ever was in the eras of jailers and pastoralists. It also became more speculative, and the gambling spirit permeated all colonies for at least a generation.

Europe and the East indirectly got gold shavings from Australian ingots. In the 1850s Australia was so rich that its population demanded necessities and luxuries which her own industries could not yet provide. In the year 1853 Australia bought 15 per cent of the total value of goods exported from Great Britain, the world's largest exporter. Britain's exports to the new gold countries of North America and Australia increased by 271 per cent in the years 1846-53; and in the same period Britain's exports to the rest of the world increased by only 21 per cent. There is evidence that California and then Australia had such purchasing power that they largely revived the sick economy of Britain, which in turn sent a chain reaction of prosperity around much of the civilized world. The goldfields were like huge public works in a depression, stimulating confidence and activity. For example, the voyage from Europe to the goldlands was so popular and so long that more ships were needed for the traffic, and shipbuilding yards in Boston and Glasgow and Tyneside prospered. The wages they paid revived trade in their own vicinity; the timber and hemp and iron they imported as raw materials spread prosperity to other industries and other countries like a chain letter. The revival of world commerce in mid-century owed much to the Sacramento and Ballarat, but it is impossible to assess the exact effects of gold on either Australia or the world. As we do not know what would have happened without the gold rushes, we cannot accurately tell the effects of the rushes.

What happened to the gold that performed these works? It was bought by Australian banks and gold buyers, minted into sovereigns for Australia and England, held in the vaults of banks in Melbourne, London, Berlin, New York, shipped to pay international debts, used as national coin and currency. Men suffocated and died in deep shafts, perished on the sealanes and goldfields tracks, died of dysentery or riotous living, to procure the yellow metal that now lies in Fort Knox, U.S.A., and great bank vaults. The metal they found, however, was not useless. Gold was valued and traded because nations did not trust one another and firms did not trust one another and townsmen did not trust neighbours with whom they traded. They gave only limited faith to promissory notes, banknotes, paper money, cheques, and contracts unless they were backed by the commodity they all trusted, the metal gold. Gold was the standard of value, the symbol of trust. The goldminer thus performed the same economic service as the soldier, the policeman, bus conductor, the manufacturer of arms and munitions, the ticket collector on railway station and theatre doors, the factory inspector, the night watchman, and all those whose trade flourished because men did not trust one another. The goldminer even today fulfils the same function as the atomic scientist, astronaut and soldier.

6

THE PROPHECY OF JOB

JACOB BRACHE was a young Prussian engineer, who was deemed slightly cranky. Near Castlemaine in 1854 he asserted that the outcrops of white quartz were richer in gold than the gravels, and if men would only help him he would prove his theory. He gathered twenty-two men from Chile and Peru, Spain, France, Italy, and England, and offered them £1 a day and a share in his venture. Braché himself had £2,000 earned as an engineer in the Americas, and he bought a steam engine and Chilean mill and set them up near his quartz reef at Chewton. His was possibly the first serious attempt in Victoria to mine quartz rock systematically, and if he succeeded a more permanent form of mining would quickly spread over the goldfields.

When bullock teams hauled in his steam engine he was hooted. Many diggers feared that machinery would enable capitalists to win more than their due share of gold, and they mimicked Braché's Prussian accent and stole water from his dam and threatened to destroy his machinery. On the outbreak of the Eureka riot most of the local policemen went to Ballarat, and his gleaming engine was unprotected. He so feared the Tasmanians who camped at what he named 'The Old Lag's Settlement' that he guarded his plant all night with two armed men. His efforts to extract gold from the hard rock were so thwarted, he said, that he lost his men and money. But he did not lose his argument that gold could be profitably wrung from the reefs, and even today the Wattle Gully mine works near the scene of his failure.

Hundreds of men saw lumps of gold as big as walnuts in the solid rock and tried to extract them with the strength of cavemen. At Castlemaine or Clunes miners could be seen cracking the rock with hammers or clumsy battering rams. A few ingenious men ground the rock between heavy grindstones, using windmills or horses. Many who tried to mine quartz scientifically were either blocked from installing machinery or confined to tiny strips of ground that could not yield enough rock to feed the machines continuously. David Syme, an intellectual Scot who later became Australia's powerful newspaper owner, worked a reef at Mount Egerton in 1854 and tried to extend his small area of ground by occupying a vacant triangle of land. His neighbour, Murphy the Irishman, jumped the ground. Syme appealed to the mining warden and was granted the land. The Irishman jumped the claim again, and won the verdict from the local magistrate and what Syme called 'his four rowdy assessors'. Syme was left with £3,000 of crushing machinery but very little to crush. The pioneers of hard-rock mining were persecuted men.

The reefs were unlocked too slowly. They were greedy for capital but company law deterred outside investors. Miners and speculators still thought mining was a treasure hunt; they were infatuated by the bonanzas in the alluvium and not by the steadier profits of mining hard rock. Above all, they were ridiculed by famous geologists. In England in 1854 Sir Roderick Murchison argued in his celebrated book *Siluria* that the experience of miners in the Urals and the Andes and many golden mountains suggested that gold would rarely be mined profitably from solid rock. He believed modern science had confirmed the 28th chapter of the Book of Job: 'Surely there is a vein for the silver. . . . The earth hath dust of gold.' While silver existed to great depths in the solid rock, most gold lay not in the rock but rather in the loose surface soil. Sir Roderick did not say dogmatically that quartz mining in Australia would be a failure but his words had enough conviction to chill the hopes of thousands of Australian miners.

Frederick McCoy, a wrestler-like Irishman who was the professor of natural sciences at the new University of Melbourne, was a disciple of Murchison. He was chairman of a crown commission that studied the mineral resources of Victoria and after touring the hot goldfields and climbing down countless shafts in January 1857 he decreed that the gold in quartz diminished rapidly with depth. The deepest shafts were then only two hundred feet but he was sure that they would rarely win payable gold. The gold in the reefs was like icing on cake; the icing was rich, the cake was indigestible. McCoy's report warned capitalists not to erect permanent buildings near the reefs, warned the government to be wary of building railways to mining towns, for the towns might quickly vanish. Murchison endorsed the report and said Australian miners were foolish to sink deep into the reefs.

The foolish miners continued to sink thousands of shafts. The Bendigo *Advertiser* denounced the prophets who were scaring its miners. The Governor of Victoria wrote despatches to London ridiculing the geologists' theory as 'a Spanish proverb of the sixteenth century'. Alfred Selwyn, the government's gifted geologist who was more familiar with goldfields than McCoy or Murchison, found in 1858 in the deepest shafts in the colony that the gold was as rich at the bottom as the top. Nevertheless, the controversy must have frightened investors.

The conditions of mining seemed to help Murchison's theory. A team of miners was rarely allotted more than half an acre, and that was hopelessly inadequate for mining a reef efficiently. If ten parties of men held sections of the same reef they could of course unite into one large syndicate, but goldminers were individualists and rarely united unless they had to unite. Fortunately for deep-alluvial mining the neighbouring parties often had to amalgamate in order to finance the expense of sinking the deep shaft so as to reach the deep gold. Unfortunately for quartz mining the parties did not have to unite. Valuable gold lay near the grassroots and was easily reached by a small party. The richer rock was busily mined, and crushed with primitive appliances. As their shaft got deeper the cost of mining the rock and hauling it up the shaft increased. Usually they could not afford pumps when their shaft reached the underground water, and so most of the small mines closed prematurely.

A crown commission visited Mariner's Reef at Maryborough in 1857 and saw in the space of two hundred yards a total of nine deep shafts where one was enough. Poor miners had spent perhaps £20,000 in sinking those shafts, and virtually all that money was misdirected. Rows of redundant shafts stood on every field, deep graves for the individualists who were victims of their own mining laws. 'Useless waste of this kind is a national misfortune,' wrote the commissioners. Handkerchiefs of ground not only wasted capital but wasted gold, for the miners rooted out the rich gold and left a huge tonnage of poorer rock which efficient companies could have converted to high profits. Victoria was slow to see that the formula for mining quartz was mass production.

The Port Phillip and Colonial Gold Mining Company was floated in London at the end of 1851 to mine Victorian gold on the large scale. For years it was thwarted by the diggers' hostility to large groups and foreign companies. Its attempt to mine gold at Black Hill in Ballarat failed. It would have succeeded in leasing ground at Fryers Creek but for cries of monopoly. On the Ovens the company got a large claim in flooded ground, pumped away the water and was about to win the gold, but diggers invaded the claim. After five years the company was still without a mine.

The wide quartz reefs at Clunes attracted Rivett Henry Bland, the superintendent of the Port Phillip Company. These reefs were on freehold land owned by the pastoralist McDonald and were therefore safe from the meddlings of local courts. Here at last was room for a large mine. 'I came up and looked at the ground,' wrote Bland. 'I was condemned by all the scientific men for taking it.' He was courageous enough to tilt at Murchison's theory. He arranged a twenty-one-years' lease of the ground from the owners, agreeing to pay them 10 per cent of all gold his company won. The agreement was illegal, for the owners of land did not own the precious metals, but the Port Phillip Company hoped the government would not interfere.

The gold glittering in the white rock on the green creekbank at Clunes had long attracted small bands of miners who believed their miner's right entitled them to the gold on private land. Bland nervously watched the bullock teams hauling his boilers and heavy cast-iron machinery on to his ground. 'At this time very wild notions were held by the miners as to their rights,' wrote C. F. Nicholls of Clunes, and as he had been an active Chartist in London and had drilled at Eureka Stockade his notions of conduct were not easily shocked. In fact few at Clunes would have been surprised if miners had invaded the paddock and smashed the machinery.

Bland forestalled some of the hostility. He arranged with Charles Kinnear, a leading Ballarat miner and rebel, to work the mine on a profit-sharing basis. Kinnear's men would mine the quartz and the Port Phillip Company would crush it and extract the gold. And so Kinnear formed a co-operative party of a hundred miners who each paid £15 for a share and worked daily in the mine. Known eventually as the Clunes Quartz Gold Mining Company, the hundred were paid alike and the laziest dragged down the others. The men sinking the first deep shaft often played cards in the shed above the shaft instead of working below, and when they dealt the cards they would cunningly strike the hammer of the shaft's knocker line to convince passers-by that rock was being busily hauled up the shaft. The co-operative was so inefficient that the mine faced ruin until piece work was enforced.

In the autumn of 1857 the winds blew the drumming sound of the stampers across the plains, summoning hundreds of men to Clunes to build tents and huts in the green hollow. At the end of the following year Clunes had 1,500 people and the alliance of London capital and Australian miners had won £88,000 of gold and paid out £17,000 in profits, half of which went to the lucky owner of the land.

On the southern boundary fence of the company's paddock, independent miners sank shafts in search of gold. Finding no gold in the shafts some secretly drove tunnels under the fence into the company's ground. They drove over a hundred feet into the neighbour's lease and apparently found gold in quartz. Kinnear and his company got word of the trespassers and sank two shafts on their boundary fence. At a depth of twelve feet one shaft broke into the drive or tunnel of the intruders, and the company barricaded that drive to prevent the miners from poaching. Another shaft intercepted the second invading drive at a depth of 35 feet. The secret entrances of the invaders were now blocked. The invaders, however, had gunpowder and threatened to blow up any miner who obstructed them; and judging by the strategic site of one of their invading drives they even planned to blow up the boiler and engine of the Clunes Company.

The mining warden, Gilbert Amos, was summoned from Creswick on 6 May 1857. He told the invaders they were trespassing and was ignored. The invaders went down their shaft, pulled down the barricade at the entrance to their drive, and broke again into the company's ground. A pitched battle raged underground, men fighting with sharp pikes, lumps of quartz, and slabs of mining timber. Up the company's shaft was hauled a miner who had been knocked insensible. Down the invaders' shaft went a miner with a loaded pistol to blast back the defenders.

All the local police could do was guard the mouth of the shafts and imprison the invaders below. Five mounted constables came the twenty miles from Ballarat and, congregating at the top of the shaft, lowered a light in a bottle to see how many men were hiding at the bottom. Underground the miners saw the bottle and smashed it. Another light was lowered in a bottle and also snuffed. The police threw down stones but the miners stood clear. Next morning the police again called on William Morgan, the leader, to come from the shaft. He climbed up and was at once charged with assault and with being an absconder at large. Eight other miners refused to come up, arguing that they had committed no offence. Finally the Ballarat warden arrived with warrants for their arrest and police descended the shaft and found the eight men, well provided with bedding and food and long pointed poles.

The company mine, and its experiment in large scale mining that was so important to Australia, was still endangered. Opinion on the goldfields was against the principle of one company holding so much valuable ground. The government was sensitive to goldfields' opinion and knew that the company had no legal right to the gold on freehold land. But if the government yielded to the clamour and drove out the company and divided the golden reef amongst many small parties, the advent of efficient quartz mining in Victoria might be long delayed.

Rivett Bland hoped the tension would disperse, but in the spring of 1857 he was perturbed to see men sinking shafts just outside his northern fence. They played the old trick, using their shaft as a base for secretly tunnelling into the company's ground. The 'Cornish' and 'Yanky' parties found to their delight a deep lead of alluvial gold and raised the washdirt at night. The company discovered the marauding drives and blocked them at the point where they entered their paddock. The outside miners were determined to raid the alluvial gold. Gold in quartz might best be worked by a company but alluvial gold, they swore, belonged by right to all men. That was the ruling credo of the goldfields, that the easily accessible gold should be divided in small lots amongst the miners.

A row of twenty-seven shafts, all capped by windlasses, were soon arranged along the northern fence. As three or four hundred men were working partners in the outside shafts the assault on the company's fort was serious. The Ballarat West syndicate was so cheeky that it drove a tunnel more than two hundred feet into the paddock.

The Clunes Quartz Mining Company sought from the Supreme Court an injunction restraining the outside miners from trespassing. They got the injunction but soon suspected that the invaders were ignoring it. They tried to send a surveyor down wild Morgan's shaft to see if his syndicate was still encroaching. Morgan boldly announced that not even the Governor could go down his shaft, and thrice he evicted the surveyor. When the surveyor requested the police to accompany him down the shaft Morgan's men set a booby trap. They chained together the two ropes that hung down the shaft, so that anyone who tried to lower himself down on the ropes would set the whim at the mouth of the shaft spinning wildly, and the whim would then 'strike with violence any person who might be standing near'. The police and surveyor evaded the trap and inspected the mine, finding that the syndicate had made 'an enormous encroachment' into the company's lease. The judge said his injunction had been defied and he sent one of Morgan's men to gaol.

The invaders petitioned the Governor that they had found the alluvial lead and had invested £10,000 in finding and working it. They complained that the company had an 'utterly unjustifiable' and illegal monopoly. Their complaint was sound. The Clunes Company in its counter petition to the Governor said it was the pioneer of systematic mining in Victoria and had taken private land simply because the mining boards virtually banned it from Crown lands. Its complaint, too, was sound. On 31 August 1858 the government ordered the Clunes Company to cease working the alluvial leads until the bill legalizing mining on private property had been passed. That bill took another twenty-six years to pass, and meanwhile the Clunes Company and the invaders declared open war.

The warriors tried to smoke out their rivals. On the afternoon of 20 October 1858 the Clunes Company's miners broke into the drives of the Sons of Freedom syndicate and set fire to rags and paper 'and other offensive and injurious substances'. Some of the Sons of Freedom were overcome with smoke and had to be lashed to ropes and hauled up a shaft. On the surface the supporters of the Sons of Freedom ran that night into the company's paddock and gathered at the main shaft. As all men had emerged from the shaft they cut down the whim, set fire to the shed, and threw timber down the shaft. Half an hour before midnight Warden Amos was in his house at Creswick, ten miles away, when a horseman arrived with a note scrawled by a justice of the peace at Clunes: 'My Dear Sir, There has been a very Serious Riot here . . . I fear this is but the beginning of Matters.' Amos sent to Ballarat for all the constabulary they could spare and then rode over the dark plains to Clunes.

The workings near the company's north boundary advanced like a rabbit warren. The All Nations men drove a tunnel only a few feet above a tunnel of the Clunes Company, and there they excavated a five-feet hole and filled it with gunpowder. When the police and company men were standing in the tunnel just below, the fuse was lit. Fortunately the powder did not fully ignite; if it had, the men below might have been killed.

Amos issued a warrant for the arrest of all men in the All Nations tunnel. Policemen with muskets stood at the top of the network of shafts to prevent escape. One bold miner climbed hand over hand 150 feet up a rope dangling in the All Nations shaft, reached the top and saw the police, grabbed the rope and slid down again. Yet somehow they all escaped.

Three of the miners were found and brought before the bench at Clunes on 2 November and charged with feloniously attempting to destroy the company's mines. And one of them, wild Morgan, was also charged with inciting riot, and was led to Ballarat for trial.

The government alone could resolve the quarrel but parliament was hopelessly divided on what to do. Eventually it refused to allow the outside miners to trespass on the company's ground but refused to legalize the mining of gold on private property. And so the company continued, like hundreds of Victorian companies on private land, to mine the gold illegally.

5 New Australasian gold mine at Creswick during the 1882 disaster
(Semmens Collection, University of Melbourne Archives)

6 The Duke United, a deep alluvial mine at Timor, Victoria
(University of Melbourne Archives)

7 The old stock exchange at Charters Towers
(Mount Isa Mines)

Most of the Clunes gold lay in the quartz rock, not in the alluvial gravels, and it was as a quartz mine that the Port Phillip led Australia. The company imported clever English engineers and invested money freely in new plant. It surmounted crises that would have crippled most mines. After six years the ore averaged less than half an ounce of gold to the ton, and the company answered the growing poverty of its ore by mining and milling on the large scale. If we crushed 400 or 500 tons of ore weekly we would lose, said the manager in 1862, but by crushing 700 tons we pay. In 1869 the company crushed and treated the astonishing total of 70,000 tons of ore. Its treatment plant was one of the largest gold plants in the world. Farmers sowing wheat out on the volcanic plains could hear the throb of its pumps and the noise of eighty stampers, each weighing half a ton and as loud as the surf. The battery and stonebreakers and Chilean mill and furnaces and the lofty engines were always open to visitors. The company was always experimenting and always generous in revealing its secrets. The company tried the chlorination process as early as 1868, and was the first Victorian mine to use Root's Blower to ventilate the workings with fresh air.

The Port Phillip mocked Murchison's prediction that deep quartz mines wouldn't pay, mocked the mining boards' creed that the small mine was ideal. If the paddock had been cut up into five or ten small mines it would have been busy for only a few years, then abandoned. As it was, the mine lasted thirty years and yielded 1·2 million tons of quartz and £481,000 profit, or more than twenty-four times the capital originally invested. The only regret for Rivett Bland was that the mine paid a royalty of £138,000 to the pastoralist who had bought the land for £50. The only regret for progressive mining men was that few other Victorian mines copied the wonderful efficiency of the Port Phillip.

The quartz industry was stronger by the early 1860s. Victoria's quartz mines were mechanized with five hundred steam engines to haul up ore and crush it in the stamp mills. Fourteen to eighteen thousand men were mining quartz and by 1872 the hard rock was producing as much gold as the deep and shallow gravels.

White reefs were hidden by winter snow in the Gippsland mountains. From the lower Goulburn River the Irish prospectors went upstream, following razorback spurs or deep ravines and finding rich reefs and pockets of alluvial. Gaffney's, Gooley's, Donnelly's, Stringer's (Walhalla) and Walsh's Creeks were opened between 1860 and 1862 and diggers and miners built huts and water-races at Happy-go-Lucky, Tubal Cain, Jericho and the Jordan, Red Jacket and

Blue Jacket, Alabama and Alhambra. Packhorses carried boiler plates and stampers up the muddy trail, strings of fifty or more horses with jangling bells. Melbourne gamblers began to hunt fortunes in the mountains and in 1866 the Woods Point district had 262 registered companies or 13 per cent of Victoria's mining companies. Several Woods Point storekeepers each used several hundred packhorses to bring beer and flour and provisions to a town that briefly had four thousand people and a ripping share market and all the cheek of a new Bendigo. Patrick Perkins ran a brewery there until the collapse of the share boom sent him to Queensland to found more-famous breweries.

The stampers drowned the song of the magpies at Matlock, 4,000 feet above the sea level. Heavy wooden waterwheels spun in the valleys, lifting the stampers. Along sixty miles of timbered mountains from Walhalla to Woods Point, hundreds of independent reef proprietors tried to profit from the surface gold or the gold in investors' purses.

Humble men made fortunes in the mountains. One man who burned charcoal for the mine smithies was known as Jim o' the Dogs—a pack of dogs was always at his heels. Chopping wood in the mountains he found a rich reef. Proudly he announced his find in a bush hotel and was ridiculed by the reef miners in the bar. He was a charcoal burner, the lowest of the low: how could he find a reef? Jim o' the Dogs eventually persuaded a storekeeper to see his reef, and the gold in sight was so rich that Jim sold for a large sum. He abandoned his charcoal and bought soap and fine clothes. In the mountain towns he threw silver coins to the children as he drove about. At the pastoral town of Sale he heard that an imported stud bull was to be auctioned, and squatters came from afar to bid for the bull, but Jim o' the Dogs outbid them all, with the astounding offer of 400 guineas. He had his bull led into the mountains and the arrival at the gold camps of such a pedigreed beast sent the rumours spinning. There was going to be a bull fight, and Rumour even fixed the day, and hundreds of people arrived at the bullyard to see the sport.

Jim o' the Dogs preferred to be a publican, not a toreador, and swapped the bull for a bush hotel. The new owner of the bull decided to set up a stud farm and sell milk and meat to miners. In thick forest and steep mountains he was troubled to find land for his farm, so he paid a boy to tend the bull on a narrow strip of grass. Before his cows could arrive over the mountains the stud bull fell sick with pleuropneumonia and died without issue. But Jim o' the Dogs kept his hotel and sat behind the bar long after gold camps began to fade.

Of all these mountain gold towns only Walhalla worked its reefs vigorously. Now a ghost town in a walled valley with a crooked narrow street and the old rotunda of the mountaineers' brass band and the old fire station on stilts above the creek, it is the most glamorous gold town to thousands of Melbourne people. Legend says it had an enormous number of hotels and breweries and twenty thousand people, but in fact its population never reached four thousand. Nevertheless large leases and good engineers made the reefs most profitable for several companies, and its Long Tunnel mine won more gold than any Victorian mine and paid £1·2 million in dividends before the end of the century. A young miner who invested £100 in Long Tunnel shares in the 1860s could have retired and lived in comfort on the dividends for the remainder of his life.

Bendigo had more quartz mines and miners than any other Australian town. It was festooned almost daily with gay flags companies flew when they paid a dividend. A score or more companies paid a dividend each fortnight and a bracket of private mines paid untold profits to owners who lived in stately houses with clipped gardens and ornamental ponds. Bendigo was independent, largely financed its own mines and kept the profits. A score of German mine magnates could be seen in the Lutheran church on Sunday mornings, seated as they sang, their thoughts straying from God to gold. Outside thirty Methodist chapels on the same morning abstemious miners chatted of shares and speculations and hidden reefs. California Gully and Sheepshead, Golden Square, Windmill Hill, Long Gully and Job's Gully, Eaglehawk, Hustler's Reef, The Unfortunate Bolle's Reef, and a hundred other reefs and suburbs formed a vigorous straggling metropolis that was Victoria's third largest city.

Bendigo was a dusty democracy in which the wealthy capitalists sometimes worked underground and humble miners owned mining shares. Hundreds of miners had no boss, working instead in the small teams known as tribute parties and sending their ore in drays to steam crushers owned by Ballerstedt and Koch and Lansell. When the tributers found a new reef or crushed a rich load of ore they would decorate the mine buildings with bunting and the flags of all nations, and celebrate in a tent with laden tables and fiddling German band. When the Governor of Victoria or an English worthy visited Bendigo he would be invited underground where carpets were laid on the rock floor and sparkling specimens of gold and champagne were arranged for the visitors.

The reefs were more widespread on Bendigo than on any other field in the eastern half of Australia. A strip of land five miles long and one mile wide was dissected by about thirty parallel lines of reef. These reefs had shed the gold that had made Bendigo probably the richest shallow alluvial field in Australia and possibly in the world. But for all the gold that had eroded from the reefs during millions of years, even more gold lay trapped in the rock.

The rocks of Bendigo resembled a neat stack of corrugated iron. The slates and sandstones had once been flat like a stack of sheet iron, but then strong pressure from the sides had created folds and undulations. Many of the quartz reefs sat on these folds or anticlines like white saddles on a horse of barren rock. They were known as saddle reefs and were so rhythmic and unusual that the American, T. A. Rickard, described them in 1890 as 'this most beautiful type of ore-deposit'. To miners the most beautiful thing about the saddle reefs was the gold, and the ugliest thing was the fact that the sides of the 'legs' of most saddle reefs were often short and sometimes barren. Many companies found the top of the quartz saddles, mined the rich gold, followed the legs down 50, 100, or 300 feet and were dismayed to see them taper from stout thighs into thin ankles. Many then sank the main shaft deeper and found a new saddle. But too often they had paid all their profits in fortnightly dividends so that when they were compelled to sink for new reefs they had to call on shareholders for capital. Given the nature of Bendigo's reefs and given the practice of dispensing the profits as soon as they were earned, Bendigo's prosperity fluctuated.

As the price of gold was fixed, gold was therefore a relatively attractive investment in bad times. Capital always flowed to gold mines when the economy was depressed or just recovering, and in 1870 Melbourne money poured into Bendigo. New money found new reefs and fired the speculators. A reef found at 400 feet in the Great Extended Hustlers was like a treasure house, and the mining journalist Macartney visiting it after much of the gold had gone saw a vaulted chamber as large as a chapel and 'spangled with glittering particles of gold'. The speculative capital increased the gold output, and in August 1871, Bendigo paid half the total dividends paid by Victorian mines. Money for share speculation was easily borrowed from banks; wool prices were rising; suddenly a Bendigo bubble appeared. Saturday, 7 October 1871 snapped all records for fortunes won on the Bendigo share exchange. The midday train from Melbourne was crowded with cigar-smoking gamblers. The speculation and jobbing at the jammed pavements and cafés was so exciting that many Melbourne men refused to return home on the evening train, hiring their own special train when the speculation had ebbed. Next Saturday the government put on a special train late in the evening, and investors reached their mansions at St Kilda and Toorak about 3 a.m.

In one week the value of shares on two of Bendigo's many lines of reefs—Hustlers and Stafford—soared by one million pounds. The Golden Fleece was the most fancied mine. J. B. Watson, a Bendigo mine owner, had paid a miner £2. 10s. in October 1867 for a share and now, four years later, the share was worth £10,000. Watson himself fanned the excitement by revealing at the height of the boom that he had a few years previously paid one shilling for another share, and that the man who sold him the share was so scornful that he spent the shilling on a drink for Watson. Thousands of investors decided that they could emulate Watson.

Such was the excitement in Victoria that Dr L. L. Smith, the physician, advertised with the prattle of a mountebank:

TO SPECULATORS!

MINING versus HEALTH!

Over-speculation, with its attendant anxieties, will produce frightful nervous effects on a frame which has been debilitated by excesses in youth, or by too strong stimulants.

THE REACTION IS SURE TO TELL

His warning was as true of mining as health but in October 1871 most sharebuyers thought they could buy shares at high prices and sell for a quick profit before the boom collapsed. Even lame men and blind caught the fever, and one man who had long sung for alms in Bendigo's streets was now seen daily pencilling his own share transactions. On dusty pavements and beneath shop verandahs the share selling sometimes continued until 2 a.m. Night visitors from Melbourne thought the quartz city was a Paris of the new world during the share mania. 'Wild was the excitement universally prevailing, whilst Hallas' excellent city band nightly animated the gay, festive, and glittering scene,' wrote W. B. Wildey, 'and all the hotel bars were brilliantly illuminated.' Anthony Trollope, the popular English novelist, visited Melbourne when the share mania touched insanity and he was told that if he walked the length of Collins Street he would hardly meet a man who had not owned a mining share or part of a share. He likened the mania for gold shares to the English railway mania of the 1840s.

More than a thousand companies owned mines at Bendigo at the close of the year, and their names were so confusing that investors who heard a tip in the noisy bar rooms must have sometimes bought shares in the wrong mine. There were three companies named Royal Standard, two New Chums, a Walter Scott and Sir Walter Scott. Hundreds of companies had the word 'Tribute' in their title. Eighteen companies began their title with the word Royal, nineteen with Extended, twenty-three with Golden and twenty-three with Great, fifty-seven North and sixty-three South. A long list of companies borrowed the names of famous men, General Moltke, Humboldt, Iron Duke, Marco Polo, Disraeli, Gladstone, and Young Gladstone. A long list took the names of nations and emblems, Rose of Italy and Rose of Denmark, Hibernia, Prussia, Cornwall, East Britain and East Little Britain.

One Saturday in November 1871 Bendigo's blackboards were white with scribbled share quotes and bands played brass music in the streets and gold cakes stood like synthetic nuggets in the banking chambers. The Melbourne investors went home after Saturday's trading, but did not return. Shares fell, everyone seemed eager to sell. Scores of wealthy speculators and working miners could not pay their debts, and Dr L. L. Smith posted them his panacea pills and potions, 'packed as to avoid observation'.

Speculators were sick but Bendigo had never been so healthy with the flood of money that went into deep shafts. Its annual gold output averaged 330,000 ounces from 1872 to 1874, an increase of 100,000 ounces on the good years of the late 1860s. Its companies in 1872 paid the astonishing sum of £683,000 in dividends but again they were too prodigal in paying away their profits. Soon they starved for capital to sink deeper and gold output slumped and the dividends in 1877 were less than one-fourth of those paid in 1872. Bendigo's mines were like all the Victorian company mines, numbers on a giant roulette wheel that span fastest when the economy was depressed.

While Bendigo fascinated gamblers in the 1870s the town of Stawell to the west of Ballarat fascinated the few overseas engineers who visited Australia. They saw the crooked street choked with men on Saturday night and saw blue and yellow flames on the heaps where the auriferous pyrites were roasted, and were variously impressed to learn that seventy saloons sold alcohol and that one mine had paid £622,000 in dividends in seven years. If they were there in 1877 they would have seen the beginning of Australia's best known footrunning race, the Stawell Gift, and would have been escorted with pride to the deep shafts that witnessed a more important race. For the Magdala and Newington and Prince Patrick were the three deepest shafts in Australia and the leader in the downward race, the Magdala, reached 2,400 feet in 1880.

Bendigo chased Stawell in sinking to the depths. Small companies with only a few acres of ground had no alternative but to sink their shafts rather than extend their drives, and the succession of saddle reefs, one on top of the other, lured them deeper. There was little water in the deep shafts to deter them and Bendigo's great Cornish managers and miners mastered easily the problems of deep-sinking. On the New Chum line of reef the shafts probed down 2,000 feet, often passing ten saddle reefs on the way. The tiny North Old Chum syndicate with its two acres sank its shaft over 2,300 feet by 1890. In 1890 the visitor could go down about 2,000 feet in a cage and then walk underground at the same depth for nearly two miles through a succession of mines, all of which had their deep shafts.

Bendigo was infatuated by depth. The Americans who went nervously down the lifts of the skyscrapers of New York and Chicago were making mere jaunts compared to the distances which the Bendigo miner travelled daily to his work. When the Bendigo field tragically collapsed during World War I it had at least fifty-three shafts that were over two thousand feet in depth and one shaft almost a mile deep. Skyscrapers reversed, they dwarfed the rising Manhattan skyline.

Professor McCoy in old age saw Bendigo crush the fragments of his theory that deep quartz mining would not pay. He and Murchison indeed had erred less in geology than economics. Their belief that quartz got poorer at depth was certainly not disproved by the deep Bendigo shafts. It was their calculation of the cost of mining and crushing that was so wrong. They had shared the widespread idea of the 1850s that an ounce of gold in a ton of rock was unpayable, but by 1900 there were Bendigo mines that made a profit by extracting only 1½ pennyweights of gold from each ton of ore. Possibly no other underground mines in the world could have lived on such mean gold.

Technical changes also upset Murchison's prophecy. The shafts that were the mines' vertical highways were enlarged, and steam engines on the surface hauled up skips of ore from below and wire rope replaced the weaker hempen rope. As mines went deeper miners went to work in fast cages or lifts with a safety catch to stop the cage from plummeting should the rope snap. No longer did miners daily have to climb down slippery vertical ladders to their working places and exhaust themselves at the end of the shift by climbing up hundreds of feet. 'A man exhausts himself as much coming up the ladders as he does doing his work,' the Port Phillip manager had noted at Clunes in 1862. The safety cage saved miners so much time and energy in the deeper mines that it almost paid for the reduction of daily working hours from ten to eight.

Mining methods overturned Murchison's theory. In the 1860s every miner used a heavy hammer to drive a sharpened rod of steel into the rock, turning the steel with his hand as the hammer struck. By sheer strength he made the holes deep enough to ram in the explosive that shattered the rock. By the 1870s Bendigo miners were fracturing the rock more easily by using, instead of black blasting powder, the dynamite invented by Alfred Nobel. At the same time the first mechanical rock drills were boring holes that miners had always bored by muscle. A surface engine sent compressed air down pipes to the heavy rock drill and pistons in the cylinders of the drill thrust to and fro, hammering at the sharp piece of steel. Hustler's Reef company at Bendigo used a locally-invented rock drill as far back as 1869, before the great Comstock field and perhaps any other United States mine drilled holes mechanically. But Australia with its host of small mines tended to be slow to adopt rock drills. They were heavy and cumbersome, had to be shifted before the holes they had drilled could be fired, and they often broke down.

The percussion rock drill and dynamite had been first tried in European engineering projects. So had the diamond drill—the miner's telescope; steam revolved a long hollow drill with a diamond edge that cut into hard rock. The diamond drill was explorer and retriever, able to collect samples of the rock through which it churned. Installed on the grass it could drill down hundreds of feet and find buried reefs or alluvial deposits. Drilling was cheaper and quicker than the old exploratory method of sinking a shaft, and by 1880 Victoria was using the diamond drill to probe for buried rivers of gold below the blanket of basalt.

Safety cage, diamond and rock drill, and dynamite were all introduced to Victorian mines between 1860 and 1880 and in time became indispensable. Nevertheless while Victoria led the continent in trying new techniques, most Victorian managers were slow to use them. When George Lansell and the mine owners of Bendigo sent Gustav Thureau to study mining methods in California and Nevada they learned in 1877 how backward were the Australian mines. Moreover, so few Australian managers bothered to read Thureau's praise of American practices that his sixty-page report is still on sale at the Government Printer's office for 2s. a copy.

The underground mines were slowly transformed but the mills that crushed the rock and extracted the mite of gold from the mountain of rock were more conservative. Victorians usually neglected metallurgy. The gold in most Victorian reefs was known as 'free gold'; the particles of gold were not intimately blended with particles of pyrite and other minerals, and even a slipshod millman could extract 90 per cent or more of the gold from rich rock. Few millmen had any knowledge of physics or chemistry. Few had enough mechanical knowledge to design more efficient mills and they let foundries design the stamp mills and the foundries adhered to the same design.

Metallurgists who tried to experiment were often thwarted by frugal directors, for Victorian gold directors really believed in directing their mines. The Lord Nelson mine at St Arnaud was managed from 1887 by a skilful Canadian with the exotic name of Zebina Lane (his son was Zebina too and also famous in mining) and Zebina decided he would extract more gold by installing Wheeler's pans. When the directors frugally halted his plan, Zebina installed them at his own expense. Within six years they had saved £30,000 of gold from the sludge.

Victorians went to manage new mines around the continent and spread their undeniable skills and prejudices. They spread the ideas that had transformed Victorian mining between 1860 and 1890 but all those techniques had been borrowed from abroad and adopted cautiously. And yet a remarkable change was coming. In world mining from about 1880 to World War I the metallurgical rather than the mining wing of the industry improved dramatically, and in that more complex and scientific arena Australia surprisingly emerged as the leader.

George Lansell was Australia's quartz king. He himself financed feats of mining which companies of a thousand shareholders would have been reluctant to try, and he won rewards which companies of a thousand shareholders would have been proud to share. He did not descend his deepest shafts, preferring to imagine the reefs his miners found. Square beard, balding head, stony eyes and tight mouth, his appearance suggested pessimism, but he had optimism and the money to give it play.

Before the gold rushes Lansell had worked as grocer and tallow chandler at Margate, Kent. On the voyage to Australia he shot an albatross, and passengers said his luck was cursed. In Bendigo he made money from tallow and candles, hay and corn, lost money in mining, then engaged experienced Cornishmen as consultants and won. He returned his large gains from the Cinderella and Advance mines into new shafts, won his first fortune from the Garden Gully mine in the early 1870s and spent £30,000 of that fortune in buying a rich mine he renamed Lansell's 180. The previous owner Theodore Ballerstedt retired to Germany and it was said that before he reached his homeland the mine had yielded the purchase price to the new owner. From one reef alone Lansell won over a ton of gold. Exhausting that rich reef he sank the shaft deeper. At a depth of 1,548 feet they found a dazzling reef, then probed still deeper. Each day he would arrive at the mouth of the shaft in a buggy drawn by two piebald horses and peer down. And the day came when he was shown gold-laced stone hauled up from one of the deepest reefs in the world.

The Victorian goldfields yielded more than half the metallic wealth won in Australia in the nineteenth century but Lansell was probably their only millionaire. Whereas most magnates retired with their first fortune to European or Australian cities, Lansell spent only seven years in London before he returned to Bendigo to live. He insisted that Bendigo should control its mines and in addition to the mines he owned he sat on thirty or forty boards. When shareholders grew impatient of paying miners to sink barren shafts Lansell rallied them, urging them to sink another hundred feet, then another, teasing them downwards with his optimism. He was one of the few Australian mining chiefs of his day who had a sense of obligation to an industry and the towns and cities that lived on it. Thousands of Bendigo families owed their livelihood to this man who, though often cunning, lifted mining above the level of piracy.

Lansell spent his last years in his rambling mansion 'Fortuna'. Surrounded by artificial lakes and lily ponds and the trees and flowers culled from five continents his house had four levels and frugal narrow corridors and stairs that led to spacious ballroom and music room, billiard room, dining room, private chapel, gymnasium, and swimming pool with deep marble baths. On pedestals and mantelpieces were bric-à-brac and specimens of gold carried from his mines, and on windows were painted homely slogans—'East and West Home is Best'. The long music room extolled in stained glass the 'Graces and the Loves which Make the Music of the March of Life', but the music he loved was the mine whistles at the changing of the shifts and the music of the stamp mills, and the important room of his mansion was not the music room with its grand piano and harmonium but the outbuilding that held his own stamp mill.

Other eastern colonies had their quartz magnates but none as rich as Lansell. B. O. Holtermann, a young German like so many quartz heroes, was a shareholder in a small syndicate that mined a series of narrow quartz veins in the switchback hills of Hill End, thirty miles north-west of Bathurst. In October 1872 the Beyer and Holtermann syndicate saw by candlelight a fantastic chute of gold in the slate rock, and with a rare sense of publicity they carefully cut away the largest mass of gold ever mined in one piece. It was not a nugget, just a lump of black slate and gold weighing nearly six hundredweight, of which one-third was gold worth about £12,000. That weekend thousands walked the packhorse track down the mountain to marvel at the treasure that stood four feet nine inches high, like a glittering anthill. Holtermann put on his best trousers and waistcoat and white shirt and stood proudly for his photograph, one hand resting on the crag of gold. Readers of newspapers later saw the portrait, advertising Holtermann's patent medicines. Holtermann himself saw it for the remainder of his life, for he had it copied in stained glass in a round window of his Sydney mansion that is now 'Shore' school.

The bonanzas enriched speculators and lawyers too. The brothers Somerville and Thomas Learmonth had pastured flocks on Ballarat's volcanic plains long before the diggers came. Rich with wool they were finally attracted to gold and paid £1,000 for the New Enterprise mine at East Ballarat in 1863. After eighteen months they sold out for £3,000 to the Llanberris Company. As the years passed the Learmonths were perturbed to see the Llanberris win £250,000 of gold—much of it from the mine they had sold. Perhaps for that reason they clung to their second mine at windswept Mount Egerton, fifteen miles east of Ballarat, long after the rich patches of gold seemed to have gone. They hoped the mine might again become rich, and read hopefully the periodical letters from their mine manager, William Bailey. By 1872, however, they were more interested in selling the mine if a buyer could be found.

Bailey the manager had lived amongst the Welsh miners on windswept Mount Egerton for seven years, a tall stooped Somerset man who, knowing sheep as well as mines, was understood and trusted by the Learmonths. Perhaps they trusted him too much. It seems that Bailey liked the mine's prospects and that instead of telling the Learmonths he may have told a mining speculator named Loughlin whom he met in Craig's Hotel in Ballarat. On 1 September 1873 Loughlin asked Somerville Learmonth if the mine were for sale. Accordingly Learmonth summoned Bailey to his homestead, Ercildoune, and asked him what he thought the mine was worth. Bailey thought £12,000. 'If I had a cartload of gold I would not give more for it', he said. Bailey was authorized to negotiate with Loughlin, and the mine was ultimately sold for £13,500.

Martin Loughlin was a handsome Irishman, with wavy hair and luxuriant moustache and a reputation as playboy and gambler in Ballarat. Once a Kilkenny baker he had worked Ballarat's deep leads, cutting his bread and meat with a clasp knife until a series of successes in his own claims and the shares market elevated him to a permanent seat at the dining table of Ballarat's most fashionable hotel. At the age of thirty-nine he spent most of his time in the bar of Craig's Hotel, or on the nearby kerb where the brokers bickered, and he had shares in many hotels and mines. He also had a large overdraft at the Union Bank, and when he decided to buy the Egerton mine he saw Williamson the bank manager and persuaded him to become a partner in the mine—possibly on condition that the bank's overdraft was extended. Another drinking crony, a Welsh-born sharebroker named Owen Edwards, also agreed to be a partner in the mine. But the unexpected member of this bar-room syndicate was William Bailey. When Bailey went to Ercildoune to farewell Learmonth and to receive his commission of £700 for selling the mine, he did not reveal that he himself had acquired one-fourth of the shares and that the mine was proving unexpectedly rich. Indeed he wept at leaving his old employer. Crocodile tears or genuine tears, they earned him the name of 'Weeping Bailey'.

The day Loughlin drove to Egerton to take over the mine was happy for the syndicate; rich gold was found in milky quartz in the Rose reef. Nothing so valuable had been seen in the Learmonths' day. Loughlin ordered the men to tear out the richer ore as quickly as possible, and more and more gold was won from the stamp mill on the hill. Whereas the yield of gold in the last week of the Learmonths' ownership had been less than 8 dwt to the ton, by the end of October it was nearly five times as rich. The syndicate known as 'The Egerton Ring' were to win profits of more than £320,000 from their small investment.

There were riotous nights in Craig's Hotel and angry nights at the Ercildoune homestead. The Learmonths, suspecting that they had been defrauded, employed a detective and solicitors. Members of the Egerton Ring argued that they had been lucky enough to buy a changeable mine a few days before it changed for the richer. The litigation went from court to court, a run of sensational hearings spiced by charges of bribery and the mysterious disappearance of the mine's books. The onus of proving fraud rested on the Learmonths; but as nearly all the evidence on both sides was circumstantial, they had little chance of proving their allegation, irrespective of its accuracy. Nevertheless the Learmonths persisted and were on the verge of appealing to the Privy Council when the case was settled out of court. The expense of the lawsuits was reputed to be £75,000, but the mine was worth much more. And the new owners kept the mine.

Loughlin and Bailey had another triumph in the 1870s. North of Ballarat were round heaving hills, bold and green, standing like nipples on a volcanic plain. Sleek horses ploughed the chocolate soil and the threshing machines were noisy on hundreds of pocket farms, and the farmers' towns of Kingston and Smeaton in the 1870s held some of Australia's largest agricutural fairs. The former government geologist, Alfred Selwyn, had once predicted that buried rivers of gold lay beneath the golden wheat and basalt bedrock, and in 1872 a golden river was found near Spring Hill.

Loughlin and Bailey rode in their buggy to see the new shafts in the wheat paddocks and decided that the buried river might flow north beneath the basalt. Forming a small syndicate of eight Ballarat and Creswick speculators they paid £36,000 for 6,000 acres of the Seven Hills Estate with its rich soil on the surface and the hope of rivers of gold below. They offered promoters the mining rights to strips of their estate in return for a royalty of 7½ per cent on all the gold mined. They helped float the Madame Ristori company, which sank a deep shaft through the basalt and broke into the river and paid out £232,000 in dividends and royalties. They helped float the West Ristori which struck the golden gutter and paid out £85,000; the Lone Hand paid £283,000, the Loughlin £70,000, the Berry Consols £393,000. But the diamond in the bracelet was Madame Berry which spent about £16,000 in penetrating the basalt to the river and paid £984,000 in royalties and dividends. In all, the gold under the Seven Hills paid over £2 million in profits, or a sum sufficient to pay Australia's entire bill for primary education in 1901. Loughlin and Bailey got much of that sum in royalties as landowners or in dividends as shareholders, and perhaps they used their inside knowledge to harvest a third source of wealth by gambling in the shares on Ballarat's Corner.

'Weeping Bailey' built an extravagant Ballarat mansion that is now St John of God Hospital, and bred merinos on the 50,000 acres of his Terrinallum station, and his racehorses twice won the Victoria Derby. Loughlin lived as a bachelor in the Esplanade Hotel at St Kilda and had his hair curled each day and his food served by his own personal servant, and his horse won the Melbourne Cup in 1885. The two gamblers founded a bank and learned in 1893 that banking was riskier than racing or mining.

Their rise to riches was unusual, but not their optimism. Most speculators think they will be personally favoured by Fortune and such mild vanity was one of the mainsprings of mining and therefore the nation's growth.

7

PACIFIC OCEAN GOLD TRAIL

HANDCARTS AND HORSECARTS, coaches and bullock waggons and leagues of leather boots had carried diggers to the Victorian rushes. Now the coastal shipowners had their turn. In the spring of 1858 ten or twelve thousand Victorians went by sea to the Tropic of Capricorn, to a goldfield known variously as Canoona, Port Curtis, Fitzroy, Rockhampton, and more blasphemous names. Forty ships left Sydney with impatient diggers, and shipowners made fortunes carrying men up the coast and fortunes carrying them back from the most disillusioning rush in Australian history.

The memory of Canoona did not halt an even larger crowd from rushing to the south island of New Zealand after a Tasmanian digger named Gabriel Read had found rich alluvial gold in 1861. In Melbourne heavy bookings for Dunedin were met by beautiful Liverpool clippers and tatty barques and brigs and tubs that had been idle for months on the Yarra River. By Christmas 1861 about 23,000 men had sailed from Melbourne and 6,000 had returned rich or grumbling. Two years later, after much coming and going, over 40,000 Australians were still in New Zealand, mostly on the goldfields.

Thousands of Victorians crossed alps or plains to new fields in southern New South Wales. That colony's gold output in the 1850s, if we prick its bloated official figures, was no more than 7 per cent of Victoria's but from 1860 to 1863 it was closer to 25 per cent. Moreover in those four years gold was more important than wool in New South Wales.

The first of new fields that enticed Victorians over the border was Kiandra. Men who pastured sheep on the summer grasses of the Snowy Mountains found gold nearly a mile above sea level, and in autumn 1860 the grasses were uprooted by the shovels and picks of at least five thousand people. Then the snow fell and gullywinds blew down tents and melting snow flooded claims in the creeks, driving most diggers and packhorse trains down the mountains. At winter's end some three thousand white and yellow men were perched on the mountains and the sixpenny *Alpine Pioneer and Kiandra Advertiser* announced with pride that the goldfield's only piano was played nightly at the Union Hotel on Broadway and the Empire Hotel was building a ballroom, and bowling saloon and drinking saloons were alive. But Cot Pulbrook, the versatile sharebroker, was selling coffee and chops—a sign that the snows had buried the lustre. The gravels in Kiandra's shallow buried rivers yielded less and less gold but today they again are glamorous, for they harden the concrete in the high damwalls of the Snowy Mountains plan.

While snow imprisoned Kiandra an American cook on a sheep run near Young washed alluvial gold in the lid of a billy. Four hundred men were digging there in September 1860 and four thousand a few months later, and Chinese swarmed down from mountain to plain to be bruised and pillaged in racial riots at Young. In the following winter there was a huge rush to a sheep run at Forbes, 240 miles west of Sydney, where miners found deep leads like those so many of them had worked at Eureka. The common estimate was 28,000 people at Forbes in 1862, but the people were so dispersed and so mobile that an accurate census was as impossible at Forbes as at most alluvial fields. Together the rushes far inland at Forbes and Young provided more than half of the £2·5 million of gold won in New South Wales in 1862, its record year.

The wave of rushes that began near Bathurst in 1851 completed a full circle north of Bathurst twenty years later in one of the last poor-men's rushes in south-eastern Australia. Fittingly the roaring days of Gulgong and Pipeclay and Home Rule in the early 1870s have been embalmed in the sharp Holtermann photos, the novels of the Gulgong gold commissioner Rolf Boldrewood, and Henry Lawson's short stories. Lawson as a lad on Pipeclay saw the generosity of luck-intoxicated diggers who handed children Chinese dolls and lollies bought in Chinese stores. He saw at night the rows of campfires and lighted tents diffusing a clear glow, saw in daylight red flags flutter above shafts that found gold in the wide buried rivers, saw tubs of gravel specked with gold and the police escort carrying away gold that in five years exceeded thirteen tons. Enthralled by the luck and lucklessness of gold mining he told of Peter McKenzie the old Ballarat digger who planted his wife and family in a St Kilda cottage and came to Gulgong. His party sank three deep shafts that were all barren, a fourth shaft that yielded each man one hundred pounds after paying expenses, a fifth shaft that they abandoned to Italians who drove a few feet and found the rich lead, a sixth shaft that was barren, and a seventh which they bravely named Nil Desperandum and won their fortune.

Lawson told of the rowdy songs in Granny Mathew's Pipeclay Inn and the rhythmic tapping of knuckles, pipebowls, and pannikins, and saw a heroic era vanish. Nostalgically he wrote that the nettles have been growing these twenty years where the bark inn stood.

Gulgong was not the end of the heroic age. From New South Wales the diggers moved north, following Queensland's mountain spine for nearly a thousand miles. When depression crippled Queensland in 1866 gold fashioned the crutches. James Nash in September 1867 found gold 116 miles north of Brisbane in timbered hills that rolled like a choppy sea, and timber splitters who had long known that some gold existed there were surprised at Gympie's riches. Queensland had not previously won £100,000 of gold in any year, but in 1868 the prize money of £600,000 enriched thousands of men. Gympie's rich white reefs were soon trailed by winding streets, fronted by shops with wooden cornices and flagstaffs like a row of fair booths, hotels with the wooden seams on the outside like Missouri riverboats and churches that welcomed all with the harmonium or the sign 'Diggers' Bethel' painted on the door. No bubbling ephemeral gold camp, Gympie's first decade was rich but its richest year was not until 1903.

Men too late to peg the reefs of Gympie rode the rough track to the riverport of Maryborough and boarded tiny steamers for new ports. Six or seven hundred miles to the north they landed on tropical beaches and crossed the ranges to perspire for gold. Ravenswood, now a ghostly town with redbrick hotels almost swallowed in a jungle of China apple bush, yielded gold in 1868. Four years later miners began to wring gold from stubborn sulphide ores at Ravenswood, the first gold escort was racing coastwards from the Etheridge, and Charters Towers was hammering gold from quartz with a quietness unusual in the first year of a goldfield of such magnificence—twenty years later it was Australia's largest.

The hurricane of discoveries towards the tip of tropical Queensland was not yet spent. Between 1873 and 1876 the Palmer River goldfield in the hinterland of Cooktown and the Hodgkinson goldfield at the back of Cairns were rushed and that fine prospector James Venture Mulligan had fair claim to be an instigator of both rushes. Coen goldfield high up Cape York Peninsula was goal for a miserable trail of optimists who went ashore at Port Stewart, and on slushy mountain tracks packhorses plodded through tropical rain to many outposts of the gold-finders. This prospecting was done quickly. Queensland in 1866 had no good goldfields but a decade later a chain of rich goldfields ran parallel to the Pacific Ocean for nearly a thousand miles.

Queensland's gold lodes were possibly as extensive and as rich as

Victoria's but were much dearer to mine in the isolation and trying climate of tropics and sub-tropics. Its alluvial gold was mean to Victorian eyes. In the nineteenth century Queensland did not match Victoria as a gold producer but its annual tally of gold far surpassed that of New South Wales from 1874 onwards. In half of the years from 1873 to 1906 Queensland's exports of gold and metals exceeded wool. And the impact of gold on Queensland's growth was powerful. The ports of Rockhampton, Townsville, Cairns, and Cooktown were made or magnified by mining fields; Brisbane and Maryborough, Port Douglas and Normanton, were enriched and enlarged by gold, and the whole colony was invigorated by the web of commerce that gold spun. In south China were hundreds of villages that fattened on the yellow diet, for numerically the Chinese dominated the Queensland goldfields in the seventies.

From Ballarat to the Palmer was halfway to the Equator, and a Chinese dragon stalked the European diggers all that way and nearly swallowed them in the steaming valleys of the Palmer. Chinese diggers had landed quietly in Victoria in the early 1850s and at first seemed just another strand in a polyglot thread. Coming from a few districts in Canton that had long sent men to Malayan tin mines and recently to Californian gold mines, they were not completely new to mining and some were skilled at sluicing for minerals. They worked hard, lived frugally, and some succeeded too well at gold digging for their own safety. Few spoke English and their contact with European diggers was slight. They were said to be exemplars of immorality, gambling, heathen practices, mysterious vices and rites, though similar failings had sometimes been observed in Welshmen or Americans or Italians on the goldfields.

The Chinese in Victoria barely exceeded two thousand early in 1854, but Bendigo thundered that there were two thousand too many. Within a year they said there were ten thousand too many, and the Victorian government restricted the number of Chinese passengers which each ship could land in Victoria and taxed also those who landed. Shipowners evaded the law by disembarking 1,500 Chinese at Port Adelaide, and with their goods in baskets or slung on long poles the Chinese walked overland to the unguarded Victorian border and reached the goldfields. The small port of Robe, being nearer the Victorian border, attracted sixteen ships from Hong Kong in April and May of 1857, and one ship, the *Young America*, landed 969 Chinese. In all 14,600 Chinese travelled from Robe to the Victorian goldfields before the South Australian government copied Victoria's anti-Chinese laws. The barrier fell too late; perhaps one quarter of Victoria's miners were Chinese in 1858. The goldfields had joss houses and Chinese barbers and stores and dens in which Chinese with pigtails lengthened by black braid smoked peacefully the pipe that had not yet been outlawed.

On most goldfields the Chinese camps were isolated and in a sense this protected them and in a sense it exposed them. In 1857, the Chinese, who had found the rich Canton lead at Ararat while walking overland from Robe, were brutally attacked and nearly deprived of their rich shafts. On 4 July 1857, day of American Independence, a small mob of white diggers on the Ovens goldfield drove the Chinese down the Buckland Valley and wrecked or burned tents and Chinese grocery and joss house. At the New South Wales goldfield of Young in the summer of 1860-1 two Chinese miners were killed and ten wounded in attacks led by armed vigilantes. The arrival of mounted troopers and artillery quelled the violence, but on the same field a few months later a rowdy mob numbering thousands marched on and destroyed a large Chinese encampment at Lambing Flat. Curiously the Chinese outnumbered the European goldminers in New South Wales by about 12,000 to 8,000 when these riots occurred; and belatedly the government in Sydney restricted Chinese immigration.

All along the Australian coast they locked the cage after the dragon escaped. When Queensland became the gold frontier the Chinese arrived on crowded ships from Australia and the Orient. Along the Palmer River in 1875 there may have been nearly 20,000 Chinese fighting flies and mosquitoes, murderous blacks and hostile Europeans. Even after Queensland had passed laws to stem the Chinese invasion the Chinese miners outnumbered Europeans. In numbers but not in wealth the Chinese were masters of the goldfields of tropical Australia from the Palmer to the Northern Territory's Pine Creek; and as this vast region was the spearhead of Australian prospecting their grasp of the goldfields seemed unbreakable.

In some years between 1858 and 1880 more than forty thousand Chinese worked on Australian mining fields, and most dug alluvial gold and a few won alluvial tin. 'They are good miners in shallow ground,' wrote R. Brough Smyth, Victoria's Secretary for Mines, in 1869. 'A Chinaman will stand for hours up to the middle in water scooping gravel from the beds of the streams, while his partners wash the stuff in cradles or boxes on the bank.' Few Chinese, however, worked in quartz mines or the company alluvial mines. They were not incapable of this form of mining—they ran the quartz mines in the Northern Territory—but most lacked capital for such mines. Above all so many Australian miners owned shares or sat on the boards of the first company-mines in Victoria and New South Wales that they could enforce a policy of employing only Europeans. When they could not enforce this policy by pressure within the company they enforced it by physical pressure. In December 1873 for example Peter Lalor, the hero of Eureka, and his fellow directors of Clunes companies, tried to break a strike by employing Chinese in their deep quartz mines, but the coachloads of Chinese they recruited in Ballarat were halted by an angry crowd on the outskirts of Clunes. This exclusion of Chinese from jobs in the company gold mines in eastern Australia was possibly more effective than all the government guards at ports of entry in reducing the strength of the Chinese on the goldfields; for the Chinese were thus prevented from working in the rising part of the mining industry.

The Chinese were even banished from new alluvial rushes in the part of the continent where they were strongest. The great rushes were moving around the continent the anti-clockwise way, and by 1873 they had reached north Queensland and the Northern Territory. For another thirteen years the spearhead of the ragged army halted in the north, and during those years the Australian colonies almost reached a common policy towards the Chinese, and when the anti-clockwise march resumed in 1886 at the tropical end of Western Australia bamboo pole and abacus were missing from the vanguard.

8

CHINATOWN ON THE ANTHILLS

WHILE CHINESE DIGGERS were filing across the steaming mountains to the Palmer, another province of China opened more than a thousand miles to the west. The lonely harbour of Darwin became a new Cooktown. Packhorses disappeared into long grass with stores for the goldfield. Ships rounded the cape with quartz captains from Bendigo. Facing the sea were the cool telegraph offices where operators tapped out messages to beckon the bold from every goldfield in Australia.

The telegraph had opened these goldfields of the Northern Territory. From the cliffs of Darwin a copper cable slipped into the sea and crossed to Java and Asia to join the telegraphic web of the northern hemisphere. From those same cliffs a wire spun south on a line of wooden posts stretching endlessly over the deserted plains and chiselled ranges of Central Australia and on to the coast of South Australia and the cities of the south. The men who spanned the continent with wire found the first payable gold in the Northern Territory. About a hundred miles south of Darwin the telegraph crossed a highly mineralized zone, and near the anthills of Pine Creek three men laying the telegraph line washed gold early in 1871; in August surveyor MacLachlan found five ounces close to the line. Only a few gangs camped long in that auriferous zone, and digging postholes in the sun few had energy or time to dig holes for gold. Most of the gold, moreover, was in quartz, locked from the grasp of men who lacked capital.

Reports of gold reached Adelaide, the distant city that governed the Northern Territory. To Adelaide came letters from Victorian miners, offering to search if well rewarded. Twelve miners wrote from Walhalla in the mountains of Gippsland, a Ballarat promoter wrote, and a Bendigo mine manager named John Westcott petitioned with influential Adelaide men. Cabinet refused Westcott's offer to report on the new gold finds, arguing that a gold rush might retard the completion of the telegraph.

Few ships sailed to Darwin. The trek overland was only fit for lunatics or heroes. A prospecting party needed heavy capital, but a few Adelaide men thought the prizes were worth the risk. Such men were E. M. Bagot, a contractor building six hundred miles of the telegraph, John Chambers, who had first pegged the copper leases that nourished the Great Northern scandal, and John Bentham Neales, who thirty years ago had floated Australia's first metal mining company to work Wheal Gawler at Glen Osmond. They formed with John Westcott of Bendigo the Northern Territory Gold Prospecting Association, and their printed prospectus invited £2,000 from the public of Adelaide and apparently got it. They argued erroneously that the Northern Territory was an extension of the rich fields of eastern Queensland, and had the dark basalt rocks they believed were essential hosts of gold.

The association waited long for a ship that would leave the sea lanes to call at Darwin. In the first week of February 1872 the brigantine *Alexandra* took on the eight horses, the arms and mining tools, and the provisions for six months and nine men. John Westcott, the leader, went on board with Captain Hulbert and J. Noltenius, his deputies, all experienced men who had dug gold in California and Victoria. A photographer stooped by his shrouded camera to record the fierce faces of men who might be famous a year hence. And then the ship disappeared down the bay, rode the swell off Cape Leeuwin in Western Australia, survived a storm in the far nor' west that shattered rudder and masts, and sailed into Darwin harbour on 11 March 1872.

Westcott went south with horses and men to find his new Bendigo. Following the trail of empty meat tins and medicine bottles along the telegraph, his men detoured into the sea of dry grass, chipping at reefs and shovelling shallow holes. It was the dry season—the climate was kind and the reefs gleamed. In August 1872, when the telegraph operators at the relay station at Pine Creek relayed the first message from Darwin to Adelaide, they found themselves in the midst of a proved goldfield.

The overland telegraph had opened the goldfields, and now was the magic instrument that raised the capital. The telegraph stations at the Shackle and Pine Creek, 121 and 151 miles from Darwin, had been built to relay the messages between England to Australia, and now they had their own messages to cable. Owners of Yam Creek mines paid dearly to send messages south offering their mines for £1,000 or more to Adelaide agents. The agent wired the goldfields warden enquiring if the claims were legally held by the prospectors. Reassured he wired the prospector at the Shackle that he would buy the claim, and then paid the purchase price into the prospectors' accounts in Adelaide banks. Owning a mineral claim he had never seen and might never want to see, he floated a company in Adelaide and recouped his outlay with part of the £2,000, £5,000, or more he raised from investors.

Miners on the field were puzzled to see the same men visiting the telegraph offices so frequently. The secretive operator at the Shackle was not puzzled, for he was said to send away £1,000 of telegrams in a week. 'In the whole history of mining in the colonies never before were claims floated by telegraph, without even a plan,' wrote a Victorian mining expert.

Northern Territory gold shares ignited Adelaide's share exchange. Sixty companies had been floated by May 1874 and £67,000 in cash had reputedly been paid to promoters and organizers alone. The stock exchange in 1873 was busy until after dark, and the adjacent café was crowded with cigar-smoke and speculation until midnight. Much money changed hands in Adelaide, small sums went north to explore the claims. The goldfields merely provided the counters for gamblers in a year of abundant money.

William Brackley Wildey, an expert, went north to report on the new fields that excited Adelaide. He had seen nineteen boxes of quartz from the Princess Louise mine at Yam Creek displayed on the counter of White's Rooms in Adelaide, and he was eager to see a mine that could yield 129 ounces of gold from less than one ton of stone. After a passage of twenty-three days he reached Darwin in the steamer *Gothenburg* and found a port without a jetty and a town without a school, and sheds of mangrove saplings for public offices. Those going on to the reefs were advised to take a copper-coloured wide-awake hat, Holloway's ointment, quinine and medicines, tent fly, hammock, mosquito net of unbleached cheese cloth, a Tranter revolver for the blacks and a breech-loading gun for the game. Heat and fever, hostile blacks, swarms of flies and mosquitoes, guarded the track to the goldfield.

A small steamer daily crossed the harbour to the Blackmore river, curving through the mangrove swamps alive with crocodiles. Southport was port for the reefs, and Wildey put his swag on a dray and walked beside it past Rum Jungle and past Dogherty's restaurant at Adelaide River, camping at night by billabongs and waterholes fringed with green grass, and marvelling by day at Gothic anthills and exotic vegetation. He was surprised like all visitors to see how the mines trailed the telegraph line as if the magic wire had some strange affinity for gold. In fifty miles the most remote mine was only twelve miles from the telegraph poles.

Coloured flags that marked the borders of mining claims blew cheerfully against a deep blue sky. Men trenched the ground in search of hidden reefs and blasted into rich gold. Drays hauled long boilers and engines from the port. Stamp mills were erected and doused with geneva. The summer storms filled the dams and the stampers at Pine Creek pounded down. Two days after Christmas 1873 the new *Northern Territory Times* published the long awaited cable: 'The Telegraph Company's first crushing has yielded three ounces three pennyweights to the ton.' The roar of the stampers blew with the wind to the lonely telegraph stations where operators cabled south incredulous news: 800 ounces of gold crushed from ten tons of stone at the Union, 609 ounces from seven and a half tons at the Bismarck.

The coloured flags on the mining claims faded. Companies ran out of money, unable to pay the price of isolation. The long sea route to Darwin inflated the price of stores and explosives and food, and for weeks in the wet season even packhorses could not get through.

There were no market gardens and no cattle runs to feed the gold camps. In November 1873 D'Arcy Uhr arrived from Normanton, Queensland, with cattle driven six hundred miles overland, giving the miners the luxury of fresh meat at luxury prices. Others followed his route across the wilderness and some arrived and some did not. William Nation lay beyond Daly Waters, his cattle dead, himself starved and exhausted, with £29 in coin and cheques and nothing to buy. Far from the telegraph he sat by his pack of bags and blankets and wrote his last message: 'Farewell, dear friends, all. I bid you an eternal farewell.'

With the mining frontier for once far beyond the pastoral frontier the miners suffered with the drovers. Many died of scurvy and unknown fevers. Their camps were littered with medicine bottles and spirits bottles. 'I found from actual experience,' said surveyor Goyder after a term in the Territory, 'that unless the men had a stimulant early in the morning that they could not eat their breakfast, and also that they could not eat their dinner unless they had first a stimulant.' Excessive alcohol, poor diet, heat and humidity weakened the miners. Companies paid them £5 a week but scores were happy to return to £2 a week at Bendigo and Charters Towers.

A few reefs were rich and all were narrow and blotchy. At Bendigo they would have paid well but at Lady Alice and Howley's they barely paid wages. Adelaide's mushroom companies had enough capital to explore near Adelaide but their capital was inadequate for the north. Their mines were burdened with secondhand machinery simply because they could afford no better after paying the sea and land freights to the goldfields. The Union company had a good reef, but the obsolete battery from Daylesford that crushed its stone was described by the chief warden as the greatest 'rattle-trap' in Australia. In 1875 the rattle-trap paid the Northern Territory's first dividend but by then the field was collapsing.

Most directors and mine managers believed that cheap coloured labour could save the mines. John Lewis, legal manager of many gold mines (and father of Essington Lewis, who was to be manager of Australia's biggest company) came down to persuade the government of the virtues of coolies. J. C. Coates, secretary of many gold companies, wrote on 24 April 1874 offering to pay £25 towards the fares of coolies if the government recruited them. Most Australians resented coolies, but South Australia's government relented. Desert separated Adelaide from the gold mines, so the coolies would neither be seen nor heard. Above all, without cheap labour, Adelaide's investment in these mines would be lost.

Captain Bloomfield Douglas chartered a ship in Singapore and escorted 186 Chinese and 10 Malayan labourers to Darwin on two-year indentures. The companies gave them daily helpings of Singapore rice and preserved fish and pay of eight dollars a month—about one-twelfth the wages paid to European miners—and complained it was too much. They complained that the coolies were lazy and stupid, did not understand orders in simple English (not surprising since they spoke no English). Darwin's mayor protested that Chinese were buried beside Christians in holy ground. When coolie Number 130 escaped, a company dutifully informed the government but was adamant that 'we do not want him back'. The coolies in turn complained, when a Chinese interpreter was about, of poor food and harsh treatment and unpaid wages. Amidst the backbiting one thing was clear: the coolies had saved the goldfields. When the two-year contract expired few of the Chinese labourers accepted the government's offer of a free passage back to Singapore. Some grew vegetables or worked mines. Alluvial diggings that were too poor for the white diggers sustained the way of life of the yellow diggers, and soon more men were washing gold than crushing gold.

The Northern Territory attracted Chinese diggers from other colonies. Queensland became alarmed at the dominance of Chinese on its goldfields and in 1877 began to harry and restrict them. The waning of the rich Palmer diggings in north Queensland likewise made Hong Kong shippers land their passengers at Darwin. In April 1879 the 3,400 Chinese in the Northern Territory outnumbered Europeans by seven to one, and the Government Resident at Darwin advised southern diggers to keep away.

At last the Territory had what its enthusiasts and sponsors had long craved—an expanding population. However, the people's skin had the wrong tint. Julian Tenison Woods, celebrated Adelaide geologist and Catholic priest, noted bitterly in 1886: 'At present the whole gold mining industry of the Northern Territory is bound hand and foot and handed over to the Chinamen. I have seen something of China and much of the Chinese, and I say we will one day regret any supremacy we give them.' From Adelaide the Minister for Education, Dr John Cockburn, denounced the Chinese miners as birds of plunder and passage who dug their gold and 'flew away to their Celestial retreat, and ended their days amidst the fumes of opium'. J. V. Parkes, Inspector of Mines, was incoherent with rage when he spoke of the Chinese: 'They are evaders of the law, liars, cheats, and everything that is bad, and as for cunning you can back them against any white man, and they become very insolent.'

The most curious allegation against the Chinese was that they mined dangerously and short-sightedly; that they honeycombed the reefs and ruined them for later companies. Pine Creek resembles a rabbit warren, wrote one Commonwealth geologist, and such strictures are common in the obituaries of Northern Territory mines. However, the main difference between the mining ways of the Chinese and Australians was that the Chinese made the reefs pay.

The Adelaide companies had called for the Chinaman in 1874 in the belief that the reefs were poor, but soon they cried that the Chinese were monopolizing and ruining rich goldfields. In 1881 a tax of £10 was imposed on all Chinese who should pass an imaginary line guarded by imaginary men a thousand miles south of Darwin. In 1886 Chinese were not allowed to work on new goldfields for the first two years after their discovery, unless by chance Chinese were the discoverers. As the Chinese were too fearful of the aborigines to venture far in prospecting, and were therefore unlikely to open new fields, they were in effect banned from new rushes.

Perhaps cheap transport could make the mines pay and enable the Territory to ban Asiatic miners. The government started to build a railway from Darwin to Pine Creek and by the middle of 1888 three thousand Chinese were navvies on the railway, and Darwin had a Chinese joss house, two Chinese shoemakers, three Chinese laundries, four Chinese cafés, five Chinese tailor shops, six Chinese gambling houses, seven Chinese brothels, and so on to thirty-nine Chinese stores. When the first toy locomotive came through the long grass with gold cakes in the guard's van, the Chinese mining princes gripped the mining industry even tighter. They owned by 1892 eight of the twelve crushing batteries in the Territory and most of the successful mines; and when the White Australia policy was made, Chinese outnumbered Europeans on the Territory fields by more than twenty to one.

The gold boom of the 1870s was a mistake and a tragedy, even if it did delight thousands of Chinese, settle the fringes of a vast area, and lay a stepping stone for the anti-clockwise march of prospectors around Australia. The Adelaide investors lost nearly all their money. Oluf Jensen won £30,000 from his own mine at Pine Creek and Michael Mullen employed Chinese and won £13,000, but most Australian miners lost money or health or life.

The Daly River copper mine tells the story of the Territory in one small vignette. A few Scandinavians, so often good prospectors in bad country, worked rich copper near Mount Hayward in 1884. On the morning of 3 September the Danish cook Schollert was pottering about the camp oven when suddenly he was attacked and killed. Landers was mining the lode with a hammer, saw a black with a spear poised, swung the hammer to defend himself but the wooden spear pierced his body. His mate Noltenius ran and was speared while running. Another prospector Roberts was dressing copper ore with four blacks and he fell insensible from a blow of hammer or club. When he awoke he removed the spears from the sides of his two mates. They began to walk towards civilization, but the two speared men were weak, and lay down to die. Roberts walked alone, hoping to meet along the track his mate Henry Houschildt, who had left a fortnight previously with packhorses to get supplies at Rum Jungle. But Houschildt had been speared beneath his mosquito net, then shot with his own revolver.

A police corporal rode from Yam Creek to investigate and punish, and returned well pleased with his new Martini-Henry rifle.

On 30 June 1886, twenty-two months after the murders, the protector of aborigines wrote a memo to the Government Resident in Darwin: 'The aboriginals sentenced to death for the murder on the Daly river are still in gaol, heavily ironed . . . Are they to be hanged or not?'

Ten years after the murders the government geologist visited the Daly River mine and heard through the bush the vibration of an engine, and found ten Chinese working the copper.

9

THE ANATOMY OF GAMBLING

HENRY JAMES O'FARRELL came from Dublin and was red-faced, excitable, and alert in his early thirties. He owned a hay and corn store in Ballarat and often in the afternoons of 1865 and 1866 he walked down Sturt Street to the open-air assembly known as the Corner, where the crowd often spilled from pavement to roadway and the jabber could be heard far down the street, and more shares were sold there than in any other share mart in Australia. O'Farrell seemed shrewd in his transactions, bought shares cheaply, sold them dearly, immersed in the daily excitement and personal challenge of a tipsy market. He speculated much in the Royal Saxon and Arrah-na-Pogue mines but in 1866 there was mild financial crisis in Victoria and most shares slumped. He lost heavily and drank heavily, got delirium tremens and lost his hay store, and drifted towards insanity. He went to Sydney and on the morning of 12 March 1868 he placed a pistol in his pocket and went with thousands to a picnic on the harbour shores to see Australia's most distinguished visitor, Alfred the Duke of Edinburgh, second son of Queen Victoria. That afternoon the mad gold speculator stepped forward from the crowd to within an arm's length of the duke and shot him in the back, the most noted attempt at murder in our history, and one sideplay of the casino that the Victorian mines inspired.

Share auctions were the nerve centre of the gold towns. Anthony Trollope noted Bendigo's vice was not drunkenness but share speculation. When the law of 1871 compelled new mining companies to list occupations of their shareholders in the *Government Gazette* it revealed to the doubters the colony's absorption in mining shares. One Inglewood mine was largely owned by Melbourne butchers. The South Extended Hustlers' Tribute Company included seventeen Bendigo sharebrokers and eleven professional speculators. Another company allotted shares in 1881 to Graham Berry, Sir Bryan O'Loghlen, and Dr L. L. Smith, three of the colony's leading politicians. But the share registers were more than a 'Who's Who' of wealth and prestige, because they listed thousands of humble carters, blacksmiths, clerks, travellers, publicans, coachmen, and miners. Share buying was the duty of all, for new mines depended on speculative money.

Australian finance is now centralized but was once dispersed. When the Linton Park Freehold Gold Mining Company was floated in 1865, investors could apply for shares to any one of twelve brokers (including J. B. Were) in any one of six towns from Happy Valley to Smythesdale and Melbourne. Towns which today haven't even a postmaster or schoolmaster had rialtos that gathered hundreds of players and spectators on moonlit nights. Ballarat and Bendigo had busier stock exchanges than Melbourne's during mining booms, and were probably always busier than Sydney until after 1880; Sydney did not even have a regular stock exchange until the Hill End gold boom of 1872. In contrast, Victoria then had at least six sharebrokers to every ten clergymen, and was not irreligious.

Gambling in shares was not surprising in a goldmining colony. Early diggers had gambled on finding gold with each swing of their pick, and when they followed other occupations they merely changed their gambling venue from a shaft lit by candles to a broker's office lit by gas. The trading banks financed mining speculation, and their ledgers from Woods Point to Golden Point were so red with overdrafts given speculators on security of their mining scrip and to companies on security of their directors' guarantee that the National Bank lamented in the mid-1860s that 'the Banks have taken the risk of working many of the Mines'.

Parliament tried to encourage and protect mining investors. Its problem was to limit the liability of each shareholder for the company's debts should the mine fail, but Ireland's and Pyke's and Frazer's acts did not answer the problem. A Melbourne merchant would buy £100 of shares in a £5,000 mining company. The company would spend its funds and then borrow from banks in the belief that by deepening the shaft it might finally strike the gutter or reef. Often the company wound up owning a barren shaft and £2,000 of debts. The creditors then moved in, only to find that most shareholders had cunningly passed on their shares to dummies—mysterious nobodies who were not worth suing or could not be found to be sued. Creditors then sued the honest shareholders, and the Melbourne merchant who owned only 2 per cent of the shares could legally be forced to pay 50 per cent or more of the defunct company's debts. Angus Mackay, a Bendigo newspaper owner and politician, claimed in 1871 that no mining investor knew where his liability ceased. Often he risked not only the money that he spent on shares but his whole estate. Thus the wealthy investor was suspicious of mining.

In the primitive state of company law, mining shares were often riskier than the mines themselves.

This trap for mining speculators was finally sealed in November 1871. Victoria passed its famous No Liability Act for mining companies, one of the most radical experiments in company law in the English-speaking world. A speculator could now lose only the money he had directly invested in shares. Once he decided the mine had no prospect of success he could forfeit his shares. This rule spurred the wealthy men to invest in mining. Likewise the No Liability companies' custom of selling their shares on time payment (they often sold £1 shares for 1s. deposit and instalments of 6d. a month) encouraged poor men to buy mining shares. The principle of No Liability was Victoria's revolutionary answer to the dearth of capital for mining, and the system spread throughout Australia. Ultimately it may have led to a dearth of capital because it encouraged mining companies to start with inadequate funds and so increased their risk of failure.

The evangelical Methodist chapels so strong in most gold towns preached against gambling but tolerated gambling in mining shares because it was vital to their towns. Mining scrip was so respectable that stock exchanges rivalled racecourses except when the Melbourne Cup drew near. 'The near approach of the Spring racing carnival,' said the report on the Bendigo stock exchange in October 1887, 'withdrew a good deal of capital.'

As the punter was more absorbed in the odds than the horses, so most mining speculators were fascinated by share prices rather than mines. Their ignorance of mining was preyed on by directors. Men fired gold from a shotgun into a barren reef to inflate shares. Directors erected crushing machinery on poor reefs to persuade investors that the reef was rich. Managers for months hoarded rich stone in dark corners and then extracted all the gold in one week and sent shares soaring. Directors of rich mines sometimes spread gloomy reports, made bear raids on the shares, and then bought them cheaply from the shareholders they had deceived. Sharebrokers employed spies in mines to glean first news of new discoveries, and on a day when a mine was expected to cut a new reef a messenger on horseback might be hiding in the scrub near the mouth of the shaft, waiting for a pre-arranged signal that would send him galloping to his master at the stock exchange. We cannot tell how common were these dodges, but they were common enough late in the century to inspire in some Victorian stock exchanges a code of ethics more courageous than those that prevail today.

Victoria's gambling edifice both reflected and encouraged an adventurous optimism not only in mining speculators but also bankers, importers, and the politicians who built state railways so freely and the Melbourne builders who erected some of the world's first skyscrapers. That Melbourne's land boom of the 1880s was wilder and more disastrous than Sydney's may partly be explained by Victorian gamblers turning from waning gold mines to a new and faster set of dice, blocks of city and suburban land. Curiously San Francisco, the rialto for the mines of California and the Rockies, began to swap mining scrip for land titles with the same abandon about the same time.

Australians owned nearly all the mines in the continent until the late 1880s. In the herculean task of financing tens of thousands of shafts and thousands of crushing mills, pumps, and smelters, Australians got little financial aid from abroad. Although Great Britain was the world's wealthiest nation and lender and a generous financier of Australian public works and companies, she rarely invested directly in Australian mines. Her more speculative investors preferred American railway shares, for they were usually safer than mining shares and yet had a tinge of the spectacular. When Britons bought mining shares they preferred mines in North America, Spain, Sweden, and their own Cornwall. The Australian mines were too far away to entice many London investors. The price of shares had to fluctuate constantly if they were to lure the speculator and, as the price of gold was fixed, gold shares could not fluctuate freely without regular news from the mines. If Australian news of rich crushings and new reefs only reached London slowly and irregularly, the shares naturally could not fluctuate enough to attract the speculator. London's share market needed quick regular news from Australia's gold mines, and the linking of Australia and England by overseas telegraph in 1872, and cheaper telegraphic rates in the following decade, at last made Australian gold shares a promising gambling counter for the British speculator.

London was now ready to float Australian gold mines into companies but first it required promising mines. Victoria was still the richest gold colony in the British Empire, and its promising gold mines easily raised their capital on the local stock exchanges. On the other hand Queensland, second to Victoria as gold producer, was hungry for capital. The cost of finding and opening reefs was higher in the tropics. The nearest city, Brisbane, was a thousand miles to the south and lacked a strong mining market. Melbourne was two thousand miles away and seemed reluctant to gamble in North Queensland mines. London's money bags therefore tempted Queensland goldmen who wanted capital for their mines.

Charters Towers was Queensland's great gold town in 1886 and when the band played Saturday nights in the warm air the streets were dusty with strolling people and rowdy with the cries of brokers and picmen. Nine thousand lived on the field, and the inflow of people and output of gold had risen steadily since the railway arrived from Townsville. Mine owners vowed gold output would soar if more money was spent in sinking deep shafts and in building mills capable of extracting more gold from the troublesome sulphide ores. The Queensland government was fanatical in its zeal to develop the land and sensed that Charters Towers' gold was sweet bait for the British capitalist.

In 1886 the Colonial and Indian Exhibition was staged in London. Trade fairs were as popular as music halls, so Queensland arranged a lavish display of its wealth. Four Charters Towers mines sold to the Queensland government over a hundred tons of golden stone, which was shipped to London and crushed in a small stamp mill in the crowded hall. Five million people saw the London exhibition and the deafening noise of the stampers attracted crowds to the Queensland court. The novel display, noted the *Mining Journal* on 11 December, 'opened the eyes of British capitalists'. It also opened their purses, and in 1886 money was cheap and plentiful in England.

Thomas Mills, rich Charters Towers mine owner, had gone to London with his town's exhibits. He was a large shareholder in the Day Dawn Block and Wyndham Gold Mining Company which had sent rich stone to be crushed in the exhibition mill and a £6,000 cake of retorted gold that stood like a crown above strolling crowds. The rich mine's publicity stirred London sharebrokers. They offered to float the mine in London on alluring terms. Mills and the Queensland shareholders would get 419,000 free £1 shares in a new company and some of the £41,000 raised by selling the remaining shares to English investors. Mills agreed, and the mine was floated in August 1886.

Many Charters Towers people were so astonished at the price the shares commanded in London that they sold their own at immense profit on their original outlay. English investors were eager to dabble in the exciting novelty of Australian gold shares and the number of shareholders in Day Dawn Block multiplied by five in three months. Englishmen bought more heavily when the company declared its first quarterly dividend at the rate of 20 per cent a year. They paid so dearly for shares that they never got adequate reward in dividends from the half-million ounces that this amazing mine was to win, but in the excitement of 1886 they were chasing quick profits on the stock exchange, not a long line of dividends. Such was the London mania for gold shares that Charters Towers promoters took their leases to London and sold worthless mines for huge sums. Twenty-seven Queensland mines were floated in London before the boom crumpled in 1887.

Charters Towers was flushed with the money that British companies were spending in the town and Brititsh investors were paying to its pioneers. After breathing optimism into British investors their own pride and optimism swelled. They built offices and shops and banks and stock exchange that seemed palatial in a remote provincial town. Their sharpest stockbrokers employed thirty or more clerks and touted business in distant cities, calling on investors to send money for shares they didn't even bother to name. D'Arcy and Company advertised weekly in the Melbourne *Australasian:*

GOLD. NOTHING BUT GOLD
The Greatest Gold City in the World
Very Soon will be Charters Towers, Queensland

It certainly became the most productive goldfield in Australia (1891-6), and the second biggest city in Queensland, but such achievements were mere signposts to the men of Charters Towers. In conversation and print they called their field 'the world'.

This was the first Australian field to entice much British capital. It proclaimed to Australian mining promoters that British sovereigns might be more profitable than gold below the ground. Australia was soon fostering a rich export trade in parchment and paper mining leases and share scrip, and London replaced Melbourne as the casino of Australian mining.

PART II

Copper Men and Silver Kings

10

GREEN MOSS

SOUTH AUSTRALIA, rich in copper, dreamed of gold. Thousands of South Australian carters and miners and labourers rushed to Victoria to dig gold in 1851 and some returned to find poor gold in scattered gullies in their own hills. Adelaide sent shiploads of flour and wheat to Melbourne to feed the diggers, exchanging loaves for gold but still envying the colony across the desert. At last it called for the soothsayer, and Edward Hammond Hargraves swaggered into Adelaide at the end of October 1863, climbed into his spring cart, and rode into the hills followed by three horsemen and a heavy dray. From the steep gullies of Cape Jervis in the south to the dry ranges beyond Port Augusta he slowly made his way, complaining his jaw was apain with rheumatism, noting the rocks, and sinking shallow holes in search of gold. His instructions were to go near the Barrier Ranges, where twenty years later the mining field of Broken Hill was born, but he refused with specious excuse. His final report had much about the grasses and his horses, and was rightly pessimistic of gold.

South Australia has proved the poorest in gold of the Australian colonies but its mining history was still written in gilt. When Victoria led the world in gold South Australia in some years mined one-tenth of the world's copper. Its Cornish captains and miners were enticed away to open mines throughout the land. Its investors financed mines thousands of miles from Adelaide and even across the seas, skimming the cream from Kalgoorlie and the milk from Broken Hill.

The long tongue of sea that poked into the hinterland of South Australia shaped much of its human history. Those gulfs floated ships to the fringe of land that was rich in copper and ideal for wheat; and copper ores and wheat need cheap transport. Furthermore, the hot, dry climate that ripened the grain on the stalk matured the deposits of copper. The original deposits of copper at Burra and Kapunda and many scattered points in the hills were poor, but through countless millennia the gentle rain and frost and heat had leached the sulphur from the ores and oxidized them, enriching copper near the surface. The richness of the copper ores and the closeness of the sea made mining pay.

The last of the Australian colonies to be settled, South Australia produced the first metal mine. This riddle can be simply explained. Metal deposits were close to the main port of Adelaide and so were soon seen and easily developed. Amongst the settlers of that paradise of civil and religious freedom were miners from Cornwall, then the world's centre of metal mining, who knew minerals and how to work them. Above all, the new colony in 1841 was depressed. The wealthy had no wealth except on paper and the workers had no work except on the roads. Therefore they had strong incentive to mine the silver-lead which had previously been ignored. There seems a consistent link between drought or depression and the opening of new mining fields.

In Adelaide at the start of 1841 the people who sat on the shaded sides of their houses on summer evenings may have seen smoke rising to the east. On a steep hill only four miles from Adelaide smoke from burning grass signalled the birth of Australian metal mining. Two Cornish miners, 'persons in the humbler walks of life', found a lode of rich galena jutting above dry grass, and set fire to the grasses to expose the hidden outcrops. Their names were Thomas and Hutchins; their Christian names have long been forgotten.

Adelaide men walked four miles to Glen Osmond to see the Cornishmen's shallow hole. Governor Gawler went to see the narrow lode of silver and lead, pleased that it was named Wheal Gawler in his honour ('wheal' was Cornish for mine). Adelaide men floated a company, issuing free shares to the discoverers and sending forty boxes of hand-picked ore by sailing ship *Cygnet* to London. Exports of mineral for the year were worth £390, an augury more than a reward, for the company and its funds vanished.

Scores of galena lodes intersected the steep hills of Glen Osmond but they were narrow and erratic, dear to mine. The ore had to be sorted to eliminate waste rock, and though ores sent abroad often carried 70 per cent lead they can have hardly paid. The field struggled until 1847, when a new London company sent out a Cornish mine captain and ten miners with 'nearly a score of chubby children' to the old Gilles mine. Forty or fifty men mined lead ore and the chubby children sorted it, returning at night to brick cottages the company had built. At nearby Wheal Watkins, one visitor vowed the lode was like a silversmith's shop. Wheal Gawler, managed by a German from the Hanoverian government's mines, was still about to prove its richness. All was vigour by 1848 with furnaces at Adelaide and the glen pouring out molten lead, and new mines opening in the hills.

South Australia had deserted villages before Victoria knew deserted

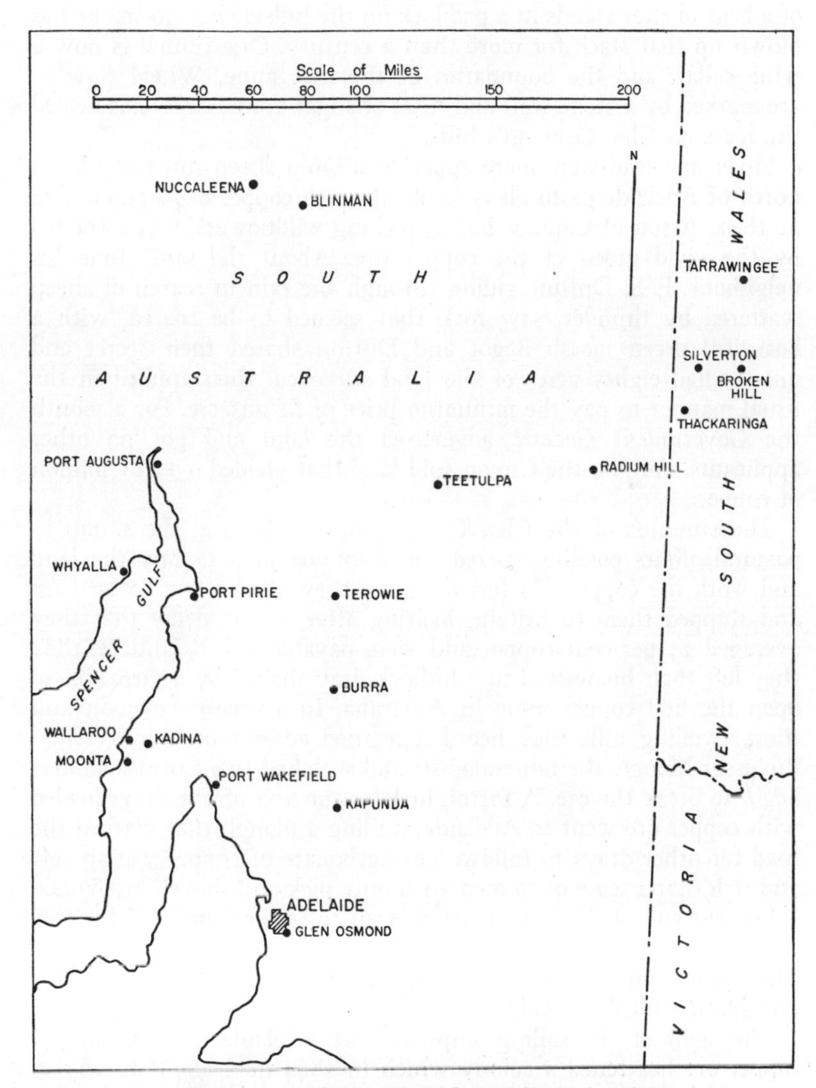

3 South Australian and Broken Hill mining fields

gold camps. Glen Osmond was one, the cradle of metal mining. It shipped its first metal in 1841 and was almost deserted in 1851. Motorists who follow the bitumen from Melbourne to Adelaide pass by the toll gate of this first mining town as they descend the hills into Adelaide, but few relics of the mines survive. The stone chimney of a lead smelter stands in a paddock on the hillside but no smoke has blown up that stack for more than a century. One tunnel is now a wine cellar, and the boundaries of the first mine, Wheal Gawler, are marked by a stone wall and rows of sugar gums. Grass has healed the scars on Glen Osmond's hills.

Silver and lead were mere appetizers. On a sheep run forty miles north of Adelaide pastoralists found the rich copper of Kapunda late in 1842. A son of Captain Bagot, picking wildflowers, was attracted by the vivid green of the copper ore. About the same time his neighbour, F. S. Dutton, riding through the rain in search of sheep scattered by thunder, saw rock that seemed to be coated 'with a beautiful green moss'. Bagot and Dutton shared their secret and quietly had eighty acres of the land surveyed, then applied in the usual manner to pay the minimum price of £1 an acre. For a month the *Government Gazette* advertised the land and got no other applicants. For £80 the Crown sold land that yielded over £1 million of copper.

The stimulus of the Glen Osmond discoveries and the slump in pastoral affairs possibly stirred the Kapunda men to buy the land and work the copper. To test their ore they filled a few dozen bags and shipped them to Britain, hearing after a long delay that they averaged 23 per cent copper and were payable. On 8 January 1844 they left their homestead in a bullock dray shaded by a tarpaulin to open the first copper mine in Australia. In a quaint ceremony on those swelling hills they heard a learned address on mining from Professor Menge, the mineralogist, and watched two Cornish miners begin to break the ore. A fortnight later the first of five drays loaded with copper ore went to Adelaide, pulling a plough that marked the road for other drays to follow. The carbonate of copper was so soft and rich that a score or so men with only pick and shovel dug £7,000 before the end of the year. Dutton went to London and sold his one-fourth share in the mine to an East India merchant house for £16,000. The other owners gave out no figures of their profits, and public imagination filled the void.

The sight of the sailing ships at Port Adelaide loading bags of copper ore heartened a colony which in 1844 held less than 20,000 people. Migration is blood and bone to new lands, and South Australia could attract few people when its only primary wealth was a cluster of farms and struggling sheep runs. Copper bred new hope. Auctions of Crown land containing copper lodes drew excited spectators and extravagant bids. Companies were floated on a froth of excitement and a slender vein of metal. Carters were alert for copper in the hills and shepherds walked with eyes on the ground.

Most adults and many boys could now recognize copper ore even if they could not pronounce such cupriferous words as chalcopyrite and malachite.

In June 1845 a shepherd named Streair came to Adelaide with handsome specimens of copper ore and offered to reveal the site to a syndicate for £6. His offer refused, he went to a shop in Rundle Street and got £8. That lode was in the wilderness eighty miles or so north of Adelaide, and the shopkeepers sent surveyor Finke to see the lode. Finke had managed a lead mine at Glen Osmond and at once he saw that the distant copper lode was infinitely more promising. The copper specimens he carried to Adelaide inspired investors, bankers, miners, surveyors, shopkeepers, merchants and the owners of the Kapunda mine to visit the new lode. There they heard from a shepherd named Pickett of another rich lode. Going north, they saw a 'bubble of copper' jutting from the hill. The bubble was Burra Burra, a copper bonanza.

Both copper lodes lay beyond the surveyed districts and therefore could not be bought at auction in the customary 80-acre lots. There remained only one way to buy the few acres of ground that encased the copper. The buyer had to select a block of 20,000 acres, or slightly over 31 square miles, and buy the lot. The price was £20,000 in gold coin. And the colony hardly possessed that coin.

Two rival syndicates plotted for weeks to find the £20,000. Named by a wag as the Nobs and the Snobs, neither could raise enough money. Finally in August 1845 they pooled their money and bought the land, cut it in two, and drew lots. The Nobs, amongst whom were Bagot and Dutton of the rich Kapunda, drew the southern half and began to explore their Princess Royal lode. Slowly they realized they had spent their huge sum on grazing land, not a mine.

The other party, the Snobs, combined with the South Australian Mining Association to work the Burra Burra lode. Their mine was rich to the grassroots. Dr von Somner, the German manager, marvelled at the size of the deposit and Dreyer, the German ore dresser, swore by its richness. Men from Cornwall were dazzled by the beauty and delicacy of the Burra ores. Adelaide shopkeepers who owned one third of the mine's shares showed in cabinets and on counters specimens of ore which the world's museums would have prized: red oxide of copper, glossy green malachite, crystals of sea-blue carbonate, and virgin copper and ruby ore. In those hot hills the miners were plundering a paradise of minerals.

After one year the Burra mine had one mile and a half of underground galleries, their rocky walls glittering in the candlelight. Rich ore was hauled up many shafts by horses circling the whim. A regiment of small boys sorted and bagged the ore for the bullock teams that arrived daily from Adelaide. At times the track to port became the busiest upcountry road in Australia, for all ore from Burra had to be carted ninety miles to Adelaide and shipped across the world to Welsh smelters.

Transport was a nightmare for the Burra company. In winter the track became a bog, impassable for heavy drays. At height of summer carters could not be hired, for they reaped on their farms or carted wool and grain. Traffic was centred on autumn and spring and Henry Ayers, the mine secretary, guessed there must have been 1,200 drays on the road at the busiest time of the busiest year. Each dray took only 2½ tons and as the return journey often took a month—or much longer if feed for bullocks was scarce or the carters fell in at wayside shanties—the haulage problem was acute. The company itself often repaired the road, clearing tree stumps and widening crossings over the creek, and even organized hotels and hay-lofts along the track. Then, of course, there was the constant task of disciplining loafing or cheating carters such as 'The Infant', who stayed so long in a wayside hotel that copper was stolen from his dray, or the carters who neglected to tie the tarpaulin over the load. Burra ore quickly absorbed rain, thus gaining weight, and the company had to pay sailing ships to take that useless moisture to Wales.

The average ton of copper ore mined at Burra contained about one-fifth copper. Steam stamps or strong men with hammers broke the lumps of ore raised from below, and old men and young boys—Ayers called them 'little bits of boys'—picked out the lumps of rich ore and bagged them. They shook the poorer pieces of ore on sieves in a flow of water; the lumps richer in copper, being heavier, tended to sink, and the poorer pieces were carried away in the stream. This ancient way of gravity concentration only eliminated a fraction of the waste in the ore. Water went so far, but fire was the great purifier of copper ores. The smelting of one ton of ore usually required more than a ton of coal, so it was cheaper usually to carry the ore to the coalfield. Cornwall shipped its copper concentrate to smelters in Wales, and South Australia in the 1840s copied Cornwall, shipping its copper ores and concentrates as ballast on the wool ships to Wales.

Early attempts to create a local copper smelting industry floundered. The Burra mine engaged Germans to build smelters at the mine, then abandoned the venture. Edward Davy, physician, chemist and journalist, who in England a decade earlier had invented the method of relaying the telegraphic message, tried to smelt copper ore with charcoal at Port Adelaide. Schneider & Co., which had smelters in Cuba, finally built smelters at Burra, and formed the strong smelting company known in turn as Patent Copper Company and the English & Australian Copper Company. From 1849 Burra had its white chimney with smoke blowing like a horse's mane on windy days as a beacon to teamsters hauling coal from Port Adelaide or to mule teams that later carried coal from the company's new port of Port Wakefield.

In the Cornish manner the South Australian Mining Association was stern godfather to its employees, and Burra became a small welfare-state. The company built rows of cottages with stone walls and earthen floors for many of its men. The company imposed compulsory social insurance, levying sixpence a week from the wages to finance a health scheme a century ahead of its time. The company's surgeon visited the sick, the company's hospital nursed them, and the mine paid the equivalent of half wages to the families of sick and disabled men. The people of Burra worshipped in chapels on land donated by the company, ate meat slaughtered in the company's abattoirs, drank water from company wells, brought their relatives from Cornwall at the company's expense, and were buried in the cemetery at whose gate the company's sexton presided. In an age of *laissez faire* Burra was the most benevolent company in the land—too benevolent in the directors' eyes. When the town in 1850 was short of chaff and the surgeon's horse was eating the company's fodder, secretary Ayers protested: 'We are not Livery Stable Keepers,' he said.

In 1851 at least five thousand people lived around the Burra mine. The visitor arriving at nightfall, seeing the crowded pavement outside the Smelters' Home or the Bushman's Home, and seeing the miners filing along the track from the mine, wondered where they lived. The main company town was Kooringa, but beyond it were the villages of Redruth, Copperhouse, Aberdeen and Hampton, and hidden from sight was a larger settlement along the creek. There nearly 1,800 people lived in caves they had excavated in the banks of the creek, with weatherboard fronts to keep out the weather, and dug-out chimneys down which foraging goats sometimes fell. Burra's cave men were washed out by floods in winter and their children died of fever in summer. 'You must not incur one farthing more expense than is absolutely necessary in affording cottages for the people washed out of the Creek,' wrote the secretary to the chief captain after a flood in 1850. Another flood in the winter of 1851 produced an ultimatum on the door of the blacksmith's shop at the mine, warning that at the end of the year the mine would employ no person who lived in the caverns.

The paternal owners of Burra were not paternal to interlopers. Their own rich mine lay on the boundary of their land, and promising deposits lay just outside the boundary. The Burra company refused to open new workings in that direction lest they find copper that excited the adjacent Bon Accord Company to search for it too. Any find in the big mine would soon be known, Ayers lamented, 'for we cannot expect to keep our miners quiet', and when he learned in September 1850 that the captain of the Bon Accord was revisiting Burra, he refused to let him go underground in the Burra mine. The Burra wanted a copper monopoly, and the sooner the Bon Accord gave up the search the better. Likewise to the south the Burra company wanted all minerals on Crown lands, and before the auction of one promising lode in 1850 it posted Captain Bryant at the land 'to keep a very sharp lookout for every arrival, to observe great secrecy himself and get all the information he can from others'. The company bid high for 800 acres and christened their mine Karkulto. All they got for their tight-lipped vigilance and £30,000 were a few tons of hungry ore.

Who could blame the Burra directors for thinking every bald hill was a treasure house? Their own mine was known throughout Australia as the Monster Mine. It had produced over £750,000 of copper before Victoria had produced a quart-pot of gold. Its dividends were huge and those who wanted them often paid over £200 for a £5 share. Clerks at the Customs House knew that Burra alone was exporting more wealth than the united tally of wool and wheat. The company's record output of 23,000 tons of ore and concentrate in 1851 would seem trifling to a modern miner, but over a thousand men and boys were employed in mining and dressing that ore and in some months another thousand smelted or carted it; no Victorian gold mine ever paid as many men as the Burra. From 1845 to 1848, recalled Ayers with a touch of exaggeration in his pride, 'you might safely say all South Australia was indirectly employed by the Burra mine'. Burra made South Australia the most prosperous of the colonies in the late 1840s, and its wealth and work attracted immigrants and ships (shipping increased ninefold in six years), bustle and optimism. Pickett, the shepherd who found the mine, now collected from the company a weekly pension, and before he died of accidental burns and was carried to burial near his mine he might have mused on the magic of metal.

The men in canvas trousers and flannel shirts who climbed down the dark ladderways were often restless. Some went to the Californian goldfields and Ayers predicted that they 'may wish themselves back at the Burra again before a year's over'. Men left for Ballarat in the wet winter of 1851, and were replaced by new men from Cornwall.

By the spring, days were warm and tracks were firm and Burra talked longingly of gold. Ayers in Adelaide anxiously wrote to the chief captain: 'How are the men affected by the late news from Melbourne?' The men were very much affected; and the company had to enforce the Cornish custom of not paying the tribute money to miners until a month after the completion of their contracts in order to keep them at the mine. At Christmas the underground captains said their farewells on the steps of the chapels, and the most profitable year in Burra's long life closed with most caves in the creeks deserted.

The wheels of the steam engine at Burra, the largest machine in the colony, ceased to spin. The water rose slowly in the deeper workings as the throb of the pumps was silenced, and the company mined only shallow ground and employed only old men and small boys in the dressing sheds. 'Thousands are now on the wing,' wrote banker Tinline in Adelaide in December 1851, and so many of them were flying east on stolen horses that more police were posted by the ferry at the river crossing of Wellington. So many thousands hurried east to dig or buy gold that South Australia was starving for coin and its main bank was in peril. In March 1852 South Australia began to escort gold from the Mount Alexander diggings four hundred miles across the desert to Adelaide where it was cast into ingots and even gold tokens to underpin the paper currency. In less than two years £1·2 million of gold was thus carted with police escort to Adelaide, and more came in the purses of returning diggers and the chests of merchants. The author of this scheme admired the gold plate given him by grateful colonists with its engraving of perplexed gentry and deserting ploughmen and hurrying gold escort, but as a banker he knew his plan was a poor alternative to the finding of payable gold in South Australia.

Men still suspected the Adelaide hills were rich in gold. After all, gold from the Montacute copper lode had been displayed in the Crystal Palace in London in 1851, and grains of gold from Balhannah had been passed around cupped hands in Adelaide. At the farming village of Echunga, men were alert for gold and a government reward in 1852, and Chapman found gold close to the Wheatsheaf Inn. His friends filled bags with washdirt and carted them by dray to ponds, where they washed an ounce of gold from each bag. When they took the gold to Adelaide in August 1852 the Colonial Secretary and fifty or sixty horsemen excitedly followed them to the scene of the find. There, however, they could see no gold. The Colonial Secretary suspected Chapman was a rogue and demanded he dig gold. Suspecting he would quietly drop gold into the dish while he swirled it, he ordered the magician to take off his coat and roll his shirt sleeves. Chapman obeyed and when finally the dirt and clay had left his pan the gleam of gold was seen. The unbelievers shouted so loudly that horses tied to nearby trees broke their bridles and galloped away. In the excitement men used pannikins and kettles and even their own hats to wash the dirt. At least a thousand men came and went, some with gold, in the last months of the year. But the diggers who scorned Echunga were wise, and one named George Lansell who carried away his swag was to win at Bendigo more gold than all South Australia won in the following fifty years.

Burra and Kapunda resumed their mining of copper, but could not last forever, no matter how sharebrokers and politicians brayed. The colony wanted new mines. The Burra company itself sent men into the red ranges north of Port Augusta in search of them. The drowsy port at the head of Spencer Gulf supplied sheep stations that stretched north into low hills and saltbush, and sometimes the wool carts brought in copper ore. But the ranges north of Port Augusta were dry and the trees offered little fuel for smelting. Long waterless stretches often prevented pastoralists from carrying rations to their own sheep stations, let alone taking ore to the wharves of Port Augusta. There was so little grass for the horse-teams that Captain Rowe, a miner since the age of eight, declared the mines could only be worked 'if Providence was to smile upon the north . . . and give us a little more feed'. Providence did not smile, and the fifties ebbed in drought.

Providence smiled instead on the shrewd. James Chambers ran thousands of sheep in this harsh country and was eager for fat that mutton could not provide. He and William Finke, the manager who had first reported on the Burra in 'forty-five, pegged an outcrop of copper ore about 130 miles north of Port Augusta and called it Nuccaleena. One lump of ore on their lease weighed a ton and with an eye to publicity they named it the Malachite Nugget. Dray after dray left the mine loaded with copper ore, and sixty tons sent to Wales assayed 27 per cent copper. Even that rich ore yielded scant profit after paying the miners, draymen, wharfmen, shipping and smelting companies.

It is usual to say that the mine wanted capital, but perhaps Chambers and Finke wanted it. Hoping to sell the mine in London they offered an Adelaide merchant and politician, Captain Hart, a commission if he sold it. After lavish banquets Hart sailed with Finke in 1859. In London they met two Adelaide mining speculators, Paxton and Baker, who bought the leases and then resold them at a profit to another syndicate. The latest owners had recently floated the North Rhine Copper Mining Company in London to work a poor copper deposit in South Australia and knew the art of beguiling British investors. The high price of copper favoured them. The bank rate in England was low, suggesting money was plentiful. These portents were not enough and the promoters wrote a vivid prospectus for their Great Northern Copper Mining Company of South Australia. They announced that John Morphett, a director of the rich Burra mine, would sit on the Adelaide board of their company; unfortunately Morphett did not know of his appointment. They announced that South Australia was inclined to build a railway to the mine; but South Australia was not inclined. They implied that a recent speech by the Governor of the colony on the 'discovery of extensive and valuable mineral deposits to the north of Port Augusta' referred to their mine; but his speech did not. And they claimed that the Burra men had offered a large sum for their mine, when in fact the Burra men had not.

In the days before the telegraph, lies were not quickly refuted. The prospectus was advertised in the London press and in three days investors paid £49,250 for 61 per cent of the shares in the mine. In three days the promoters collected £25,000 of this sum for their own pockets and a packet of shares that were worth even more.

Months went by before Adelaide heard that the name of its Governor, its parliament, its richest mine and one of its respected citizens, had been taken in vain. In November 1859, Governor Sir Richard MacDonnell decided to see the mines that had produced the famous 'Malachite Nugget'. In the heat of summer he rode through dusty ranges to the mines and found them deserted. A shaft sunk through clay had no trace of ore. A few shallow holes had given up rock with faint traces of copper. The puzzled Governor informed the Duke of Newcastle that the shares were not worth taking as a gift. For shareholders in London the news was too late.

However, the month of the Great Northern scandal marked a new era for the copper seekers. South Australia's largest copper field was found by a burrowing wombat and a shrewd shepherd within a day's coach ride of Adelaide.

11

NEW CORNWALLS

FORTY MILES across the water from Adelaide a leg of land jutted south. At the top of the leg the landscape was flat and dreary with hillocks of sand rippling the surface. No river or creek cut through the plains to the sea. In winter the grass was green and sweet and kangaroos grazed and lizards scuttled over the stony earth, but in summer there were no pools or waterholes for the sheep. It was poor land for the pastoralist and mean land for the prospector. A blanket of earth and rubble covered the ancient rocks and minerals of Yorke Peninsula.

Captain Walter Watson Hughes ran his sheep across the thigh of the leg of land. From his homestead near the beach at Wallaroo he saw the placid sea, and to an old sea captain the view was perhaps nostalgic. Hughes was a Scot and as a boy had been apprenticed to a cooper and had made barrels for whale oil, spending much of his youth whaling in the Arctic. He preferred sailing to carpentering and became ship's mate and then owner of the brig *Hero*. He was familiar with most ports and winds of the East, and for years he carried opium between China and Calcutta.

In his late thirties Hughes left the incense of the Orient and joined the Adelaide mercantile firm of Bunce and Thomson. That was the firm which the Burra shepherd approached in 1845 with rich copper specimens, but Hughes by then had left the firm and bought his flocks. He tried to redeem his misfortune in leaving that lucky firm by speculating in copper lands, and he took up many leases. One of his sheep runs was at Watervale, south of Burra, and he sought copper there and on his sheep run at the top of Yorke Peninsula. Sea captains are observant and therefore are sometimes good prospectors, and Hughes eventually found copper ore near the beach at Wallaroo. It was 1851, not an apt year to open a copper mine, and eventually he forfeited his mineral lease of sixty acres. He possibly could not find the lode from which the ore had eroded but his observation that the tree roots of the area sometimes burnt with green flames convinced him that strong lodes must exist. He told his shepherds to look.

James Boor, a shepherd, was walking with his sheep about five miles from the coast when he found, amongst the earth and pebbles thrown up from a wombat's hole, a green stone as small as a pea. He showed it to Hughes who identified it as carbonate of copper. The summer of 1859-60 was poor for the sheepmasters and Hughes, probably in debt, was anxious for another income. He engaged four miners from Burra and they chose a site for a shaft in the old Cornish manner by swinging a pick around their heads and digging where it landed. As the pick couldn't fly far, their superstition was tempered by common sense; they sank a shaft by the wombat's burrow.

The shaft found rich copper ore and Hughes quickly pegged mining leases. The mine needed capital, so Hughes invited Scottish merchants in Adelaide, Thomas Elder, Robert Barr Smith, and Edward Stirling, to join him in a small syndicate. The money they put into the mine was repaid so handsomely that it was a gilt prop of the great pastoral house of Elder, Smith. The old opium trader himself won more money than any other mine magnate in Australia had hitherto made.

Hughes' sheep station was known as Walla-Waroo, a native name for wallaby's urine, and his mines became known as Wallaroo. Five miles across the flats the hot beach became Port Wallaroo, and sailing ships anchored offshore to take bags of copper ore from barges. By 1861 the brick chimney of the copper smelters was a beacon to ships that brought coal from Newcastle, and Cornish miners and Welsh smeltermen. The previous year had been miserable for South Australia. Once again copper revived the colony.

One of Hughes' shepherds on the limestone plains south of the new mine was an illiterate Irishman, Patrick Ryan. Passing the new smelters on his way to the homestead with his flock he saw the green copper ore and so he searched for similar ore where his sheep grazed. He too picked up ore and instead of taking it to the captain he showed it to cronies at the hotel at Port Wakefield, forty miles to the east. The publican was excited and hurried the shepherd to Adelaide to apply for a mining lease. The publican's syndicate applied at the Lands Office for the lease but learned that no lease could be granted until they accurately specified the land. And so Ryan rode more than a hundred miles back to his sheepfold at Moonta to show the syndicate his lode.

Captain Hughes heard of his shepherd's find and also persuaded him to say where it was. The fickle Irishman had already promised to support the publican's syndicate but he agreed to enter a syndicate with Captain Hughes and Stirling, the Adelaide merchant. The Hughes syndicate was now behind in the race. The first party had surveyed the land they wanted for the lease and had started on horseback for Adelaide to register it. Hughes sent his messenger on the same mission, providing fresh relays of horses along the way. His man rode for eighteen hours and reached the Lands Office in Adelaide a few minutes after it opened. The other syndicate was already there, but the clerk in the office interviewed Hughes' messenger first and accepted his application for the mining leases. The other syndicate protested all the way to the Privy Council, but Hughes retained the rich mine.

Ryan the shepherd got £6 a week until the Moonta mine should pay dividends on his one-tenth interest, and he drank away his success in pub and shanty. He died less than a year after finding a mine which became the first in Australia to pay a million in dividends.

Captain Hughes and the Adelaide Scots were the largest shareholders of both the Wallaroo and Moonta mines, and they resolved to smelt all ores at the Wallaroo Company's smelters by the sea. Visitors who came by ship on a stormy evening could see across the water the glare of furnaces reflected in the clouds and white smoke trailing across a black sky. Morning did not soften the landscape. Commodore James Goodenough, Commander of Her Majesty's Australian Fleet, sailed into Wallaroo seven months before he was to die from arrow wounds in the Pacific. 'This is the barest, driest spot conceivable,' he wrote in 1875. No gardens or trees lined the dusty crooked streets. The solace of most smeltermen after their twelve-hour shifts was the Sunday sermon in their own Welsh language, and the town heard Welsh sermons and hymns for over forty years. The commodore saw little drunkenness, 'but, as in Wales, a good deal of keeping company of boys and girls, &c.' As the boys of the town worked at the smelters from five in the morning to five at night, the wonder is that they could snatch time for keeping company.

Overlooking the sea was the main furnace house, 700 feet long, massive stone arches at each end and no walls. Sea winds tempered the heat from the row of roasting furnaces that were fed with ore from a long overhead tramway. Counting calcining and refining furnaces, Wallaroo had thirty-six furnaces by 1866 and they burned about one-tenth of all the coal shipped from Newcastle (N.S.W.). A small fleet of colliers sailed on the thousand-mile run from Newcastle, carrying back sand as ballast until the Wallaroo company sensibly chose in 1867 to fill them with poorer ores. Henceforth the richest ore was smelted at Wallaroo and the poor ore averaging about 7 per cent copper was shipped as ballast to Newcastle where the company's Hunter River Copper Works were a smoking landmark for twenty-five years.

The triangle of copper towns on the peninsula lay in 'perhaps the most wretched tract of country to be found in any inhabited part of the colonies', wrote one of the Kennedy family singers who visited the mines in 1874 during a world tour. Along the tramline between the port and Moonta they saw a hotel advertising water for sale and saw tanks sunk in the earth to trap the precious rain. Moonta men paid twopence a bucket for mineralized water condensed in stills fed by the mine pumps.

While the towns were thirsty the mines were often flooded. As early as 1862 water in the mines perturbed the companies and at the depth of ten fathoms the whim horses drawing barrels by rope could no longer master it. Eight engines were pumping when the Wallaroo mines were barely two years old, going 'like the old tread mills of India', said Captain Hughes. On the plains the water pumped from the mines lay in long lakes, the minerals in the water destroying the grass and scrub so that strips of country were 'as black as this table', said the captain vividly. There were no creeks to drain the water and it lay in stagnant pools, harming men's health in summer.

Moonta and Kadina became the largest towns outside Adelaide. Miners built their own cottages from sun-dried bricks with low streamlined chimneys above the wide Cornish hearth. At Christmas they whitewashed the cottages until they glared in the sun. Inside, Cornish pasties were cooking in the oven and saffron cake for Sunday, and the tub of warm water awaited the miner home from his shift with his billycan and the white calico bag which carried his pasty.

These were possibly the largest Cornish communities beyond Land's End. At the New Year carnivals the game was Cornish wrestling and at week-ends the favoured game was cock-fighting—until the law and religious revivals tossed the aged birds into the boiling pots. The towns celebrated Midsummer Eve with fires and fireworks, and as 22 June was midwinter the custom gained by crossing the equator. For a generation the mines gave all men a holiday on 'Midsummer Eve', deliberately ignoring the Queen's birthday.

If the miners were to die far from Cornwall they preferred to die near the end of the week and be buried on the Sunday near the whispering trees and stone walls of Moonta's burial ground. The coffin rested on chairs at the cottage gate and the crowd stood in black at a distance, singing slowly the verses of a hymn of Isaac Watts to a tune composed by a Moonta fiddler. On reaching the third stanza, 'What e'er we do, where e'er we be, We are travelling to the grave', the men carried the coffin to the hearse. If Sunday were fine hundreds of mourners fell behind in pairs, the men in black frock coats with a woman on their arm, and marched along the sandy road, singing the slow burial tune that had been sung at miners' burials in Cornwall for unmeasured years:

> Sing from the chamber to the grave,
> I hear the dying miner say;
> A sound of melody I crave
> Upon my burial day.

On windy days the strong slow singing could be heard far from that stone-walled graveyard where Methuselah Tregonning and Ephraim Major lay with a thousand Cornish dead.

Broken Hill has long had unjustified reputation as the wildest city in the land. The towns which most influenced Broken Hill were Moonta and Kadina, and in their day they were known as the most religious towns in Australia. The man who climbed to the balcony of Prankerd's engine house at Moonta could see sixteen churches. That was not surprising in a town of over five thousand people. What was surprising was that thirteen were Methodist churches—Wesleyan, Primitive Methodist, or Bible Christian. All three were convulsed by religious revivals, and one miner recalled standing one week night at a place equally distant from three large chapels and hearing from each the cries of penitents. At the 1874-5 revival miners returning from work late at night walked to the doors of chapels to learn how many were saved that night. In 1875 three or four hundred people left a Bible Christian church at sunset one Friday night and marched down the street singing in loud voices, 'Hold the fort'. They marched through crowded streets, hundreds joining the procession, all singing to the beat of hobnailed boots, 'My God, the Spring of all my joys'. Next evening a larger procession marched the streets singing with remarkable vigour before returning to cram into a huge Wesleyan chapel 'which was lighted with gas and presented a brilliant spectacle'.

The chapel flavoured every activity. In 1874 the copper field had a dignified strike and miners called an urgent meeting on Sunday, 12 April to discuss the latest negotiations, and several thousand Kadina miners formed their customary circle in the open air and began their union meeting with the Wesleyan hymn 'And are we yet alive', and closed it with the doxology and benediction. When the strike was settled four thousand people formed a ring in the open ground at Moonta and held a thanksgiving service.

Miners preached in Moonta's chapels with strong oratory and voices of Boanerges. Two preachers from the Primitive Methodist chapels became Labor premiers. James Jeffrey, stump preacher from the Victorian gold diggings, worked in the Yelta mine in the 1870s and his quaint phrases were laughed at and revered: 'We thank thee Looard for what little spark of grace we d'possess,' he preached. 'Oh, water that spark.' Church musicians were as eccentric as the preachers, and in a crowded chapel Billy Mitchell, bearded Cornish miner, with big bass viol, was heard to whisper aloud: 'Gimme that rosin and I'll soon show 'ee who's the King of Glory.'

In his realm at Moonta, Henry Richard Hancock was king of glory. He managed the mine on which the town depended and on whose leases most houses stood. He belonged to the dominant church and race, and though not strictly Cornish because he was born five miles the other side of the Tamar, the miners conceded he was 'near nuff to wan'. His unique place at Moonta made him sally for legend and joke, and his fame was spread by migrating miners and by the cartoons of Oswald Pryor so that he became Australia's legendary mine manager. More than six feet tall, skilled in assaying and ore-dressing, Captain Hancock had migrated to the colony about 1859 to manage the small Wheal Ellen lead mine in the Adelaide hills. He was at Yelta mine on the north boundary of Moonta when he was offered at the age of twenty-eight the big Moonta mine, and with him went five captains from Yelta. He could survey and assay and think and plan, and his mine became possibly the most progressive in Australia. He mechanized the mine and mill to a degree which staid Cornishmen could not appreciate, and the Hancock jig for ore-dressing was as celebrated as the captain. Not altogether accepted as a preacher in his own church, he had a pulpit built in his office and from the heights he dictated letters and reproved his captains, though later he blew a whistle and spoke to them through tubes installed in the walls. He had three office boys, recalls Oswald Pryor in his fine book *Australia's Little Cornwall*, and he rang a different handbell to summon each boy. The captain mechanized the giving of orders just as he mechanized the works. However, the reign of this fatherly captain with the white beard and the grey belltopper was warm and personal. He watched over his employees in work and sickness and leisure, set up night-school for the illiterate boys of the dressing sheds, giving each a metal token for each night at school, and insisting that he hand the captain of the dressing floor four tokens before starting work each Monday morning.

The Cornish custom of calling all managers and foremen 'captains' survived until Captain Hancock retired in 1898. Some of the captains relished the title so dearly that in bad times, if told that they must either forgo the title or forgo part of their pay, they preferred to lose the pay. Thus Captain Hancock played with their pride. The captains in the mill were known as floor captains and sometimes as grass captains, but the real captains managed the underground. At Moonta the day shift of miners went down the shaft at 7 a.m. but the captains did not descend until 9.30. Dressed in white duck trousers and coat, blücher boots, and a hard miner's hat over a white skullcap, they walked with dignity to the cage and went into the darkness with a candle stuck by wet clay to the front of the hat. After inspecting the workings they came up at 2 p.m. and bathed, cleaning their finger nails with sharpened match sticks which an attendant provided. They were the elect of Moonta from the ringing of the loud bell atop the engine house at six in the morning until the bell called the night shift an hour before midnight. Some captains were not pleased when the Salvation Army marched into Moonta and usurped their titles.

The captains had their dignity and so did the miners. The Cornish tribute system ruled in most South Australian mines and gave the miners some hope and freedom. Virtually all the ore at Moonta and Kadina was mined on tribute. Miners selected an area down the mine and worked in only that area for a term of two months. They were not paid wages and they were not paid solely by piece-work. Instead they were paid a share of the value of the ore they mined.

The day on which miners selected their underground area or pitch was known as Survey Day, and it came every ninth Saturday. For days previously the captains were busy underground, surveying and pricing the pitches that would be put up for auction. If the pitch was in hard rock and the copper seemed poor, the company might offer the men ten shillings for each £1 of ore they mined; if the ground was rich and easy to work they might offer only two shillings in the pound. After lunch on Survey Day hundreds of men waited outside the mine office. At one o'clock Captain Hancock mounted the platform with his clerks and frock-coated captains. Loudly he called the number of each pitch and the terms on which it could be worked. If two groups of men wanted the same ground they bid, pushing up the amount of tribute to the company. If the tribute seemed unfair, and no men bid, the company often cut the price.

The gamble of the system pleased most miners. They usually received subsistence money or 'sist' during the term of the contract, and if their ground was unexpectedly poor they could abandon it. If it was rich and the price of copper soared the company could not deprive them of their ground. When the day known as Measuring Day came, and all the tributers' ore lay in neat stacks at the surface, the captains would sample and measure it and work out the sum the company owed the men. And if Captain Philp were there, that masterly judge of copper who could estimate the assay value of the ore without the trouble of a chemical assay, the admiring miners would accept his word that the pile averaged 16 per cent or 12 per cent copper. The last day of the nine-week cycle was Settlement Day, and the men were early at the office to collect their money, the leader of each team counting it into equal lots with slow precision. Some men made enough money from one small pitch to retire. They told alluring stories at Moonta of the 'wallows' of rich copper in the Prince Alfred shaft that gave tributers of 1870 remarkable fortunes at the low price of tenpence in the £1. Commodore Goodenough heard of two miners who earned in two months what the average labourer would not earn in two years.

For the company the tribute system had advantages. The men worked hard, often enthusiastically; they mined the ore frugally because they paid for their own candles, blasting powder, and mining tools; they mined very little barren rock because they were paid for copper, not mullock. As A. K. H. Jenkin, the Cornish mining historian, explained, the company with good tributers had in effect many practical geologists whose knowledge of the lodes of the mine was unrivalled. The tributer had to study rocks because he was bidding for a block of ground which, exposed on only one side, might prove unexpectedly rich or poor when broken into. In Cornwall and Australia alert tributers had saved many mines from early extinction by their ability to find new lodes or old lodes that were lost. The tributers were aristocrats amongst miners. They worked in a company mine but were their own bosses, entering the mine early or leaving it early, even working all night when the pitch was rich and their time was running out. There were Cornish miners who worked thirty years in South Australian copper mines or Victorian gold mines and never worked for fixed wages.

While tributing usually made for industrial concord it had abuses. In some mines tributers worked for months and actually made a loss while the company that owned the mine profited by their labour. Moreover, a skilled captain could deceive the tributer, and the cunning tributer tricked the captain. If he found rich ore near the end of his term he tried to hide it in the hope of getting his pitch again on favourable terms next Survey Day. ''Tis no sin to take cappen in,' said the old Cornish precept. Some tributers broke the stern rule that they were not to mine ore outside their own pitch or territory, and at times the captains of Burra and Moonta left rich patches of ore in remote parts of the mines as a reserve for bad days ahead, only to find years later the areas honeycombed with the illegal burrowings of tributers straying from their pitch. One old

Moonta tributer was asked if any of the miners stole such ore, for theft it was. 'All av'em,' he cried, 'local preachers 'n all.' The spiritual Cousin Jack, born of Cornish smugglers, was renowned for his dual morality on the Australian mining fields. In some mines they called the double-headed hammer 'The Wesleyan', though there was more banter than truth in such barbs.

The men who sank the vertical shafts and drove the horizontal levels through barren rock were known as tutworkers. Tutwork was piece-work, and the men were paid by the fathoms of ground they mined, and not by its value. The cautious miners did the tutwork, preferring a more even income. They explored the ground for the tributers to come after them and in the good mines they were often two years ahead of the tributers. The essence of sound mining is to develop the orebody and establish reserves of ore, and this foresight distinguished the copper captains from the Victorian gold managers. Thus in 1865 the Wallaroo and Moonta mines had about five years of ore in sight at their current pace of mining, and this reserve was constantly being augmented by teams of tutworkers. When the falling price of copper clipped the company's profits, the company could afford to reduce the essential tutwork solely through its foresight in the good years. In South Australia copper mining was moving from a gambling system to an industrial system, an essential step that many leading Australian fields did not make for another half century. Imagine a Victorian manager of 1860 refusing to gouge the rich ore near the surface because he had not first explored the deposit. Yet in 1860 most South Australian captains would agree indignantly with Captain Prisk: 'It is not doing the property justice.' To the Cornishmen it was essential to win the most wealth from every deposit of mineral, and the sophisticated idea of ore reserves was as essential to a mine as a haystack to a farm.

The thigh of Yorke Peninsula was a second county of Cornwall. The Cornish pumping engine, the man-engine, and the Cornish boiler were carried to Moonta and Wallaroo and improved by local artisans. The Cornish welfare policy towards the men, the system of tribute and tutwork, the idea of ore reserves, all the pithy mining vocabulary from crib to skimps—everything except the Cornish practice of employing girls or 'bal maidens' as ore dressers—came to South Australian mines. By the 1870s South Australia had replaced Cornwall as the largest copper region of the British Empire.

Early in 1872 Captain Walter Hughes, the copper king, was amazed to read in the latest English newspapers that copper had risen dramatically towards the close of the previous year. By April 1872 the London price of South Australian copper, which always carried a premium because of its purity, had jumped to £102. In May it averaged £105, and in June it was £112. Over five months it averaged £107, an average not to be attained again until World War I devoured copper in munitions. The copper prices added immensely to Hughes' fortune, and he had so much money that he donated the £20,000 that led to the founding of the University of Adelaide. With Moonta's chairman, Thomas Elder, he gave another large sum to equip Colonel Warburton for exploring the deserts of Western Australia. Elder himself, benefiting too from the high copper prices, gave £20,000 to the University in 1874.

It was not only Adelaide's dignified North Terrace that shared in the buoyant times. The copper fields shared too. Captain Hancock's mine employed over 1,400 men and boys in 1873, more than any other Australian mine until the rise of the rich Broken Hill Proprietary silver-lead mine. The Wallaroo company's mines, which usually employed only half as many men as Moonta, reached their peak of a thousand men in 1872. The Wallaroo smelters and port had never been busier. The three main copper towns had 11,500 people at the census of 1871 and in the next few years, as copper prices leaped, they may have had nearly 20,000 people or about 2 per cent of Australia's population. From the balconies of the stone engine-houses the engine drivers could see across the plains the steam rising from the engines of many little mines that were reopened in the boom. To the colony copper was like a huge flywheel, generating wealth and activity.

While the mines by the blue sea and sand dunes of Yorke Peninsula prospered, the early copper mines of Kapunda and Burra sank. Exhausting their best ore by the 1860s, they scavenged the poor ore cheaply by excavating Australia's first large open cuts.

A Scottish syndicate briefly revived Kapunda, taking the mine on royalty in 1865 in the hope that modern metallurgy and civil engineering and a new manager would make poor copper pay. And so the old manager, Captain Oldham, retired after a versatile career in which he simultaneously was manager of the mine, post office, and National Bank, pastor of the Congregational Church, captain of the Mines Rifles and conductor of the philharmonic choir.

Kapunda needed specialists and new methods. The Scottish syndicate created large quarries in the upper levels of the old workings, sluicing with water the green veins of ore that permeated the white clay until the old timbered drives and workings were visible in the craters. In place of costly smelting the Scots tried a novel leaching process and spent £80,000 or more on possibly the largest chemical works in Australia. Rows of furnaces roasted the poor sulphide ore mined underground, and the sulphur fumes were turned to sulphuric acid in a vast lead reservoir. From the acid and from salt harvested in distant lakes they made hydrochloric acid; that acid ran into vats of crushed ore from the open cut and leached the copper liquors from the ore. The copper liquor flowed to more vats filled with scrap iron and precipitated itself as metallic copper on the scrap iron.

The Kapunda farmers who gazed across their wheatfields to the gleaming white mounds of salt, the railway trucks running up the steep quarry, the smoking chimneys, the vats and condensers, marvelled that copper could be extracted with the simple ingredients of salt, waste rock, and scrap iron. In Scotland Sir Charles Tennant did not marvel. Despite the skill of his syndicate's alchemy and their courage in trying to profit from 0·5 per cent copper ores they made a profit in only the boom year of 1872.

Fifty miles to the north, amongst bare rounded hills, Burra was also dying. The mine and dressing sheds alone had employed the amazing total of 1,208 men and boys in September 1859. 'I was much struck with the splendid physique of the Cornish miners; they were splendid-looking fellows, active as cats, and strong and brave as lions,' wrote engineer West. Within ten years the lions had gone to Moonta, the mine was idle, and Australia's veteran manager, Captain Roach, had been retired. The company which had paid £800,000 in dividends was still optimistic, and as its secretary, Sir Henry Ayers, was now premier of the colony he persuaded his own government to build a railway to the mine. John Darlington, an English engineer, came to design a long open cut that would salvage the poorer ore which underground miners had rejected, and by 1871, 300 men and boys were again employed and the captain said that 'the Burra Burra will still make a lasting and profitable mine'. Why he said it will never be known, for the company was merely eating the crumbs that the high price of copper made edible.

Burra had yielded 51,600 tons of copper worth £4,700,000 when the men queued for their last pay-day in September 1877. Even after the horses and stores and machinery had been auctioned, many miners waited years around the windswept town in the hope the mine would reopen. Kapunda lasted eighteen months longer, and again the miners lingered in hope, nurturing the belief that only an inrush of water had closed the mine. One of the historians' problems comes from these myths that mist the finding and closing of great mines.

After the gold rushes South Australia became the granary of the continent, but as late as 1868 copper exports could still surpass those of wheat and flour. In the high copper prices ruling from 1872 to

1875 mineral exports still averaged over £750,000 yearly but by 1880 their value had halved. The price of copper had fallen. In hollows of the Mount Lofty Ranges fires were drawn in the copper furnaces and blind horses led from the mines. Nearly every mine had exhausted its rich ores and now depended on the sulphide ores that were poorer and were dearer to mine. The deep Wallaroo mine was idle for two years, and Moonta tried to cut costs by expanding output. Captain Hancock himself busily improved his famous jig that mechanically separated poor ore from barren rock in the mill, and underground he experimented with a cumbersome piston rock drill worked by four men. None the less the profits were thin, and in 1880 the old narcotics runner, Captain Hughes, living in luxury on his Surrey estate in England, had the pleasure of being knighted but the sadness of receiving no dividends from his famous copper mines.

The older Cornish miners were superstitious. They believed dwarfs known as 'knackers' lived down the mines, and whenever they heard the sound of the knackers' hammers and picks they imagined that rich ore lay near. They left scraps of food for the knackers, were careful not to offend them; they heard their massed choirs echoing through the mine on Christmas Eve and heard their warnings of danger. On Yorke Peninsula in the 1880s old miners wondered if they had offended the knackers, for they seemed to begrudge their copper.

Cornishmen also believed that Jack o' Lanterns sometimes shone above rich lodes that lay buried a few feet below the grasses. At Wallaroo and Moonta in the 1880s many Cornish men and women walking the plains at night watched for the 'Jacky Lanterns' that might point to the hidden lodes now so desperately wanted. As late as 1960, when the fields had long been dead, an old farmer wrote to the government of South Australia to unburden something that had long been on his mind. He had seen as a farm boy in the 1890s, when lightning was flickering, the Jack o' Lantern clearly crossing the dark Wallaroo plain like a figure eight, travelling half a mile or more, disappearing and then faintly resuming its path to vanish into the sea. The government sent a geologist to examine the path of the lantern, but he saw no sign of copper in the rocks below.

12

MOFFAT'S METAL

SOUTH AUSTRALIA ceased to produce all Australia's base metals and bake most of its Cornish pasties. Moonta miners and captains were enticed away to hundreds of tin and copper mines in the east, and the wide Cornish fireplaces could be seen from Tasmania to the tropics by the mid-1870s.

Peak Downs in central Queensland was the first successful copper mine outside South Australia and also the first rich mine in the Australian tropics. The copper-stained outcrop jutting ten feet above the ground was found by a gold digger in 1861 and Thomas Mort and Sydney friends opened the mine and imported men from Burra to stoke furnaces and work underground in dry heat. Carters carried the copper over ranges to the beach at Broadsound, and so rough was the road and so mean the port that the copper often spent a year reaching the London market. The mine was rich enough to triumph over isolation and heat and high wages, but not the optimism of its owners. During the copper boom of 1872 and 1873 the company imported two hundred Cornish miners and their families in the hope of mining on the grand scale. In a frenzy of optimism the company paid £100,000 in dividends in half a year; but it had not even sold the copper from which it had borrowed the dividends, and when the copper reached England it fell onto a tumbling market.

The company now lacked money to finance the search for new underground lodes, and it exhausted its good ore. It had paid shareholders eight times the money they had invested and with less greed could have paid them far more. With copper prices still slipping the auctioneer came to Peak Downs with its death warrant in 1877. Many other small Queensland copper mines from Cloncurry to Mount Perry died without benediction from auctioneer or creditors, for they were so remote that there was nothing to salvage.

Queensland copper was tarnished until the end of the century and New South Wales became the hope of the industry. On the dry western plains the decomposing copper ores coloured rocks in many places with the sapphire blue of azurite and the dark green of chalcocite and the turquoise blue of malachite, and the colours invited the curious and the Cornish. Out toward the River Darling and its paddle steamers marooned by drought, well-sinkers were busy near the bare pastoral homesteads in the late 1860s. Hartmann and Campbell, two Danish well-sinkers who had once mined gold at Bendigo, saw blue and green splashes of copper on the rocky sides of a blacks' waterhole, and gathered specimens. Going south they met a woman who had once been a 'bal girl' in the Cornish mines, and she assured them it was copper. They returned to the waterhole at Cobar in 1870 and prepared to send a dray of copper ore to the river ports and thence by slow steamer to South Australia's copper smelters.

From Moonta came captain and miners and from Peak Downs smeltermen and from Wales came thousands of furnace fire-bricks that reached the mine at a cost of 2s. 6d. each. Furnaces were lit in 1875 and hundreds of people came in the stage coach and built huts of whitewashed hessian and bag walls and sat on boxes outside in the evening cool and watched the molten slag trickling through the darkness. In the wilderness of fuchsia and cassia and wild lemon that blazed colour in the spring the Cobar axemen chopped over 50,000 tons of wood a year for boilers and furnaces, laying waste the country for the rabbit plague and then for the summer winds that spiralled pillars of red dust across the plains. Cobar had 2,500 people in 1882 and must have been the largest of all Australian towns lying more than four hundred miles from the coast. Moreover it became the heart of a wide copper region. The Danish discoverers of Cobar found the green and blue stains in another native well at Girilambone to the north-east and opened a rich mine. One of Cobar's richest shareholders, Russell Barton, floated a copper company to mine at Nymagee to the south-east, and between the rivers Darling and Bogan in the early 1880s were many squat chimneys and smelting sheds and mounds of rich ore. Near the dividing range to the east were Burraga and Peelwood and small copper towns with slag-pots burning like yellow flares in the winter snow.

The dry interior of New South Wales and Queensland promised to support a great copper industry, but the promise was empty. Most of the mines were far from railways, and the cost of carting copper by horse and bullocks to the nearest port or railhead was high. As the mines were in dry country the cartage of copper was impossible in the long droughts, for grass withered on the plains and mud baked in the waterholes. Companies had to pay miners one and a half times the wages of Ballarat to entice them to Cobar or Mount Hope, and dear labour meant high costs of producing copper. They sold their blister copper on a falling market from the 1870s and often ore of fabulous richness barely paid a profit. In 1883 and 1884 New South Wales produced an average of over £400,000 of copper but a decade later the annual output was barely one quarter of that sum. Drought, isolation, high wages, falling copper prices, and improvident mining almost crippled an industry blessed with mountains of fair ore.

Cornwall's mineral wealth was in copper and tin, but the Cornishmen in Australia were quick to mine copper and slow to mine tin. Australia's tin was sparser than the copper, confined to a few areas, and its few outcrops lacked the plumes of green and blue that made copper conspicuous. Nevertheless the price of tin was usually higher than copper, for tin was used in pewter and solder, bronze and typemetal and tinplate, and was mined in relatively few countries. And as most of the world's tin came and still comes from the alluvium, it could be mined like gold by men with scant capital and experience. At the Ovens in Victoria and on goldfields in northern New South Wales, diggers sometimes found the grains of heavy black tin in sluicebox ripples that had been designed to catch the heavier gold. They sluiced occasional bags of tin in the same manner as gold, but their success was erratic. Australia's richest deposits of tin were not near the goldfields.

On the New England tableland in northern New South Wales, on land that sloped west to the dry plains, granite rocks containing sparse tin had been eroded for millions of years. The tin oxide or cassiterite was scoured by the rains into the creeks and lay buried in the sand and gravel and rubble of old and new watercourses. In some areas volcanoes had erupted, and the basalt buried rivers of tin just as it buried Victoria's rivers of gold. In some slopes or valleys the black grains of tin were sprinkled like small gunshot or turnip seed on the surface of the earth.

Near the pastoral town of Inverell a shepherd named Joseph Wills saw the black grains, and when he felt them their heaviness aroused his curiosity. In an Inverell hotel he sold samples to a commercial traveller, and in Sydney the tin was passed from hand to hand until it attracted C. S. McGlew, a mining adventurer. McGlew went by train and coach to the northern tablelands in 1871, found the eccentric shepherd and was led to the gentle hill at Elsmore where the tin was exposed.

McGlew started a rush to take up rich tinlands around Newstead and Tingha and Inverell. Gullies of alluvial tin were found near Glen Innes and Tenterfield. The tin at Vegetable Creek was so rich that thousands of Australians and Chinese camped in the creek and they churned 150 acres of flats so vigorously that they got 15,000 tons of stream tin, making the canvas town of Emmaville one of the largest and wildest in the north. Like the miners of Ballarat they traced the layers of mineral under the basalt and sank rows of deep shafts for 2½ miles along the lead and won another 6,000 tons of tin.

McGlew had publicized the shepherd's tin near Inverell and, a year later, through the same attention to gossip he opened another rich tinfield at Stanthorpe, just across the Queensland border. He heard from a police magistrate that the Quart Pot Hotel used a heavy sandbag to stop the wind from whistling under the door. The magistrate said the bag was so heavy for its size that it might contain tin, so McGlew went to the hotel in 1872 and found the sand had come from an adjacent creek that was rich in tin. So this hotel in the granite ranges got more customers in a month than it had served in all its previous years, and by September 1873 the sign 'BAR' was scratched or painted on scores of boards and canvas flaps down Stanthorpe's long gutterless street, and hucksters were selling oysters and Chinamen selling tea in a town that was said to be catering for eight thousand miners in the vicinity.

From Stanthorpe in Queensland south to Tingha in New South Wales ran a hundred miles of tinlands with miners in high boots shovelling the gravels and tributers pushing wheelbarrows along muddy planks or stabbing the sluiceboxes with eight-pronged forks. In 1873 the Border tin region alone produced more tin than any other region or country in the world, and Australia was the world's largest tin producer in the decade 1873-82. But the discoverer of the great tinlands, Joseph Wills, the shepherd, got no reward, as his tombstone in Newstead churchyard proclaims:

> Here lieth poor Wills who found out tin,
> But very little did he win;
> He paved the way for others' gains,
> And died neglected for his pains.

John Moffat, a tall Scot with chiselled face and short sight, was in his early thirties when he joined the tin rushes. He had been a shepherd near Roma and a grocer near Brisbane, and when Stanthorpe became a rich tin town he set up a general store under the name of Love & Moffat and also bought tin concentrate for a Brisbane smelting works. Sensing that smelting works on the Border tinfields would make profits he joined with John Reid in building a tin smelter at Tent Hill near Emmaville in 1874, and spent heavily on reverberatory furnaces, portable engine, brickworks, stables, carpenters' and blacksmith's shops, and workmen's cottages. With a rare eye for neatness he painted every building white and, being a puritan, he refused to allow a hotel in his smelting town. He was a model employer, honest with his smeltermen and the miners who sold him tin. But his furnaces had metallurgical troubles and the falling price of tin increased his monetary troubles and Moffat & Reid's tin smelters had to borrow from Love & Moffat's general stores. In 1879 the stores slid into debt and creditors forced them to liquidate. Moffat went to Tent Hill and gave all his energy to the smelting.

Two of Moffat's shop assistants who had lost their jobs at Stanthorpe decided to go a thousand miles north to the new Herberton tinfield in north Queensland. There they found a big outcrop that seemed rich. They sent a sample to be assayed at Tent Hill and when Moffat sensed the value of their lode he caught the steamer to Cairns and bought a one-fifth interest in the lode and later a majority interest.

Herberton was an unusual tinfield, for its lodes were more valuable than the alluvium, and the lodes gave mining a stability that the Border tinfields lacked. Moffat increased Herberton's stability by building a large and efficient tin mill and smelter at Irvinebank and spending much of his smelting profits in the search for new lodes. He became not only the magnate of Queensland's largest tinfields but the most generous financier of new mining ventures in Queensland. He built dams and tramways and railways, opened mines of many metals, searched for oil, and developed the copper-silver-lead mines of the Chillagoe field and the Mount Elliott mine near Cloncurry.

Moffat's kindness to his employees was so spontaneous that legend says hundreds of children in north Queensland were taught to mouth his name in bedside prayers. He himself was a Swedenborgian, but he respected all beliefs and had such a sincere manner that people were captured in his presence. His honesty and sense of honour were famous. After his storekeeping firm of Love & Moffat had gone insolvent and he was no longer legally liable for the debts, he patiently saved until he had enough to call creditors together and pay them with interest. When his Chillagoe mines were floated into a huge Melbourne company and his own free shares commanded a high price on the stock exchange, he alone of the directors refused to unload them when they had inside knowledge that the mines were over-valued. He lost a paper fortune but he saved hundreds of share-buyers from losing.

His memorial carries the unusual plaque: 'In memory of charitable and philanthropic John Moffat . . . The children ended their prayers each night with "God bless John Moffat".'

Moffat's working life spanned and sponsored the spread of copper and tin mining in eastern Australia. When he reached Australia in the early 1860s South Australia was the only centre of copper mining and there was no centre of tin mining. When he died in 1918 Queensland, New South Wales, Tasmania and South Australia had each produced close to £30 million of the Cornish metals—copper and tin—and in that achievement no man played a greater part than Moffat.

13

THE STAR OF SILVERTON

SEVEN HUNDRED MILES west of Sydney the Barrier Ranges stood on the western border of New South Wales like whales stranded on a beach. The surrounding plains were hot for much of the year and sometimes barren. Nine inches of rain fell in the average year, but sky and earth were hot and no creek lived more than a few weeks in the blazing west. The one living stream was the Darling, which drew water from the high coastal ranges of New South Wales and Queensland and flowed gracefully by shaded banks to join the Murray's long slow glide to the ocean. Paddle steamers tugged barges up the river with fencing wire, corrugated iron, flour and sugar and all the provisions for the sheep stations. When the river dried the steamers lay months in the mud and when the river flooded they sometimes wandered far from the banks, to be marooned again in mud when floods vanished.

East of the Barrier Ranges the only two towns worthy of the name hugged the thin oasis of river. Their wharves unloaded supplies for stations hundreds of miles away, and in their courthouse on drowsy days the magistrate punished the fighting shearers and remanded outback murderers. From the telegraph office the operator tapped news of cattle movements, grounding of paddle steamers, and the price of wool clips. Wilcannia was the larger of these towns with its two breweries pumping water from the river, and its saddlers stitching in the sweet aroma of leather and its five blacksmiths' forges a hell from November to March as they shoed the horses of drovers who each year swam tens of thousands of cattle and sheep across the river on their way from western Queensland to abattoirs and saleyards in the south. The paddle steamer and the horse-drawn coach and the bullock team were the lifelines of the plains. Wool prices and rain-clouds ruled the plains, and seemed likely to rule them forever.

Across the plains to the east the copper mines of Cobar in the early 1870s sent loads of rich copper ore by bullock team to the riverport at Bourke, and the river steamers called at Wilcannia on the long journey to South Australia with the beautiful copper ores. Many Wilcannia men invested in the Cobar mine, paying £250 a share in 1872 and selling it for more in the bars and cool verandahs of the hotels. Sheep owners riding back to their homesteads began to examine the rocks for the familiar stains of copper, revisited places where they had once observed minerals. In hot hills north-west of Wilcannia two neighbours, Quin and Kennedy, found copper on Nuntherungie station and formed a small syndicate. With the price of copper soaring above £100 a ton in 1872 they had the incentive to mine a few tons and send it by horse team seventy-two miles to Wilcannia as loading for the river steamers. For a goodly fee Captain John Warren came from Moonta in the winter of 1872 and inspected their mines; and he must have seen that the green stains of copper covered huge blocks of useless sandstone.

'Few men,' wrote a New South Wales inspector of mines, 'have done more for the advancement of Australia than the so-called back-blockers of these wild, hot, dusty, and arid regions.' The men of Nuntherungie made nothing from their copper mine but they were humble pioneers of one of the great mineral regions of the world.

The wealth of the Barrier Ranges was not in copper but all early hopes were fixed on copper. Nearly three hundred miles to the east Cobar had copper and two hundred miles to the south-west the dying Burra was still renowned for copper. Once the railway reached Burra in 1870 many stations near the Barrier Ranges forsook the river ports and sent their wool by bullock team to Burra. Some of their bullock drivers and carters had once carted Burra copper or had worked in the vast open cut, and in the ranges their eyes were alert for stains of copper. Any bloodshot eye could see that the Barrier Ranges were mineralized. White quartz littered slopes like bread-crumbs. Fool's gold glinted in the sunlight. Green carbonates splashed the rocks on sheep runs. And so when the price of copper was high, many prospecting pits and holes were dug and blasted from Nuntherungie in the north to the Broken Hill in the south.

In 1875 and 1876 drought captured the plains and ranges. Sheep died in thousands on the cracked mud of empty waterholes; without grass to hold the soil the red storms whirled across the plains. Pastoralists called for well-sinkers and water-diviners, and Captain Sanders of the Burra mine complained that good miners left him to make high wages sinking wells in the outback. So indirectly the drought spurred on prospecting. And at Elder, Smith's pastoral station of Thackaringa on the western edge of the Barrier Ranges two men, Julius Charles Nickel and his mate McLean, found veins of silver while sinking a well in 1876.

Thackaringa homestead lay on the western edge of the Barrier Ranges, and the road from Burra to the sheep stations upcountry passed close by. On the dusty road John Stokie had a small bush hotel where he sold spirits to passing teamsters and retailed gossip culled from a radius of a hundred miles or more. He soon heard of Nickel's discovery.

Stokie sent word of the silver find by mailcoach to Menindee, a riverport a hundred miles to the south-east, where one of the general stores was kept by Paddy Green. Like many outback storekeepers, red-bearded Green supplied free stores to prospectors and miners in return for a share in what they found. Hearing of the silver of Thackaringa, he came on Maiden's coach and walked with Stokie across the brown hills to examine the small lodes of galena in the brown soil. He sent away samples for assay and the report pleased him. In June 1876 he applied for forty acres of mineral lease and Stokie applied for eighty. The drought spurred their efforts to open a new source of wealth. On 19 October the Wilcannia newspaper reported that thirty tons of silver-lead ore lay at the mine awaiting the teamsters, and already nine tons were at Menindee awaiting the paddle steamer.

Newspapers were soon silent about the silver. A reporter of the distant *Burra News* travelled in the winter of 1877 to Thackaringa and saw Stokie's pub, saw cattle and sheep pads on the sandy graveyard, but saw no sign of mines. Walking forty miles from Thackaringa to Mount Gipps he saw more men searching for a job than he had seen in two months in South Australia: all walking with waterbag close to the Broken Hill with its vast treasure protruding above the rocky soil, their campfires the lonely lights in a district which soon would glow with scores of furnaces and the lamps of an inland city. But the reporter heard no talk of minerals on his long journey; if he had, he would have told it to his readers in Burra, where the famous copper mine closed and threw three hundred out of work in the month he was away.

Why had Thackaringa's mines collapsed so quickly? Patrick Green had sent his bags of silver-lead down the Darling on the first leg of a journey to overseas smelters. On the voyage to Europe the ship apparently ran into stormy seas and the bags of ore were jettisoned. Yet it seems unwise to explain, as so many writers have explained, that the silver field collapsed because the cargo was lost. It was insured for £270, so Green can hardly have lost his money. The delay in receiving a report from a smelter on the value of the ore would not have retarded the working of the mines, for they had ceased work before any report could have been received. Thackaringa appears to have died simply because the ore was poor and few investors would touch it.

In 1881 there were strong rumours of gold near the stony desert where the borders of Queensland and New South Wales and South Australia met. Past Stokie's pub at Thackaringa and German Charley's shanty at Mount Gipps, a straggle of men with spring carts and horses or nothing but strong boots pushed north nearly two hundred miles to the rush at Mount Browne. At the riverports of Bourke and Wilcannia small processions left the wide streets and disappeared into the mirage. The largest newspaper in Australia, the Melbourne *Age*, sent a reporter a thousand miles overland in April 1881, and from the railhead at Deniliquin he spent three days in the coach jolting to Wilcannia where all was excitement and the banks of the Darling were lined with tents and waggons of men halting briefly to rest horses or buy supplies. At the riverport he bought blanket and waterbag and flyveil and with a neatly-rolled swag boarded the coach for the last dreary leg of the journey, eating biscuits and tinned fish in the crowded coach and stopping at wayside shanties where the horses fed and passengers drank adulterated spirits known as 'chain lightning'. Already a thousand people were on the goldfields and there was barely enough fresh water for man and horse, let alone to wash the auriferous dirt. Men were still coming in from trips of endurance, and the reporter met one who was so mad with thirst that he had gashed his wrists and drunk the blood, and others who had abandoned fagged horses and heavy swags on the Bourke track in order to reach the diggings alive.

Melbourne newspapers denounced the rush as a trick of the local shanty-keepers, but hundreds who had battled to reach the field did not leave without testing it. The quartz reefs seemed promising enough to float companies in Adelaide in the cheap-money times of 1881, and the alluvial gold in the clay seemed rich enough to pay wages if only there was rain to wash the dirt. Throughout eastern Australia there was recession in the cities in 1881 and publicans arrived and built hotels of iron or thick stone and storekeepers came with goods in camel waggons and the Commercial Banking Company of Sydney opened two branches at Milparinka and Tibooburra. Everything came except the rain so essential to extract gold from alluvial or hard rock. Men built dams and tanks to catch the rain that fell, but within a month sun dried the dams. One company sent up machinery from the distant coast and never bothered to erect it. Men piled paddocks high with washdirt and won gold by dry-blowing the soil in the wind. The warden at Milparinka daily read his thermometer in what shade he could find and announced that for six months of the year the average daily temperature was over 100 degrees. At the end of summer men died of typhoid and dysentery, and many lay in darkened rooms in the agony of sandy blight, and a team of camels hauled a waggon of sick men to find shade and medicine in Wilcannia.

As a goldfield it might have proved payable if it had been near the coast and if it had had a gentle climate, plentiful water, and cheap food, but it had none of these advantages. In three years it produced about £80,000 of gold at an alarming cost in money and life and health, and there were men who camped without paying in the stone rooms of the deserted hotels and mumbled that the field should never have been opened. They were not quite right. The gold attracted so many men and so much capital from mining companies and business houses because 1881 was a year of recession, and men and money were idle. In such times a poor goldfield, no matter how far away, seems better than none. Significantly all but a few of Australia's main metal fields were opened during, or soon after, recessions.

Men who retreated from Mount Browne in 1882 found the silver-field in the Barrier Ranges stirring again. At Thackaringa men turned windlasses and hauled bullock-hide buckets of silver-lead ore from narrow shafts, men sewed the bags of hand-picked ore and loaded them on waggons, and newcomers hammered in pegs of new mining leases. The Gipsy Girl, Goat Hill, Homeward Bound and a dozen other little syndicates hewed ore from shallow shafts and trenches. Returning diggers from Mount Browne, and well-sinkers and stockmen from the parched sheep runs, worked their own little mines, pig-rooting the richest ore. A man with little capital could almost make a living at Thackaringa by selling rich ore to ore buyers who sent it to the railhead two hundred miles away and shipped it overseas to smelters in Wales and Germany.

Eighteen miles up the track to Wilcannia another silver-lead mine was opened. Its name was Umberumberka, and Stokie the publican was said to be its finder. A mile away was a beautiful place for a town with permanent water in wells and lofty gum trees in the dry watercourses and white cockatoos screeching from the branches. The town became Silverton, an oasis in the ranges and a base for men who rode or walked north and east to find more lodes of silver and lead. In 1882 and 1883 Silverton was excited by the finding of slugs of cerargyrite, amazingly rich in silver. An Adelaide assayer came with his chemicals to set up shop in Silverton and was astonished at the richness of the specimens he was paid to assay. Stone from the Chanticleer mine went 10,000 ounces to the ton, Jo' the Marine assayed 10,724 ounces to the ton, and Hen and Chickens 3,400. As silver was worth nearly four shillings an ounce, these assays were seductive. Without silver in the bank it had been impossible to mine the silver in the ground, but Adelaide stock exchange was enjoying a mining boom in 1882 and investors subscribed capital to the Umberumberka Silver Mining Company and the Pioneer and several smaller mines in the ranges.

The waterholes and soaks were so dry in 1883 that prospectors were confined to their camps and Thackaringa itself was almost closed. Drinking-water there was almost as dear as cold beer in the cities. Swagmen walking to the stations or the silverfields were found dead of thirst or sunstroke, and on the River Darling ships were stranded below Menindee for over a year, baked mud below them and clouds of red sand above them. Sheep died by the thousand and some pastoralists employed men to go around cutting wool from dead sheep. 'There is nothing but death and desolation,' wrote the *South Australian Advertiser* on 16 May 1884, but it predicted that once rain came all Australia would see the value of the silverfields. In fact, unknown to Adelaide, the rains had already fallen, and horsemen were crossing the flooded plains to the Menindee telegraph station with news that three years of drought were over.

Waterholes in the ranges were brimming with water and prospecting revived. Grass grew on the long track to the nearest railway station at Terowie, where the line from Adelaide ended. Teamsters returned and their low rates enabled scores of little mines to send away loads of their choicest silver. In Terowie at daybreak each morning eight horses pulled the coach to the door of the hotels, and men and sometimes women crowded aboard and drove into the saltbush country, halting to change horses at wayside camps where flies clung to the sweating faces of the passengers until the coach once more wheeled into the ribbon of red dust. Big Jim Wilkinson, a Canadian with an auburn goatee on his long jaw, drove one of the coaches, shouting aloud the poems of Adam Lindsay Gordon as he held the ribbon through the night, and like hundreds of his passengers he was sucked into the eddy of speculation and lease-selling that almost swallowed the silverfield. The population of perhaps seven hundred at the breaking of the drought had swelled to nearly five thousand one year later. Leadville and Purnamoota and The Pinnacles and Broken Hill became busy mining camps, sending strings of waggons loaded with rich ore for Europe and daily messages to buy and sell shares at Silverton's busy stock exchange.

Some companies, certain of the field's future, decided to smelt the ore which had hitherto cost upwards of £8 a ton to send overseas. An American metallurgist, la Monte, built small smelters at the Day Dream, and his chimney still rises like a gun barrel in the ranges today. At The Pinnacles a Melbourne company built small furnaces and employed a manager who not only ran mine and smelter, but helped unload the firewood; for all their optimism the small silver companies were wary of experts. W. H. Corbould went out to the smelter as assayer for £6 a week, working by day in a hot shed and sleeping at night on cornsacks stretched taut between two poles; and being fresh from Ballarat's good school of mines and knowing the rudiments of metallurgy he marvelled at the smeltermen's inability to work furnaces. Riding to Silverton to collect the mail and the pay-bags he wondered like many quiet observers whether the silver-fields would last.

While all the mines were infants and none had been explored at depth, who could tell whether many would pay? Meanwhile, investors bought shares for quick profit, and mine managers showed visitors the richest parts of their lodes and the most beautiful specimens of silver, and directors in Adelaide and Melbourne pronounced optimistically so that their companies could call for more capital when needed. And so all conspired to spread optimism about a field which was still unproved. (It is easy to forget from the remoteness of the years that mining fields can never be created without optimism, even if such optimism seems absurd after a field has failed.)

A touch of the optimism on the silverfields came from the famous play by Henry Jones, *The Silver King*, a melodrama of an Englishman making his fortune in an American silver camp. Tens of thousands of Australians were moved by that play in city theatres. Even in Silverton, where stock exchange and ten hotels and crowded streets were the only theatre, men dreamed they would be as rich as Denver, the Silver King. Silver lodes in the hills were sold for high sums. Silver kings were throned and dethroned. Harry Meaney, young and earnest, won a fortune from the Day Dream mine and for a few months they called him the Silver King. At the height of fame he chose to marry, and it was long remembered how after the marriage in the Catholic church on the hill the bridal pair paraded the busy streets in a drag drawn by four grey horses with the town auctioneer riding in front as bellringer and Apple Jack riding a white charger in the rear. At one corner the brass band stood on a horse drag and played bridal marches, and at another the hotel balconies were packed to see him ride out the dusty track towards the Day Dream mine for afternoon tea with his bride amongst the silver. In the town that night the Nevada Hall was ablaze with light, and twenty cases of champagne as cool as Silverton could make them awaited the wedding feast, and the musicians played until they lost all harmony.

The town in the valley on the edge of the great plain prospered throughout 1885. At least three thousand people lived in this new Nevada and several thousand more came to town on festive days from the outlying mines. Its stock exchange that year helped float a company which today is Australia's blue ribbon stock, and another company which vied for that honour at the turn of the century. Its nine or ten hotels earned fortunes for such men as German Charley, red-bearded and coarse-faced, and John de Baun of giant frame and little voice, who later built famous hotels in Coolgardie and Perth. Not all was gambling and drinking in this thirsty ephemeral town, and guests sitting on the verandahs of the silent hotels on Sunday evenings could hear in the distance the voices of Cornish men and women singing hymns and part-songs as they filed over the hill from the Bible Christian chapel to the Umberumberka mine.

All the silver kings wanted to assure their fortunes was a railway to the coast, and in 1885 the South Australian government began to extend the railway two hundred miles north-east of Peterborough. The government, however, could not extend the railway beyond its own border at Cockburn, so the bold promoters of Silverton and silver investors in Melbourne decided to bridge that narrow gap between Silverton and the border. They floated the Silverton Tramway Company in May 1885, envisaging great profits from a railway that tapped so rich a mining field. They could not foresee that when their first train arrived, Silverton and all its mines would be dying.

Silverton today is quiet. Near the beautiful river gums and the wooded range, churches and public buildings and houses of pink rubble stone are scattered about the overgrown streets. The massive walls of one old hotel still stand, but there are no customers and no roof. No court meets in the large courthouse and no mayor sits in the municipal chambers, and no prisoners await trial behind the tall walls of the gaol. Of the few score people who live near the town one is the son of German Charley, the famous publican. A man in his eighties with kind face and white Edwardian moustache, he lives in an old stone house and muses over the silver kings, believes still in their mines, and presided until recently over one of the town's last institutions, the Star of Silverton Rechabite Lodge.

14

THE BROKEN HILL

EAST OF SILVERTON lay the Mount Gipps sheep station with stone homestead and stone shearing shed and all the stores and huts of a self-contained village. Not far away was a grog shanty, where the station hands could cash their cheques, and a police station where the constable could supervise the shanty. In a small cemetery in the red sand, wooden tablets marked the graves of swagmen and station hands who had died of too much alcohol or too little water. Glassy pieces of quartz littered the ridges like shining salt and signified to many travellers that the country was mineralized.

Scattered hills eased the monotony of the undulating plains. One bald hill was known as Round Hill. To the south-west more hills rose like pimples and were called The Pinnacles. In between was a long low hill with jagged crest as if the top of the hill had been broken. They called it the Broken Hill. The crests and slopes of the hill were black and red and brown with what Cornishmen called 'gossan', great boulders and blocks of manganese and iron oxides. Gossan had been formed by the slow oxidizing over the ages of the top of the original mineral deposit, and scores of Australian deposits had similar 'iron hats'. But the iron hat on the Broken Hill was enormous, stretching over four thousand feet along the surface. Just as the iron hat was enormous, so was the deposit of mineral below. On most Australian fields a lode five feet wide was said to be large, but on Broken Hill the lode's width exceeded five hundred feet.

A giant beside the pigmies of Silverton, its wealth was not evident to the men who had often passed or climbed it. Boundary riders had climbed it for nearly twenty years to scan the plains for straying flocks. Surveyors had camped near the hill and had erected a survey trig on the crest. The men who built the wire fence that divided Mount Gipps from Kinchega station had dug postholes at the foot of the hill, and the fence actually crossed the outcrop. And when Silverton and Thackaringa were engrossed in the treasure hunt, prospectors had chipped samples of the rock from the hill and sent them for assay, only to be disappointed by the seeming poverty of the ores. Seven years after well-sinkers found silver by the dusty track at Thackaringa the lode on the Broken Hill was still unwanted.

When eventually Broken Hill proved to be greater than even Ballarat and Bendigo, many prospectors in the ranges could not understand why they had neglected the jagged hill. Writers later echoed their puzzlement. Roy Bridges, in his book *From Silver to Steel*, was the first of a dozen who hinted that the aggressive manager of Mount Gipps station, George McCulloch, wanted no silver miners disturbing his flocks and therefore scared the prospectors away from the Broken Hill. But it was shielded from the prospector not by McCulloch's oaths but by less obvious forces. It was farther from travelled roads than Silverton and Thackaringa, and far from waterholes or wells. Its lode differed from Silverton's in appearance and in geological environs, and was much poorer on the surface. Even after Silverton became an excited mining camp few of its silver hunters rode towards Broken Hill, for it was isolated from Silverton's myriad of lodes by an alluvial plain on which few rocks were exposed. So it remained for the sheepmen who were ignorant of mines to develop Australia's richest single lode.

It was shearing season at Mount Gipps in September 1883, and twenty shearers worked in the long stone shed and the boundary riders trotted in with the sheep from distant paddocks. One of the riders was Charles Rasp, a lean German with small balding head and squirrel eyes. Thirty-seven years old, he was said to have trained as a chemist in edible oils in Hamburg before weakness in the lungs made him sail to a warm climate. His mind was sharp and curious. When Silverton was simmering he chanced to take a holiday in Adelaide and bought a guidebook on prospecting. Reading in his hut at night he gleaned some knowledge of the colour and weight of the common minerals and whenever he rode around the huge paddocks he was alert for them. On 5 September he had to ride twelve miles to the south end of the run to bring in sheep and as he had often seen the Broken Hill he decided to examine it. Chipping at the rocks he gathered samples and perused his guidebook. They were black, heavy for their size; perhaps they were tin. He took samples to the camp of two contractors, David James and Jim Poole, whose bullocks were excavating a large dam on the dry station. They handled the black rock and were also impressed. Banding together with Rasp they each put in a few pounds and applied for a mining lease over part of the hill. Sending samples to an assayer in Adelaide because they did not trust the Silverton assayers, they were told that their ore was carbonate of lead with a little silver. As silver, not lead, was the prized metal on the field they were disappointed. Still they hoped for more silver at depth and they paid a miner to sink a narrow hole to find it.

At the homestead George McCulloch, the manager, heard of

Rasp's discovery. A heavy Scot, pompous and jocular, thirty-five years old, he had no experience of mining, but he knew that mines meant money at Silverton, and he wanted money because, at the end of the long drought, his small share of the profits the Melbourne company made from the station must have been meagre. He therefore persuaded Rasp that more partners should join his syndicate and finance the search. Rasp, Poole and James agreed, and the syndicate was enlarged to seven. McCulloch himself took one of the seven shares, and so did George Lind, who kept the station's store and books, George Urquhart, the sheep overseer, and Philip Charley, a young jackeroo from Melbourne.

At Mount Gipps the gulf between the men who ate their meals in the huts and the men who dined at the manager's table was not so wide as on most pastoral stations, and the ability of both bosses and men to unite in the syndicate possibly had profound effects on Australia. If they had refused to unite, the most valuable part of the lode would have been split amongst several groups, and later amongst several companies. As it was, they united to form one strong syndicate that became one strong company, and that company gained enormous wealth and influence from the large area it acquired. Each of the seven partners invested £70 in the syndicate. Rasp could now afford not just the one lease of 40 acres, but seven leases totalling 297 acres. Sixteen days after the first discovery he formally applied at the warden's office in Silverton for the seventh lease. The syndicate thus got two miles (less ten chains) along the line of a lode that was to prove richer, for each foot of lode, than any metal lode in the continent.

The Broken Hill syndicate had acted wisely and swiftly. For all their ignorance of minerals they grasped a deposit that experienced men had overlooked. Ignorant of mining law, they simply took a treasury of a size such as no Australian legislator had ever envisaged should or would fall to one small band of men.

The lucky seven still had no idea that they were lucky. Gossip in the bars of Silverton damned the Broken Hill. 'For twelve months,' Rasp later confessed, 'it was really doubtful whether we would make anything out of it.' His darting, squirrel eyes saw scant silver below that iron hat. His partners were optimists, but there was no real base to their cocktail of dreams and hopes. They knew no geology, no mineralogy. They paid their weekly pound to the common purse and engaged a miner to sink a narrow shaft, but the best silver he found assayed only ten or twelve ounces to the ton. Hopefully they awaited a rich discovery or the offer of a few thousand pounds for their leases, but no offer came from the promoters who were fêted in Silverton.

Charles Smith Wilkinson, government geologist, visited Silverton in July 1884 and, sparing only half an hour at Broken Hill, was not quite sure what to think of the craggy outcrop. On Victorian goldfields until inflammation of the lungs drove him north, he was inclined to predict that much gold might be found but he finally predicted much silver and lead. Next month a surveyor measured out the leases and he too oscillated between hot and cold. A month later another geologist, Norman Taylor, walked the hills and stroked his mutton chop whiskers in puzzlement. 'The ridge contains the most extraordinary and largest lode I have ever seen on the Barrier Ranges silver field, or, in fact anywhere.' Timidly he concluded that it was only *one* of the best mines on the field.

Shallow shafts and potholes had found plenty of lead but little silver. Perhaps the silver was deeper. An advertisement in the *Silver Age* newspaper at Silverton drew a team of Cobar miners to deepen the Rasp shaft for 45s. a foot. This was the deepest shaft but by poor luck or poor judgment it skirted the massive lode. The ore was poor in lead and very poor in silver, and as the shaft deepened both metals vanished. A horse of barren rock appeared in the bottom of the shaft and the miners continued to sink, hammering away at the sharp steel while the shareholders shivered with fear. After four or five feet the lode reappeared in the bottom of the shaft. At least the lode was massive if poor. What consolation was that? Everyone knows only rich mines paid in the outback.

The syndicate had got from the government the usual twenty-year lease over their mine, paying five shillings an acre or an annual fee of £74. If, however, they had been more confident they could have bought the freehold and the mineral rights of all the ground they wanted for £2 an acre. Many Silverton companies had been optimistic and had bought the freehold, but the owners of the largest mine of all missed their chance. In 1884 the state virtually ceased to sell the freehold of mineral lands. Thus when the twenty years of their lease expired, the owners of Broken Hill were at the mercy of the government and, for the privilege of renewing their lease, paid millions of pounds in royalties.

The seven partners could not afford to buy the freehold of Broken Hill in the first year of struggle. Some were paying more than half their monthly wage simply to explore the mine and could not afford it. The storekeeper and the sheep overseer each sold their share at a loss. The contractor Poole was short of money while excavating a dam and he sold a one-fourteenth interest to a passing cattle dealer in return for ten old bullocks worth about £40. McCulloch sometimes sold and sometimes bought, and on one famous night he played a game of euchre with a visitor to determine whether he would sell a one-fourteenth interest for £120 or £200; the visitor won and paid £120 for a share certificate that was soon to be worth more than a million.

Wealthier men visited Mount Gipps and bought interests cheaply from the station hands. William Jamieson, a big bearded Scot versed in theology and theodolites, was sent west by the government to survey Silverton leases early in 1884, and the worry of getting the waggons through the sand drifts to the field was rewarded by the chance of buying cheap shares in Broken Hill. W. R. Wilson, a burly Ulsterman, was managing the Day Dream mine and smelters and became the first experienced mining man to invest in Broken Hill. From the stone homestead of Corona station with its 3 million acres and 80,000 sheep rode Harvey Patterson to buy into Broken Hill. From nearby Poolamacca with its long shearing shed and its string of wool waggons rumbling down to Adelaide, Kenneth Brodribb called on McCulloch and was persuaded to invest in a certainty. One of the state's rabbit inspectors, out west in a vain effort to halt the invasion of rabbits, called at Mount Gipps and was astonished at the promise of the lode. Some of these driving young men were to become the real silver kings and the nation's great industrialists, but their immediate gift to Broken Hill was money to sink shafts deeper into the puzzling lode.

The Silverton men remained suspicious despite this buying and selling. Wilson was the one local mining engineer who thought the mine worth a gamble. Even pastoralists who paid a hundred or two for a share were more interested in selling at a profit than waiting for the mine to pay dividends. Sidney Kidman, a cattle dealer and later a cattle millionaire, was the man who bought an interest from Poole in exchange for old steers, and he had no wish to cling to the share. Travelling in the coach from Terowie to Silverton with a sharebroker named Harris he told him to sell his one-fourteenth share in Broken Hill for £150. Harris held the share in Silverton for some time until there strode into his office a huge sandy-bearded Irishman named Bowes Kelly, once a station manager near Wilcannia and now an investor. Kelly bought the share without even examining the Broken Hill and with no intention of ever examining the mine that would transform him into a magnate of many mines. In strong circling handwriting he wrote to his brother on 18 October 1884 confiding that he would sell the share at large profit when the fever for silver shares came again, 'which I am morally sure it will or I am greatly mistaken'. Back at his sea front hotel at St Kilda near Melbourne he waited for news of his many silver shares. At last he heard good news and wrote of Broken Hill: 'They are no doubt the daddie of the field.'

Philip Charley, jackeroo on Mount Gipps, rode to the hill one morning to see what the shaft was finding. Fossicking about the dump of ore hauled from below he cracked a stone here and there as all mine owners do. One stone he broke carried grey specks. Showing them to the miner he exclaimed, 'Chloride of silver.' The miner disagreed: 'Carbonate of copper.' Nevertheless, the lad took samples of ore to the homestead, and from there it went to the Silverton assayer who reported 700 ounces to the ton. A shareholder was lowered down the shaft in a boatswain's chair to find the place where the chlorides had been intersected. He chipped more specimens and they were just as rich. About the same time men found chlorides of silver on the surface. Broken Hill had the rich silver of the Silverton mines and was infinitely larger.

The shareholders spoke now of floating a company. Their Broken Hill Mining Company was not registered as a company and thus was fragile in the eyes of the law. The shares had been so split and decimated that it was uncertain who owned them. The shareholders were tired of paying calls to explore the mine and felt entitled to pass on the burden to others. Above all, the floating of a company would swell the market for the shares, enabling the gamblers to sell out handsomely.

Four leaders of the syndicate—the Scots McCulloch and Jamieson and the Irishmen Kelly and Wilson—met in June 1885 to organize the company. All in their thirties, all over six feet tall, they argued out and wrote a strange prospectus of a company that was to be Australia's largest, the Broken Hill Proprietary Company Limited. Their aim was to raise £18,000 from investors; £3,000 of that sum would repay them for all the money they had spent on the mine, and the remainder would explore the mine and perhaps equip it with tiny smelters. In addition to retrieving the money they had invested in the mine, old shareholders would get 14,000 of the 16,000 shares in the new company. Wilson explained this in the prospectus, inviting the public 'to become partners in what appears to be one of the most valuable mining properties in Australia'.

Curiously, the pamphlet enclosing the prospectus gave little evidence that the mine was valuable. The only report on the mine was written by men who owned shares. The only evidence of the value of the silver came from samples taken by the syndicate and tested by their assayer. There was no attempt to test the tonnage of ore available. Not one lead assay was mentioned in the prospectus, and yet if the mine were to prove valuable lead would be a vital source of profit.

Disguised by the crude manifesto, the magnificent lode was hawked on the stock exchanges. Five hundred shares were offered in Silverton at £9 each and rushed. Jamieson boarded the coach, wearing one of Wilson's suits and carrying a bundle of printed prospectuses, and went to Adelaide where solicitors sneered at the document and persuaded him to rewrite it. The five hundred shares offered in Adelaide were then quickly sold. A similar parcel of shares was offered in Sydney and sold more slowly, and Melbourne's allotment of shares was nearly shunned. At last, in August 1885, the B.H.P. was floated and registered, a name amongst hundreds in the share lists.

Some Melbourne speculators marvelled at the company's impudence in asking so much for shares in a mine that hadn't sold enough lead to fill a pencil or enough silver to mint a florin. At the mine the old owners grumbled that they had sold shares too cheaply. Soon after Jamieson had gone to the coast to sell the shares his black-boy Harry Campbell found rich chloride of silver at a point of the hill which no man had prospected. The silver lay in soft white kaolin, the mineral from which pottery is made, and assays proved it was richer here than at any part of the mine hitherto opened. A shaft was begun, the Knox shaft, and it penetrated ore that made the Rasp shaft seem a waste of energy. After two years the company had begun to prick the heart of the mighty lode.

During the struggle to explore the lode, McCulloch's syndicate had not sold the better ore found in the shafts, and maybe it was too poor to pay freight to the coast. Now, with richer ore, they could safely load the bullock waggons. In the Melbourne suburb of Spotswood the Intercolonial Smelting and Refining Company smelted the first 48 tons of ore for the astonishing return of 35,600 ounces of silver. The glittering bars earned B.H.P. nearly £7,500 in cash and roused the curiosity of hundreds of mining speculators who saw the silver on display in that popular mining house, the City of Melbourne bank. The fact that the ore was easy to smelt was more heartening than the fruits of the smelting. The company now planned its own furnaces at the mine, and Kelly and Jamieson in August went from Melbourne to Sydney and then over the Blue Mountains to Sunny Corner to inspect a silver mine and smelters, which were then Australia's largest. They decided to buy two Nevada furnaces capable of smelting one hundred tons of ore a day.

The furnaces took months to reach the mine and the company won revenue meantime by sending a thousand tons of rich ore across the range to the nearby Day Dream smelter. Income from bullion produced at Spotswood and Day Dream was so high that the company in effect had no need for the money raised in floating the company. The mine paid its own way and paid for the company's new smelters.

Hundreds of men walked to Broken Hill and won a regular job. From the ironstone crest they looked down on the baked town through which spun whirlwinds of red sand, and each week they saw a dozen tents or a row of shops fill gaps on the plain. There was a Silver King hotel to compete with Delamore's hotel hauled from a deserted camp at the back of Silverton, and W. R. Thomas remembered in his small book *In the Early Days* how John John came from Silverton to erect the first boarding house and paid as ground rent a dozen eggs, nine of which were rotten. A preacher of the Bible Christian Church drove his buggy from Silverton to sing Sankey's songs in an iron hut by Rasp's shaft and quickly his flock followed him east, for Silverton, body and soul, was about to be reincarnated in the city of Broken Hill.

The Day Dream smelters that had been the hope of the Silverton field closed in April 1886, and the new B.H.P. smelters opened a month later. Then dividends began. The short Silverton railway was continued to Broken Hill by the Silverton Tramway Company, and on 12 January 1888 the Duke of Manchester stood on a carriage at Broken Hill and opened the railway to the coast by smashing a bottle of wine over iron buffers that were untouchable in the heat. Thousands came in railway trains from South Australia, the most depressed of the six colonies, which was suffering the hangover of a boom that had toppled the feckless Commercial Bank of South Australia with which the Broken Hill Proprietary had first banked at Silverton. The flow of unemployed men from South Australia united with cheap transport to lower the costs of mining at Broken Hill. They swelled the profits of the Proprietary mine and the hopes of smaller companies along the line of lode. All those smaller companies now wanted was capital.

In the 1880s British capital surged into Australia, and much of it built cities and some of it speculated in suburban lands and distant mines. The value of mines on the Melbourne stock exchange increased by about £10 million in the last half of 1887, and Broken Hill mines were the market hares. But when the stock exchange closed for Christmas and brokers went to their seaside mansions there was no sign of the excitement that would break in the new year.

Broken Hill Proprietary shares each had a nominal value of £19, and the last sales for 1887 were at the startling price of £174. 10s. When stout brokers in frock coats entered the Melbourne exchange at the start of 1888 they heard a slightly hysterical sound when the list of silver shares was read. Messengers hurried in with bundles of telegrams from all over Australia, from New Zealand and Fiji, ordering shares in any mine at Broken Hill. B.H.P. shares led the rise, jumping from £175 to £275 in nineteen days. As the price of each share was beyond the budget of those small buyers who were so prominent in booms, some brokers thought shares had reached the peak. They had not; by 2 February they were £380 and on the 24th there were sales at £409. That gave the mine a market value of £6·5 million, thirty times the sum it had paid in dividends.

The high price of B.H.P. shares made speculators think adjacent mines might prove as rich. Shares in the Central were 3s. at the start of 1887 and 185s. thirteen months later. Junctions, Souths, Norths, and a ring of other Broken Hill shares rose and fell like puppets on strings. In one week South Broken Hill doubled, then halved, then rose by 50 per cent. On the same day they were selling at £9. 15s. in Adelaide, 13 guineas in Melbourne, and more in Sydney. Here was the first national mining boom.

The clamour for silver shares was strongest in Melbourne. Speculators crowded the entrance to the two stock exchanges. Brokers worked night and day, their clerks so weary that their errors cost brokers thousands. Brokers could afford them, for it was estimated that eighty brokers had made a quarter million in brokerage from silver shares alone. On several days more money possibly changed hands through silver shares than through all shares in any day to the end of 1886. Even the great gold booms seemed insipid by contrast, and brokers were abandoning Bendigo for Melbourne. The popular Governor of Victoria, Sir Henry Loch, was gambling in silver shares, and at fashionable balls and church tea-meetings all talk was silver. (And when this boom fizzled, businessmen in the insolvency court and bank clerks in the criminal court told fascinating tales of big gains and bigger losses in Broken Hill shares.)

Most silver investors probably did not care if the mines were poor or rich. Any share was worth buying so long as it could be sold at a profit a week later. And so miners who held leases five miles from Broken Hill and even five hundred miles away sold them, promoters hired an engineer to report quickly on ground holding a few squibs of lead, and a new company advertised its prospectus. In the Melbourne *Age* eighteen new silver companies advertised their prospectuses in eighteen days. Mines at Thackaringa and Silverton that had long been derided and abandoned were floated as Melbourne com-

8 Miners of the Day Dawn Blcck & Wyndham, Charters Towers, 1890

9 An underground forest: square-set timbering at Broken Hill in the 1890s

10 George McCulloch and Charles Rasp, Broken Hill

panies with their own office in Queen Street, and £5,000 or more in their bank account. Men who had never descended a mine were persuaded to direct them; Professor Elkington took time from his chair of history to take a chair on Wheal Byjerkno—a tin mining company, but what did that matter?—it was only forty miles from Broken Hill. Melbourne merchants floated one silver lease merely by printing an anonymous letter praising it. Zebina Lane, soon to be mayor of Broken Hill, hurried to a hungry lease 4½ miles to the north-west and declaimed with that assurance that was to win him a fortune at Kalgoorlie: 'In my opinion this company will pay from its very inception . . . the very first requirement of the mine will be ore bags to take the ore away to market.' The first need of the mine was something to mine.

Despite the excesses of the boom it stimulated silver mines. The boom provided money to test lodes far from Broken Hill. Melbourne, Ballarat and Sydney blew money into the new silverfield of Cordillera in the mountains south of Bathurst, N.S.W., and the name was magic for a month, 5s. shares in one mine selling in Melbourne for 77s. 6d. before machinery had arrived. Adelaide money revived the silver mines of Glen Osmond, the oldest in the land. Melbourne and Ballarat capital opened silver mines in Tasmania. Wilson, Jamieson and the giants of B.H.P. floated the Commodore Vanderbilt, raising £20,000 in a few days to develop Captain's Flat near the present city of Canberra. And along the line of lode at Broken Hill two of the huge companies still mining today, the North and South, grasped money they needed.

The Broken Hill Proprietary even floated mines, severing more than half its ground and splitting it amongst three new companies. It originally held seven blocks of ground numbered 10 to 16, and in February 1887 it severed block 14 and floated a new company of 100,000 £5 shares; 80,000 shares were handed as a gift to B.H.P. shareholders, and the rest were sold at £4. 10s. a share in Melbourne, Sydney, Adelaide, and Silverton. Of the money raised £40,000 went to B.H.P. and £50,000 to finance the new company known as Broken Hill Proprietary Block 14. There are few parallels for this decision of a rich mining company to sell valuable ground; presumably the shareholders were pressing the directors for higher dividends and a bonus issue of shares in a new company was the directors' shrewd reply.

The big pastoralists on the board were so pleased that they decided to sell their two northerly blocks 15 and 16, and after spending a fortune in commissions and publicity in London they floated the British Broken Hill Proprietary in November 1887. It was an astonishing deal. The promoters created 240,000 shares of £5 each, gave a third as a gift to the lucky shareholders in B.H.P. and sold the other two-thirds to British investors for £800,000. For the shareholders of B.H.P. it was a wonderful bonus, for they not only got their free shares in the new company but £576,000 of the money raised by floating the company in London. For the Cambridge widows and retired colonels of Brighton who had bought the new shares to provide for their retirement, it was a sad transaction. The British mine was not worth the price.

At the peak of the boom B.H.P. floated a third company, the Broken Hill Proprietary Block 10, on terms almost as favourable to its own shareholders and almost as cheerless for those who jostled to buy the shares. From these three deals the B.H.P. shareholders got in cash and shares a gift worth at least £3 million at the zenith of the boom. That was a mere appetizer to the treasure they got from their company's remaining leases.

Broken Hill the town savoured its own sudden fame. Spreading below the long dark ridge it amazed reporters and investors who arrived by railway. In the red dust of Argent Street the oyster vendors cried, and drunks lay on the footpaths, and the tide of people halted at the windows of iron shops where delicate specimens of silver gleamed by kerosene lamps and prospectuses of new syndicates were pasted on walls. All talk was shares. Professional men, illiterate labourers, boys and braggarts, talked shares and bought shares. Beneath the balcony of the Denver City Hotel men gathered in knots all day, brokers shouting aloud that they would sell Rockwells or Pinnacles, clerks chalking on blackboards, the hubbub growing as the hot sun went down. No shares could be ignored when promoters from Melbourne were paying absurd prices for small mines they hoped to launch on city stock exchanges.

'Argent Street would disgrace a Hottentot village,' wrote the *Silver Age*. So would all Broken Hill, it had grown so fast. Streets were coated in dust to the ankles, and a sudden wind turned them into a Sahara. Water carted by horse three miles from the government dam was too dear to be wasted in washing gutters or perspiring bodies. Houses were built without outhouses and flies were a plague, and scores of strong men in a town of strong men died from typhoid and were buried the same day in coffins of packing cases.

It was freely predicted that the field would have 100,000 people, and such boasts seemed sane in January 1888 when the silver fever raged and Australia celebrated its centenary. The town then had only six or seven thousand people, but many new furnaces were blowing smoke across the sky and many new mines were calling for men and high wages drew in thousands more. At the 1891 census Broken Hill had nearly twenty thousand and still it grew.

The first impact of the field on Australia was not closely observed, being spread over three colonies and blurred in their statistics. While Broken Hill was the third largest city in New South Wales it got most of its supplies from South Australia. The mines gave work to South Australia's wharves, railways, farms, factories, warehouses, and silver smelters. Broken Hill was part of South Australia to all but map-drawers and politicians, and it must have had in the late 1880s an effect akin to a vast programme of public works in a sick economy. For every eleven breadwinners in South Australia there was one in Broken Hill and nearly all he spent or saved returned to Adelaide. The Proprietary alone paid over a million in dividends in both 1890-1 and 1891-2, and much of it went to Adelaide. Indeed, 1891 was Broken Hill's first peak of output, and it contributed nearly all the £3·6 million of silver-lead exported from New South Wales. That year the narrow ridge earned more than all Victoria's gold mines. That year it earned a sum equal to one-fifth of all the wealth earned from agriculture in Australia.

The Proprietary mine was directed from Melbourne and that city gained much from the field. The young pastoralists had chosen Melbourne because they came from that city or spent their holidays there or chose to retire there as magnates and gentlemen. They had planned a main board of three directors in Melbourne and a local board of four directors in Silverton, but such was the exodus of directors to build their mansions in Melbourne that the Silverton board does not seem to have met.

Every member of that first board of a company that was to dominate in turn the base-metal and the steel industry was essentially a wool man. A. R. Blackwood, Duncan McBryde, Harvey Patterson, Bowes Kelly, Kenneth Brodribb, William MacGregor, and George McCulloch knew little about the complex industry they were creating. Spending much of their time in Melbourne they were isolated by an 800-mile trip by train and coach from the mine. They therefore needed skilled managers and at first lacked them. William Jamieson was first general manager, strong in build and flexible in mind, chewing at his pipe as he mused over the mine's problems but not always able to solve them because he was by training a surveyor. He gave way to Samuel Wilson, brother of one of the large shareholders, and this nepotic appointment was unwise. Sam Wilson was a mining engineer but he had got most of his experience on small, hungry mines, the Phoenix gold mine at Mount Browne and the Day

Dream near Silverton. Luckily his brother, William Wilson, joined the board in 1886, the first director who knew mining, and Wilson persuaded the board that there was no man in Australia with skill to manage the gigantic lode.

The directors sat around a dark table in a small room in Collins Street, Melbourne, in July 1886 and made perhaps the most momentous decision in Australia's industrial history. They decided to send Wilson abroad to sign the best man money could sign. And they sent him not to Cornwall, which had long been the crumbling home of Australia's mining skills, but to the United States, which was probably the most advanced mining country in the world.

From San Francisco, Wilson went to Virginia City in Nevada, where the famous Comstock lode was waning. The Comstock had long boasted that it was the greatest mining camp in the world, and if this was an extravagant claim for a camp that never had as many people as Ballarat, it was still close to truth. Its sidewalks were said to have restaurants as fine as Paris's and drinking saloons more lavish than Broadway's. Its miners worked in great heat at great depths, retreating often from the heat of the working face to underground chambers lined with ice sent daily from the surface. The silver and gold they mined was astonishing, the great Consolidated Virginia alone mining sixty-one million dollars of precious metal in nine years. In the 1870s it was probably the world's most productive mine, and its successive superintendents were the promiscuous James G. Fair, later United States Senator, and a versatile engineer and metallurgist named William H. Patton.

Wilson met Patton in 1886 and was impressed with him and his methods. He heard that Patton had designed Comstock's second largest stamp mill in 1874, that he had later played a part in conquering the blend of bad ventilation, high temperatures, underground fires, and shifting ground that made the Comstock one of the most difficult lodes to mine. With the consent of the board in Melbourne he induced Patton to become general manager at Broken Hill for the huge annual salary of £4,000—twice the allowance the highest-paid Australian politician received.

Wilson crossed the Rocky Mountains to Colorado. Whereas the Comstock led in mining engineering, Colorado led in metallurgy, for its ores were harder to treat than those older fields to the west and that challenge was met by metallurgists who were schooled in Welsh smelters and German academies. In Colorado's grimy smelting towns one metallurgist in this new tradition who appealed to Wilson was Herman Schlapp. He was aged thirty-two, native of Iowa, lean and generous with cropped hair and small spectacles. His schooling at the Royal School of Mines at Freiburg in Saxony had been applied with success as metallurgist of Pueblo Smelting and Refining Company and as manager of Gunnison Smelting Company. For a high fee Schlapp came to Broken Hill as metallurgist, the first of many metallurgists, mine managers, and skilled miners to come from the Rockies.

Patton, arriving five months after Schlapp, came in time to face a crisis. The lode in the Proprietary leases was oxidized, heavy and friable. Australians were inexperienced in mining such a lode. The quest for early profits and the difficulty of supplying strong timber to support the underground workings before the railway reached Broken Hill made them tear out the rich ore without adequate care for the safety of the mine. The native gum and the New Zealand pine they used sparingly in the mine soon decayed. Slabs of soft crumbling ore often collapsed underground. Shafts which passed through the lode twisted shape with the slow movement of the ground. The foundations of buildings above the lode cracked as the ground subsided. The wise practice of filling the abandoned underground chambers with rock sent down from the surface was rarely followed. The Proprietary mine was soon like a series of catacombs, the walls often as weak as sodden paper, the roof pressed down by the weight of pillars of heavy ore and the wide ironstone outcrop. The canny Cornish miners could minimize the risk to life by mining carefully and by leaving an area at first sign of danger, but they could not prevent huge blocks of ground from settling and subsiding.

Patton went underground and thought square-set mining was the answer. A young German engineer, Philipp Deidesheimer, had devised it on the Comstock to mine a lode of clay and decomposed rock that swelled and shifted when exposed to air from shafts and drives. In an underground stope his square-set timbering resembled, in cross-section, the scaffolding of a steel-frame building. Heavy logs of sawn Oregon formed strong stagings that supported roof and walls of the stope and shielded men from a fall of rock as they mined into the walls. Slow and expensive, it enabled miners to work blocks of ground which otherwise were hazardous to touch.

In 1888 miners came from the Comstock and turned Broken Hill miners into builders, and several thousand people in the forests and seaboard of North America gained their bread, directly or indirectly, from hewing and milling and shipping forests of Oregon to the vast city of the underground.

The Comstock way of mining did not entirely suit the wide lode at Broken Hill. Even after many local mines had adapted and varied it, the sheer pressure of the lode occasionally crushed these scaffolds of timber in parts of the mine, and the heavy use of timber in the dry oxidized zone of the Proprietary mine led to serious fires.

On 21 July 1895 fire broke out in an abandoned area where the crushed, splintered timbers dipped vertically for two hundred feet. Though the area was sealed the fire burned slowly for years. Smoke was again seen underground on 12 September 1897, a small fire in the timber, but its fumes killed three men. Five years later that fire revived, and thirty-seven men had to be carried to safety, and lofty walls of brick were erected far underground to prevent the fire from spreading. As late as 1906 the town was shocked by news of smoke and fumes drifting along galleries and drives in the Proprietary mine, and one miner was killed by fumes and one rescuer was killed after falling down a shaft.

The low range was stalked by disasters in these years. Even when mines entered the sulphide zone where the lode was more stable, the troubles continued. In the Broken Hill South mine on 18 July 1895 a slow slipping of rock made the manager order twenty men from a stope. Nine men deemed it safe to sit around on the level nearly two hundred feet from the stope, and as they sat there a great slab of ground fell in the stope and bellowed a tremendous concussion of air along the narrow drive, lifting them 'off their feet like mere feathers' and smashing them to pieces.

The Broken Hill lode was widest at the south end and there that slow movement of ground which miners call a 'creep' was widespread. In the Central mine in 1902 a creep crushed two men, and their bodies were not found. The rock soon settled, the mines inspector no longer heard the rumbling and whispering of moving ground, and work went on with normal caution. But the ground began to talk again in July 1905, and the men took the warning and left the mine. On 27 July there was a drastic movement of ground such as no field in Australia had witnessed. On the surface about 300,000 square feet of ground subsided, and the mine buildings and most of the mill had to be abandoned. Underground most of the workings between the 400 foot and 600 foot levels were destroyed or distorted. After the vast slide of rock had settled, men found it eerie to walk along a drive and suddenly find it blocked by a mass of rock that had slid down from far above. In time they mined out much of the ore but for years the ground moved intermittently. On some levels of the mine men would mine all the ore, fill the cavities with waste rock, and abandon it. Returning years later they would find a fresh block of ore there, having slipped from above. It was as if a tenant of a city office had caught the lift to the eighth floor, walked along the passage towards his rooms, and found in their place the remnants of the rooms from eight storeys higher. To many miners from the narrow, stable lodes of Moonta or Bendigo, the field was frightening.

Managers took long to realize the extent of these movements of rock. They built mills and smelters and offices on the hanging wall of the lode, and masons built beautiful walls and foundations of dark local stone and thought they would last forever. But the creep of the ore affected the strong overhanging rock near the surface, amazing Patton and Retallick and men who had managed Comstock mines for a decade or more. Soon after Patton arrived, the Proprietary built a mill near the lode and had to abandon it five years later because the moving foundations had set all machinery out of line. Block 10 and Central mines had the same costly lesson. Perhaps £100,000 was lost on the surface of the lode through trust in its stability; much more was lost in the darkness below.

Most Broken Hill companies found their smelters a heaven beside their mines. The Proprietary led in smelting as in mining, and Schlapp from Colorado created the largest metal smelters in the land with fifteen furnaces and nearly a thousand men. Smaller companies built nests of furnaces at the Hill, and visitors delighted to walk at night through the town and see the red rivers of slag trickling in the darkness. Although the British Broken Hill company preferred to send its rich ore to its own smelters amongst the mangroves of distant Port Pirie, and the busy Junction mine had its own smelters at Port Adelaide, most companies smelted their ores at Broken Hill. In drought they had to cart water from Silverton and at all times they had to rail ironstone from tiny quarries over the South Australian border. And to provide the furnaces with the essential limestone as flux, a public company built forty miles of railway north to Tarrawingee and carved a quarry on a windswept hill. But first the small smelters closed and then the great Proprietary smelters moved to the coast, and Tarrawingee shares, famous on every stock exchange in Australia, fell from pounds to pence, the railway useless, birds nesting in the silent quarry, and a township ruined overnight.

Most speculators had long seen a halo around Broken Hill but in the early 1890s the halo faded. They had exaggerated the treasure under the hill, they had exaggerated the wealth of all Australia. The whole field at Christmas 1892 was valued by the public at only two-thirds the price they had placed five years earlier on the Proprietary mine. The price of lead had fallen slowly since 1886, and the price of silver had receded ever since the field was found. Silver was a political metal and now was fiercely challenged as a sister to gold in international currency. The Indian mints closed, the United

States government ceased buying. A bar of silver worth £1,000 in 1884 was worth only £572 in 1894. Worse still, Broken Hill was now mainly mining the deeper sulphide ores, and the metals in them were not only poorer, but harder to extract. And so in 1896 the city said amen to the prayer of the Rocky Mountain silver states and their presidential candidate, William Jennings Bryan, that 'you shall not crucify mankind upon a cross of gold'. And when Bryan and his Democrats, and his plea for a silver standard and dearer silver, were defeated at the American election of 1896, Broken Hill had to solve its own problems.

PART III

Resurgence

15

INDIAN OCEAN GOLD TRAIL

THE LONG SLOW DECLINE of goldmining ceased about 1886, and that year the lodestars shone in northern Australia. Charters Towers gold was exhibited to thousands in London, and Englishmen invested heavily in its mines. Three hundred miles north-west of Charters Towers, in the heat of January 1886, Croydon was proclaimed a goldfield by the government and by thousands who rode or walked overland from the east or from the wharves of Normanton on the Gulf of Carpentaria. Far west on the goldfields of the Northern Territory the Chinese coolies were building the railway from Darwin, which all the optimists of 1886 swore would transform old gashes in the earth into a rich goldfield. Further west the amazing rush to the Kimberley district began in 1886. Such was the glamour of this first Western Australian goldfield that hundreds left Charters Towers at the peak of its share boom and followed the sun for nearly 1,800 miles.

The late birth of goldmining in Western Australia puzzled many men but not those who knew the land and its vastness. The gold regions lay in some of the hottest and driest parts of the continent, and so pastoralists long ignored the grasses that grew above the gold. Without sheep and cattle stations to supply lifelines for prospecting, few men ventured into the hot inland in search of gold. Those explorers and land-seekers who did venture sometimes saw gold but signs were few. There were few waterholes where they could wash the soils for gold, and even if there had been more water most of the colony's auriferous areas lacked the splendour of Victoria's surface alluvial gold. Moreover, long, undulating plains of soil and stones concealed like a blanket most of the solid rock that was host to the gold. This was Australia's Pre-Cambrian shield, formed long before the golden lands of eastern Australia, and the ancient rocks that did outcrop were unfamiliar to many travellers.

The northern part of Western Australia was known as the Kimberley district. Its tumbled mountains rose in the path of the monsoons, and sand dunes and desert isolated it from the pastures of the south. Alexander Forrest, the surveyor, crossed this north-west corner of Australia in 1879 and vowed that the valleys of the Ord and Fitzroy would each nourish a million sheep, and a few pastoralists came from east and south to call him liar. Forrest and his geologist Fenton also saw gold in the desolate ranges, and when the weary party reached the overland telegraph line in the Northern Territory and rode north to Pine Creek and spread their news in that waning goldfield there was excitement. Adam Johns, a mine manager at Pine Creek, and his mate Phil Saunders had had poor luck and they decided to risk their savings in the expensive search for gold in the Kimberleys. Few ships sailed from Darwin so they hired a cutter in August 1881 to take them to the small pearling port of Cossack, seventeen days' sailing to the south-west. After prospecting 250 miles of the Ashburton River without finding gold they rode far north to Yeeda station on the Fitzroy River and waited two months for their provisions to arrive by sea. They had already been a year away from Pine Creek when they left the Fitzroy River to prospect in the ranges to the east, and the expenses of their trip, the chartering of ships, the buying of horses and stores, were so high that one can understand why the desolate parts of Western Australia defied for so long the skilled prospectors. Only rich miners, skilled with horses and used to rough terrain, could even attempt the search for gold. With few ships trading the coast and few sheep stations inland, their task was herculean.

Johns, Saunders, a white man and a Darwin blackboy left the new sheep stations on the Fitzroy River with twenty horses and eight months' provisions. A forlorn band, they rode into the ranges early in August 1882. They tried for days to cross sheer sandstone cliffs of the King Leopold Range. The natives menaced them. The horses found little grass and lacerated their feet in the stony country. Johns rode in pain, barely able to see or walk, ailing with a paralysis he thought was rheumatism. In tributaries of the Ord they found gold in many places but, being the dry season, they had no water to wash the soil thoroughly. With Johns still in pain they rode into the Northern Territory, reached the overland telegraph at Katherine, and got to the gold diggings at Yam Creek in time for a welcome banquet on 28 October 1882. No men more deserved their welcome. They were the first experienced prospectors to find gold in the west, and, but for the expense of working poor gold in wild country that had no line of supply or communication, they would have returned after resting. They did return four years later, when the spread of the pastoralists had eased the opening of the goldfield. Saunders in time followed far south the chain reaction of rushes which the Kimberley gold exploded. He was a gold man to the last, dying at the age of ninety-three after falling into his campfire at the gold town of Menzies.

In the steps of the Pine Creek men came the geologist, E. T. Hardman, leading an official party well supplied with men, stores, horses, and medicinal rum. They could stay in the ranges longer than Johns and Saunders and they found more gold. Often Hardman counted nearly thirty quartz reefs in a mile. Hammering quartz to powder with an axehead, he observed specks of gold. At night in his tent with rum bottle and monocle, he wrote glowing reports of two thousand square miles of auriferous country. How payable the gold would be he could not tell, but when late in 1884 he returned to the civilized south and brandy and soda he spoke of grains of gold in most dishes he had washed. A visit to Bendigo and a study of its reefs and quartz convinced him that the Kimberleys were rich.

Hardman had failed to find the heavy nugget or the dazzling reef which alone could make men run to the Kimberleys. But his report inspired Charles Hall, a frugal bushman who saved money from sinking wells and making fences near Roebourne, to gather a party and search for gold. Hall and John Slattery led a team of six to Derby, the new port for the Kimberleys, where they borrowed ten horses and pack saddles from the government. On their first trip east they found ten ounces of gold. On a second trip they got eighty-one ounces including a nineteen-ounce nugget. Within months Hall's Creek was the magnet of the longest overland gold trail Australia had seen.

The Kimberleys were so isolated by land or sea that diggers spent hundreds of pounds reaching the field. Some rode up the coast from the pearling port of Cossack, crossing scorched desert. Some hired pearling luggers and landed at Derby, a tiny town on King Sound where the spring tide rose over thirty feet and receding left the town encircled by mud. Into the mud came a steamer from Sydney and a schooner from Darwin, loaded with horses and diggers. On 17 May 1886 Perth got a message sent by ships to Cossack and then by telegraph, 'The rush to the fields has already commenced.' A fortnight earlier a cable had reached Perth from Auckland announcing that a large party was about to sail and wanting to know where the Kimberleys were. 'Great excitement' were the last two words in the cable and they echoed through Australia.

Some ships jettisoned the diggers at Wyndham, where not even a jetty stood in the mud, and some passengers looked south and saw the bastion of mountains like the wall of China and went no further. A few drove horse and cart into the mountains and realized that only packhorses could reach Hall's Creek. Most diggers, however, landed at Derby and followed the easier trek three hundred miles east. At the nightly camps along the track, waggons, barrows and handcarts lay by muddy waterholes with baggage strewn beside them. Horse teams were overloaded, and as the horses had eaten hay on the sea trip and now ate coarse grass they could not pull the loads. Along the way goods were unloaded and concealed in marked graves. No water lay on one stage of thirty-seven miles between the Margaret River and Soda Springs, and the rotting carcases of horses and the tinned foods tossed from heavy swags became mileposts on the track. Tough men with fit horses who set out at four in the morning did not usually complete the waterless stretch until midnight.

This was the track where Russian Jack the digger won a place in Australian folklore. He had left Derby pushing a barrow with shafts seven feet long and a wooden wheel so wide that it did not sink far into the sand, and on the track he overtook two old men who were too tired to carry their swags. He loaded the swags on his barrow and pushed them to the next waterhole with the old men walking beside him. Another day he pushed a sick digger in the barrow to the nearest shade and water. Stumpy in body, grizzled of face, unable to write but able to curse, a gentleman to females and a lout to gentlemen, Russian Jack had loyalty. For years he was to mine gold for himself or for wages, and his loyalty to men he worked with became a proverb. Working years later on a Mount Morgan (W.A.) mine, he fell down a jagged open cut while walking through the bush in the darkness. Three days later they found him seventy feet below, cut and fly-blown, and his one comment was, 'I've missed a shift.' Russian Jack had reached Hall's Creek in 1886 with his barrow loaded high with water cask, tools, blankets, and food for months. Hundreds carrying half his load turned back without seeing the goldfield.

Far to the east the excitement was intense. The Government Resident at Darwin saw the ships calling with farmers, labourers, clerks, townsmen, bound for Wyndham and Derby and thought it was the greatest rush since Port Curtis and likely to be as disastrous. Nearly every European miner in the Territory who could afford horses and provisions took steamer from Darwin or rode into the wilderness, abandoning alluvial claims yielding £12 a week for the lure and uncertainty of the Kimberleys. Even the Chinese would have hurried west but for the new law preventing them from settling on a Western Australian goldfield until five years after it had been proclaimed.

Hundreds of men crossed the continent to join the rush. Saddlers in Charters Towers and blacksmiths in Croydon saw them preparing for the long journey. Towns in Western Queensland saw them ride through. The little port of Borroloola, a few iron buildings on a crocodile river near the Gulf of Carpentaria, provisioned them with potatoes and painkiller. The police magistrate counted hundreds going through the town. He saw one man of seventy pass on foot and a woman ride side-saddle with a child on her knees. A thousand miles lay ahead of them before they reached those shining hills.

The ragged cavalcade crossed under the telegraph line, calling at Katherine telegraph station for directions or food and water. From there to Kimberley three or four homesteads were the only habitations in 570 miles, and these homesteads lost countless horses and cattle. The manager of Spring Vale reported that 'great numbers of men from Queensland have passed by, some of them very undesirable characters, who prefer picking their own beef and horseflesh'. They faked the brands on their stolen horses with any piece of iron they could find, and at Kimberley one could see horses from nearly every pastoral run in Northern Australia.

Others rode up the telegraph line from South Australia, then pushed west. One party comprised two famous Moonta miners, George and John Brown (known better as Wonoka Jack), another whose only name was Tommy the Rag, a runaway sailor from Port Augusta named Hugh Campbell, and three others. At Newcastle Waters the seven joined six men who had come overland from Queensland, and 'The Ragged Thirteen' rode north to the telegraph station at the Katherine and then out west. On the way they fought bare fists with shanty-keepers, stole horses and cattle, desperadoes living off the land. Near Victoria River Downs their unshod horses bruised their feet; so the leader Tom Nugent, alias Tom Holmes, called at the station and pretended to be a Queensland pastoralist looking for land. He played crib that night with the storekeepers while his mates prised slabs of wood from the store and borrowed flour and sugar and four hundredweight of horseshoes.

Thirteen men riding together could protect themselves, but some went alone or in pairs; and when eight years later the South Australian geologist, H. Y. L. Brown, visited Willeroo station on a tributary of the Victoria River he heard of many diggers going or returning who had been killed by natives and he saw the natives using iron instead of stone in spears and tomahawks. The springs of broken drays and discarded strips of wire from the passing rush had brought the iron age to the district.

Near Hall's Creek the three treks met: the men from Wyndham, from Derby, and from far overland. Culled by endurance, they could still marvel at the feats of those who continued to arrive from the east. One man walked from Queensland with a bag of oatmeal and a small tent fly, eating his gruel by the fire at night with no gun to shoot into the moving shadows. A tattered group of six Afghans arrived with a woman who called herself The Mountain Maid, and she made money from the field though how she won it was never said.

And now they were there, men from everywhere and The Mountain Maid. The first warden counted about two thousand in September 1886 and perhaps there were never more than that. But so many came and went and so many did not reach the field that in the first year six thousand people possibly set out for the Kimberleys. The rush was small, but then the vast colony held only 36,000 people.

Some men had spent £500 on fares and equipment and, on the warden's estimate, the average man spent £100 just to reach the field. Few took that much gold away. The layer of auriferous soil on grassy flats and in the bars of creeks was so thin that large areas were denuded of gold in weeks. New Zealand diggers were dismayed to find no defined runs of alluvium and no shallow leads. Victorian diggers hurried away from ground where the nearest water was miles away. In the dry season the only way of winning the gold was dry-blowing, a method used in North Queensland, in western New South Wales, and far back in antiquity. Some diggers had two dishes, filled one with alluvial material, held it above their heads and let the dust and stone fall into the other dish on the ground. The dust blew away in the wind, leaving the pebbles and heavy specks of gold in the dish. Enterprising diggers scavenged amongst wrecked drays and abandoned harness and they improvised shaking boxes which they rocked in the wind on the principle of the winnowing machine. Dry-blowing was grimy and inefficient; and there were few rich patches of gold.

Men fled the field in their hundreds, gold diggers became grave diggers. Late in 1886 one digger counted eight graves on the diggings, eight near Wyndham, and three on the road between. Fevers and ague and dysentery were so rife that Warden Price wrote pathetically to the government: 'Great numbers were stricken down, in a dying condition, helpless, destitute of either money, food, or covering, and without mates or friends, simply lying down to die.'

Those who stayed to dig conquered despair. Cammilleri and Maitland, young seamen, spent Christmas of 1886 by a dray bogged to the axle. Maitland mixed a plum duff in the prospecting dish with a lacing of brandy and tied the duff in a clean shirt and sat till midnight by the boiling pot. On Christmas morning he curried some tinned tongue and produced the duff in a tin dish afire with brandy.

11 and 12 The smokestacks of the B.H.P. dominated Broken Hill in the 1880s

13 Broken Hill Proprietary's steelworks at Port Kembla

They gorged that tongue and duff until they were light in the head, and so Cammilleri wrote, 'We forgot our troubles, talked of old times at sea and in London, our friends and relatives and the girls we had met in many ports.' Others anointed their sorrow in the stifling heat of Christmas with whisky bought for a shilling a nobbler, the nobbler being a Holloway's ointment pot, the shilling being a lot of money. Ernest Giles, a brave explorer who had turned to digging, was one of many on the Mary River who celebrated by drinking a rum so deadly that he exclaimed to his mate Carr-Boyd: 'Carr, dear boy, it would kill the devil.'

In time the messengers of civilization came, a telegraph line from Perth, and a Catholic priest who was feasted with johnny cakes and honoured with the intent hearing of a packed hotel that drank no liquor while he preached.

The patches of auriferous country spread ninety miles almost parallel to the border of Western Australia and the Territory, but after one year they held only four hundred diggers with perhaps another two hundred teamsters, packers, and miners along the tracks to port. The reefs which Hardman had seen through tinted monocle still lay untouched. William Carr-Boyd, a tall buccaneer and raconteur, believed they were rich and sailed for Melbourne with half a ton of quartz as proof. One trial crushing assayed 162 ounces to the ton, and a delirious report on the mine from two Victorian engineers was enough to float the Jackson's Reef Gold Mining Company in June 1887. While shares soared, a Melbourne broker sent a man all the way to Derby to report on the mine, but he was not allowed to enter it. The stamp mill shipped around the coast did not reach the mine. The manager did and denounced it as a fraud. More Melbourne money came to work the reefs and erect eight stamp mills. A battery manager was speared by blacks, and the boom companies did not long outlive him. The reefs were patchy, and greedy for capital in that remote region. The warden telegraphed that the future of the field as one of the nation's richest was assured, but his message faded as finer fields were opened to the south.

Kimberley's influence was not measured by its meagre output and discredited name. It drew to Western Australia tough and alert prospectors who had learned elsewhere how to survive on scorched earth. They were the men whose petitions and resolutions helped persuade the colony to learn from the experience of Queensland and the Territory and ban Asiatic aliens from new fields. And they were the men who found rich gold in so many places down the coast, for Kimberley began a lightning run of finds that in seven years stretched in a long arc one and a half thousand miles from the Timor

Sea to the Great Australian Bight. No large mining region in the world was opened so rapidly.

Two years after the Kimberley rush diggers rushed for gold at the Pilbara. The nearest towns to Pilbara were the pearling port of Cossack and the inland town of Roebourne, the capital of the north-west where a Government Resident was ruler and a judge held quarter sessions when the town was not devastated by hurricanes. Seventy miles east of Roebourne, Withnells ran a sheep station near swampy coast, and their fifteen-year-old lad Jimmy was fascinated by gold, because Hall and Slattery had stayed at his father's homestead before they found the gold of Hall's Creek. On a sweltering day he was cutting slabs for a hut when he saw a crow sitting on his tuckerbox and picked a stone to throw at it. The stone gleamed with gold. Excitedly he went to a big outcrop of quartz nearby and found similar stones carrying free gold. His father filled the buggy with rich stone and drove to Roebourne to report his find to Colonel Angelo, the Government Resident. Angelo gazed on the gold and sent a hasty, incomprehensible telegram to his superior in Perth: 'Jimmy Withnell picked up a stone to shy at a crow.' The Colonial Secretary in Perth is said to have replied: 'What happened to the crow?'

Nothing happened to the crow but the sheep run was soon invaded by diggers. On Pilbara Creek to the south-east Ben Christie found rich, jagged gold, often adhering to pebbles of quartz, and quietly gathered a harvest simply by picking the gold from the red earth. He worked alone for six weeks until one moonlit night at a waterhole he shot his own mouse-coloured horse in mistake for a kangaroo and had to walk to Mallina for another mount and more supplies. He was followed back to Pilbara, and from there diggers pushed over the hot tablelands and the dry gorges to Coongan, Marble Bar, Nullagine, Robber's Gully, Paddy's Market and the Ashburton River, 15,000 ounces being won in half a year on that river alone. Rich nuggets were picked from the ground, and Villars found a 127-ounce nugget at Pilbara Creek in November 1888, the first of many culminating in Doyle's 'Little Hero' nugget of 1895, a beautiful, pale lump weighing 333 ounces, and Archie Clive's 'Bobby Dazzler' nugget of 487 ounces found in 1899.

Kimberley was powerless to hold its dusty men. It had inspired the legends, Russian Jack and the Ragged Thirteen, and men who were to be revered, impersonated, or misted by folklore on the richer fields to the south; but those legends were of hardship, not wealth. In contrast Pilbara had wealth, and diggers hurried to Derby and went in luggers and small steamers to Cossack, where the new lighthouse guided in the ships and the Pearler's Rest hotel had every inch of floorboard taken by sleeping diggers. Others rode eight hundred miles overland by coastal sands, drinking brackish water from wells where pearlers called, eating salted beef which dried hard in the pack bags, and plagued with flies by day and mosquitoes by night. On the way many diggers passed the French Trappist missionaries riding atop bullock waggons on their slow way north to found the mission at Beagle Bay, and at night the Trappists would set up an altar beneath a tarpaulin and in white robes hold divine service for the diggers. 'Them fellows,' said one digger's blackboy, 'look all same cockatoo.'

More than a thousand men scattered across the tablelands of the Pilbara, men of every rank and occupation. Warden W. L. Owen, who settled the diggers' disputes, met a broken-down baronet, a former parson, a man who said he'd been captain of the Cunard Line, a forty-niner from California, and a London lawyer who camped under a tree—the dirtiest man on the field and perhaps the poorest, for he won only fourteen ounces in a year. Statistics are too flimsy to tell the real richness of the Pilbara, but those who saw diggers pull out gold in the bank at Roebourne, or saw them, in hessian grog shanties, boastfully pat their chests and withdraw a chamois cloth bulging with gold, glimpsed Pilbara's richness. Out at Marble Bar and Nullagine the heat danced from bare hills and 160 successive days of century heat were once recorded, but men gladly endured the hottest part of the continent for gold in such profusion.

The march of prospectors and camp followers down the western coast of Australia quickened in 1891. Seven hundred miles down the telegraph line from Cossack and Roebourne was the wool port of Geraldton. Lead and copper mines had long been worked near Geraldton, but the mines were struggling. The port wanted gold in the dry hinterland where sheep and cattle roamed sparsely by salt lake and crumbling tableland. Certainly gold was there, and isolated specimens were found by station hands.

On Sunday, 13 July 1890, J. F. Connelly found a rich reef on a sheep run about three hundred miles by track from Geraldton. He was an observant and scholarly fellow of twenty-nine, native of the Victorian gold town of Inglewood, a champion pedestrian, a trained draughtsman in the New South Wales railways, and an amateur ethnologist and mineralogist. He had toured the Kimberley and Pilbara goldfields, observing them sharply. In 1890 he was paid by F. C. Monger to lead a team of three men with blackboys to find gold in the Murchison district beyond Geraldton. In the glorious winter days he reached the salt lake near Annean sheep station and learned that station hands had recently found faint colours of gold. Shown the place, he looked carefully and found slugs of gold in a big white blow of quartz. His trained eye had found the first rich gold of the Murchison field, one of the greatest of Australian fields.

Connelly reported the find to the little police station at Mount Gould, but he decided the reef was too patchy. Restlessly he pushed on to find gold elsewhere. His party split, and two of the men lost their way and perished of thirst. When at last Connelly returned to Nannine, the scene of the original find, others who had read reports in the Geraldton newspaper were busy picking up pieces of alluvial gold. Gilles McPherson, tall Scot with chiselled face and the toughness of granite, arrived first with his mate Peterkin and found a thousand ounces in weeks. The gold was so rich, the water plentiful, meat everywhere for the killing, that no digger had need to return to civilization, there to spill the news of the field's richness.

At last a brawl occurred and an injured man returned to Geraldton, arriving with his waterbags heavy with gold. The news was cabled up and down the coast, passed by word of mouth from Roebourne to Marble Bar and the Ashburton diggings. Men shoed their horses in the sheds attached to bush stores and rode in haste to the coast to catch the first steamer going south. Some had travelled on horseback and ship a thousand miles when they landed at Geraldton, and the diggings were still three hundred miles to the north-east.

For farmers' sons and labourers around the capital city of Perth the news from the Murchison was more compelling. The journey there was far cheaper than to the tropical fields of Pilbara and Kimberley, the climate was kinder, food was cheaper, and they heard that the gold was incomparably richer. They landed at Geraldton, paid £5 to a teamster to carry their tools and heavy load of provisions, and walked beside the waggon to Nannine. Those with horses and packhorses raced ahead, nursing that nervous fear of all rushing diggers that the finest gold would be taken a day before they arrived. But Murchison could quell all nervous fears.

Connelly, the finder of Nannine, leased six acres over a reef with veins of gold as thick as his fingers. He broke off much gold for himself, sold the lease to an Adelaide syndicate for 3,100 sovereigns, and rode away with the sovereigns under armed escort. The reefs were fit not only for city syndicates but for brawny miners, the gold so rich in patches that miners could dolly the hard rock and extract the visible gold at high profit. The alluvial gold was shallow and soon scarce, but before one place was worked out a new one was ready for the rushing regiments. Steadman found gold near Mount Magnet while looking for a lost swag and with a station owner and natives picked up two hundred ounces in a few weeks. Tom Cue found the field that honours his name, the first tents were pitched at Day Dawn, and the riches around Lake Austin enabled diggers to indulge in long drinking sprees with enough money remaining to search for new fields. At the height of the rushes, fevers killed as many men on the Murchison as they had killed on the Kimberley. Enteric fever was widespread in the tent hospital at Cue. On some days at The Island on Lake Austin four or five men died, were wrapped in their blankets and buried in coffins hammered from packing cases labelled 'Coleman's Mustard' and 'Milk Maid Brand'.

Within five years of the Kimberley rush, gold was being mined on four main fronts in Western Australia, each front with its own isolated ports. In the far north was the Kimberley goldfield, with its tidal ports of Wyndham and Derby. Farther south was the Pilbara, shipping its gold through Cossack and the long muddy beach of Condon. Far south of the Tropic of Capricorn the Murchison obtained its provisions and medicines from Geraldton. And lower down the long coast Perth, the largest town in the colony with nearly 10,000 people, had its own goldfield out by the low hills and dazzling salt lakes of the Yilgarn. With each successive rush the centre of gravity had jumped south. Now the Yilgarn, east of Perth, became the spearhead for the most important of the long line of Western Australian discoveries.

The first gold of the Yilgarn, like the first gold of Pilbara and Murchison, had been found on the fringe of pastoral settlement by men of the sheep runs. Their awareness of gold was heightened by the rumoured riches of Kimberley in 1886. In the following year three men, Anstey, a mineralogist, Payne, and Greaves, went east on behalf of a Perth syndicate to inspect rumoured finds of gold in the distant hills. A smart native on a sheep station led them to a native well to point out the white quartz his tribe called 'yilgarn', and in the quartz they found gold. They employed two labourers to blast a trench along the reef, but the men wandered and died from thirst. Others came to inspect the reef, were not impressed, and went eleven miles south to find the gold of Golden Valley. New parties pushed further south, finding richer reefs in January 1888, and naming them the Southern Cross. Each discovery was a guidepost, a supply base, a stepping stone to new reefs. After 1888 there seemed no more steps to take on the Yilgarn.

Companies were floated in Perth in 1888 to work the reefs, and the Perth stock exchange was born to speculate in the shares. At

Southern Cross six stamp mills were soon pounding the rock, and several hundred miners worked for wages underground. Lying 250 miles from Perth, Southern Cross became one of those quiet dead-end towns to which men escaped from wives and debts and murders. In 1891 Alfred Deeming came to Fraser's mine as engineer after quietly cementing the bodies of two wives and four children in the floors of Melbourne and London houses, and even after he had been led away to face the most publicized murder trial Melbourne had seen, Southern Cross still failed to stir the imagination of visitors who had seen the individualist mining camps of the north. It seemed a tame town; it had resident clergymen, a bank, and the amenities of social life. No one envisaged that it would be the starting place for the last of the huge gold stampedes.

East of Southern Cross swept an undulating plain with millions of spaced gum trees on the pink and red soils. The miners of Southern Cross called it desert, and desert it was in so far as it was deserted and parched. It had repelled pastoralists who had hoped to find watered grazing land. It had disappointed and endangered some of the government explorers who pushed through—Lefroy, Hunt, Ernest Giles and John Forrest. It was well-known that these explorers had found no gold, but then well-fitted parties had crossed most of Australia's mineral fields without observing them.

These plains were a barrier as real as mountains to the men of Southern Cross. If there had been rich patches of gold twenty or thirty miles apart, prospectors could have reached one goldfield, and used that as a stepping stone to reach the next; but the nearest rich gold lay nearly 120 miles to the east. If there had been pastoral stations to the east there would have been a supply line for new prospectors and, moreover, some of the station hands in their rides over the country would have observed the gold; but east of Southern Cross was no hut or telegraph station. Water was scarce, blacks were said to be hostile. The blanket of soil on the plains concealed most of the gold-bearing rocks. And there was no certain prospect that the easily-won alluvial gold so vital to the prospector lay in the east, for there was virtually none in the Southern Cross area. No wonder the eastward drive of the gold-seekers had halted at Southern Cross in 1888.

The break through the desert was one of the hardest tasks in the history of Australian prospecting. It called for the stamp of prospector that tropical Australia had tested in the 1880s. The men from the dry regions of North Queensland and the Northern Territory who had rushed to Kimberley and Pilbara and the Murchison were in fact a distinct breed from the prospectors who had blazed the gold trail along the mountains of eastern Australia. These men employed blackboys to tend their horses, to guide them to water, and to look for gold; the honour of finding many valuable fields no doubt belonged to, though was never credited to, blackboys. This new breed of prospector could endure dust and intense heat; he often preferred them. He was adept at prospecting without water, a master of dry-blowing and specking. Often he liked native women and so he prospected for long periods in the one area. He succeeded where other gold-seekers failed.

From 1888 searchers probed the plains beyond Southern Cross, and not all returned. Gilles McPherson, the experienced bushman who was later to win a fortune at the Murchison, rode about 140 miles east; a blackboy was his guide and a Fremantle doctor his financier. He carried insufficient food on the packhorses and his water ran low. On the hot plains he nearly died from thirst, but his life was spared for another decade, to be snuffed in the Klondike snow. Southern Cross townsmen financed another party, Jim Speakman, Erickson and Day, and they nearly died from thirst in 1891 at a place about 150 miles north-east that later became the Ularring goldfield. These parties found gold but not enough to tempt them back.

Bayley and Ford were the team that succeeded. Arthur Bayley was only twenty-seven years old but there were few finer prospectors. He had magnificent physique, won footrunning races at meetings from Southern Cross to Roebourne, could throw a hammer further than anyone he met. He liked to gamble, would risk his money or his life, but had strength and stamina to minimize the risk when his life was at stake in arid places. Born at Newbridge on the Loddon, on the famous Victorian nugget grounds, he had mined gold at Croydon and the Palmer in North Queensland and at every field in the West except the Kimberley. He could almost smell gold; he found a sixty-ounce slug in the wilderness by the Ashburton River; he was one of the few who found much gold on the beach at the Nickol, and with one mate he won over a thousand ounces at the Murchison and thought he was perhaps the luckiest of all on that rich field. Generous and reckless with money, he interspersed each rich find with a drinking bout. Impetuous, he sometimes fought barefisted in camps and hotels, and at the Murchison he allegedly hit one man with a shovel. He could punch like a kicking colt in the opinion of William Ford, who had first seen Bayley fighting in the streets of Croydon, Queensland, and who met him again halfway around the continent and became his closest mate. Ford was wiry, full-bearded, much older and more cautious, a Victorian who had sought mines or worked in them from Broken Hill to Southern Cross. He was Bayley's partner in the trip they planned.

Bayley saved enough of his Murchison gold to buy ten horses and food for months, and he was better equipped for a long search than any others who had left Southern Cross. In the winter of 1892 he and Ford rode into the rising sun. They rode about 160 miles to the east and then turned back, for their waterbags were light. Near the present town of Coolgardie they found a small rockhole of water and decided to camp a while and prospect. Bayley recalled in his first recorded interview that he was rounding up horses before breakfast when he saw a nugget weighing half an ounce. They gulped breakfast and sought more nuggets. By dinnertime they had twenty ounces. Recent showers had washed the fine red dust from the slugs of gold on the ground, and walking about with eyes on the ground—a mode of prospecting known as 'specking'—they found many specks and bits. After they had scoured the surface they skimmed the topsoil and got more gold by dry-blowing. In a month they had about £800 of gold. Short of rations, they rode 120 miles to Southern Cross and loaded their packhorses and then quietly left the town. Men in the town saw them leave so quickly that they suspected they had found gold. As the miners were then on strike three men decided to follow Bayley and Ford, but owning only one riding horse and three packhorses their pace was slow. Despite the defection of their native guide they eventually found Bayley and Ford working in a hollow like a saucer on the plain. Then 'we found gold galore, in fact gold all around us. We could see it glittering in the sunlight, for at least 20 yards in front of us,' recalled Tommy Talbot. 'I think we were off our heads for quite a few minutes.'

On the ridge was a strong white reef which Bayley may not have been the first to see but which he certainly claimed. One Sunday evening he got fifty ounces from the reef. Next day he broke with a hammer another three hundred ounces. The newcomers took several hundred ounces for themselves before Bayley warned them to leave his claim. Bayley now had to register his claim with the warden at Southern Cross if he and Ford were to keep out the newcomers, and he reached the town on 17 September 1892, with 554 ounces of gold worth at least £2,200.

In the bank the gold was magnificent to see. Experienced miners gaped at the beauty of the quartz-studded gold that Bayley drew from his saddle bag. Men packed swags, tried to buy or beg or steal packhorses, in their haste to rush east. The morning after Bayley arrived, men rode into the rain with loaded horses, men slushed along the track with heavy swags on their backs and prospecting tools in their hands. A week later the cavalcade from Perth and coastal towns passed wearily through the town: camel teams, horse teams, human teams. They were followed a week or more later by the leaders of the exodus from the Murchison, men who had followed the rushes south from the Kimberleys and found each field richer than the last. And those hardened men picked up gold as easily as mushrooms when they reached Coolgardie and the fields beyond.

Bayley and Ford went on breaking the rock in a narrow trench. The gold was so thick that Ford had another 528 ounces two days after Bayley returned from registering his claim in Southern Cross. Bayley made another trip with the gold and, a fortnight after returning, another 642 ounces were waiting to be escorted to the bank in Southern Cross. All this gold had come from stone that had been pounded in a simple pestle and mortar worked by hand, and the waste rock from this simple crushing weighed six hundredweight and later yielded 278 ounces of gold. In three months the two men won gold worth £8,000—that money equals £50,000 in our inflated currency—and the gold was still gleaming in the bottom of the trench.

Silver speculators from Broken Hill bought Bayley's Reward mine and were lucky to buy it. Sylvester Browne was a wealthy Victorian pastoralist who had gambled in Broken Hill shares and directed some of its mines, but the Barrier boom was over and silver prices were slipping and with his nephew Everard Browne and Gordon Lyon, young assayers from Broken Hill, he was travelling round the coast to Geraldton to see the Murchison mines. Their steamer called at Albany for an hour or two and hearing of Coolgardie they left the ship on an impulse and went by train and coach to Southern Cross. They reached Coolgardie in style, driving their own light waggonette they had taken from the ship, and in one hour they had seen Bayley's Reward and offered £6,000 for a five-sixths share in the mine. They paid £100 deposit to clinch the deal. The remainder of the purchase price they won from gold.

Sylvester Browne floated the mine into a public company in Melbourne in January 1893, and apart from the Brownes and the discoverers the main shareholders were a few Melbourne and Sydney gentlemen, barristers, and merchants. One-twelfth of the shares was held by George McCulloch, who had already minted one fortune in silver from his old station at Broken Hill and now was to mint another in gold.

Bayley's Reward was a bonanza on the top. A trench, eight feet deep and eight feet wide along sixty-four feet of the outcrop, gave up a third of a ton of gold. In the company's first year the mine treated a meagre 48 tons of ore for 25,872 ounces of gold. Soon after an antiquated stamp mill was unloaded at the mine, the little Melbourne company sold out to a new London company at fantastic profit. Arthur Bayley, prize-fighting, tenderhearted ruffian of the gold frontier, became a gentleman with his own sheep station on the green plains of Avenel, Victoria, and assets worth £30,000. He died aged thirty-one when the goldfields he had found were becoming superior to all but the Transvaal.

Bayley's mine still yields rich gold. I saw in the safe of Gold Mines of Kalgoorlie a sugar bag holding beautiful samples of free gold in quartz, like ornamental gold on white satin, mined in 1961 from 900 feet underground at Bayley's. One can understand the astonishment that even richer samples once created amongst the diamonds and amethysts and finger rings in the jewellers' shops in Perth, Melbourne and London.

Dr E. D. Peters, an American mine expert, saw Bayley's specimens on display in 1893 in a jeweller's window in Melbourne and thought such stone was never seen in the gilded age of California's quartz mines. The wide streets of Coolgardie seemed paved with such stone in the imagination of men.

16

COOLGARDIE

THE LAD ALBERT GASTON worked in a sawmill at the small town of York when news of Bayley's gold came on the wires. Driving the mill engine, he was so excited he forgot to draw the fire and nearly blew the boiler. He left work that day, rolled a small tent and spare shirt and pair of blankets into a swag, and next morning walked east with tin billy and gallon waterbag. York was one of the nearest railway stations to Coolgardie, a mere three hundred miles away, and he was one of the first of perhaps a hundred thousand men who walked to Coolgardie in the four years before the steam train reached the goldfields.

Fifteen miles from York the fences and ploughed paddocks ceased. Thirty miles farther and water became scarce and the mileposts seemed more than a mile apart. With sore shoulder and blistered feet Gaston walked thirty-eight miles one hot day to reach the next waterhole. At Southern Cross he bought tools and stores and walked out into the hot white sand carrying miner's dish, shovel, pick, saw and twenty pounds of provisions—the same old flour, oatmeal, tinned meat, sugar, tea and tobacco. After half a day he threw away coat and blanket, a pick handle and walls of the tent. The flies were such a plague that he tied to his hat a few corks that jigged as he walked. So many men camped the night at the first waterhole that he waited long for his turn at baling water to boil for tea. His days on the track were exhausting and dreary, but thousands shared them.

Under the desert stars the campfires burned at a score of points along that straight Coolgardie track; men cooking damper, washing clothes, chattering of gold, with the bells of grazing camels and horses murmuring from the darkness. F. Cammilleri, an alert prospector making for the new Kalgoorlie rush after chasing gold south from Kimberley, wrote his diary most nights in the winter of 1893. He had paid a teamster a shilling a pound to carry his swag and he walked unimpeded, one of thirty-five men walking with the waggons. The horses' pace was slow and one teamster was so drunk he had to be lashed to his load. Cammilleri had that nervous haste to be at the new rush and with a mate he pressed ahead of the rumbling teams, anxiously questioning swagmen passing the other way for news of the latest rush. 'Met three teams and a good number of men and several swagmen, all of whom say there is no tucker on the field and that half the men were starving.' His main complaint, apart from weariness, was that 'we have had nothing but tinned meats since leaving York'. But he was grateful for the winter rain that soaked clothes by day and sometimes blankets by night.

Water was so scarce along the track in the first two summers that Warden Finnerty warned diggers 'to move back to Southern Cross—preferably in small groups—so as to avoid over-drawing at the few wells and soaks'. In time, surface dams scooped out by government navvies collected water, and a Cobb and Co. coach could make the journey in two days and a pony and spring dray in four. With at least six hundred horse teams and many trains of camels taking goods to Coolgardie in the third year of the rush, iron stores and hessian shanties arose to refresh the travellers, and one South Australian baker counted twelve small bakeries selling loaves in bush clearings on 120 miles of road.

The end of the journey for some and the start for others was Bayley's Coolgardie with Everard Browne rooting out the rich gold on the red slopes, crushing it with a wooden pole clad with iron, and sitting dirty in his camp at night, talking of Broken Hill or sheep or his father, Rolf Boldrewood, who was Australia's popular novelist and was to write his last book, *The Last Chance*, on the golden west. Those who had spent weeks in the steerage of a swaying steamer and the heat of the track expected to find an oasis at Coolgardie, but it was an oasis without water. Their first sight on reaching the camp was fine red dust from the dry-blowing, and the first sound was the peculiar rattle of gravel falling on tin dishes. Water was so precious that men 'washed' their clothes by belting them with a stick, and carried their waterbag as if it was their wallet. When rainclouds gathered, a man would dismantle his tent and lay it as a tarpaulin on pegs to catch the rain. After a shower every man in the camp walked about with hands in pockets and eyes on the ground as if in contemplation, for the rain washed away the red dust that disguised slugs of gold on the surface of the ground. After the rare showers Coolgardie was beautiful with the air crisp and the sky soft blue and the gum tips shining above the smooth trunks of gimlet and salmon gum that gleamed like oiled wrestlers.

In less than a year Coolgardie had a dozen stores and hotels of hessian and bush poles with earthen floors and a room or lean-to at the rear where the owner ate and slept. Scattered tents and shelters, the warden's camp and the police camp, completed the settlement. The warden settled the daily disputes over ownership of mining claims, but the police rarely were called to detect the petty thefts so common in a city. Men left gold in a bottle in an empty tent for weeks and, returning from an outback trip, found it undisturbed.

Coolgardie was often full of deserted camps in 1893, for it was the base for reconnaissances and rushes to a score of new Coolgardies. Old Charlie Grainger wandered into the hills three miles away and returned at night with his billycan heavy with specimens. A crowd raced out next day and shovelled sand and scoured the ground and found almost nothing. The rush ended in a day, this billycan rush. Courageous men loaded their horses and went north, finding water and gold at the 25-mile, then the 40-mile, a hop, step and a jump to the 90-mile, later known as Goongarrie. Forty men hired an Afghan to take their swags the ninety miles on his string of camels, and he charged them a shilling a pound for their swags, and gave each man a gallon of water a day. Albert Gaston found only two pennyweights of gold in a week of sweat and jettisoned most of his goods and walked back the ninety miles, drinking the red muddy water from Canegrass Swamp and seeing on the way the familiar sight of men digging a grave for someone who had died on the track.

Gold to the north, gold to the east. Rumour of gold over the hills at Mount Youle drew out the horse waggons with men trudging beside them in the clear air of winter. Patrick Hannan was edgy to leave but saw no point in walking with waggons and stopping when they said 'whoa' and going when they said 'go'. He had prospected long enough to know the value of being independent; he was almost fifty, lean with small eyes and flowing beard, and he could talk of deep mining at Ballarat, rushes in the rain forests of New Zealand and Western Tasmania, dry rushes to Temora and Teetulpa, and rushes innumerable. Hannan, Flanagan and Shea, three Irishmen, left Coolgardie on 7 June 1893 with their swags on a horse and followed the wheel marks of the waggons towards the new rush. Days behind, they still dawdled, cautious not to move far from the waterholes in the granite outcrops, careful to inspect the ground away from the waggon trail. On the side of the low range of hills they saw a slug of gold, then gold in quartz, and they shovelled and winnowed the soil until they were red as rust. They had found Kalgoorlie, the most productive goldfield in the continent, and today a bronze statue of Hannan and his waterbag stands near the scene of his discovery.

Under Western Australian mining law those who found gold more than twenty miles from the nearest workings could take a reward claim of twenty acres. The Irishmen were twenty-four miles from Coolgardie, and being anxious to get their twenty acres they sent Hannan on horseback to the warden's camp to register his discovery. He rode away on 17 June 1893, skinny and dishevelled, with two quarts of water and eight or so pounds of gold. When they saw his gold in Coolgardie on that cold Saturday evening they prepared for the exodus, storming the stores for provisions, queueing at the bakehouse for flour until not a pinch remained. The sky poured rain and the water ran through the shops and tents, gold and rain the same day. Before daybreak next morning the fires were flickering along the flats, and men packed swags and sodden tents and the race began. The track was soggy after rain and the older and heavier-laden men were soon straggling, and when the sun pushed through the clouds those in the rear saw far ahead a string of men 'carrying loads that were more suited to camels'.

Two thousand men in a week went to Hannan's Rush. The early comers found hundreds of ounces of gold gleaming on the wet soil. Others worked on hands and knees, their knives 'lousing' gold from the few inches of loam that covered the bedrock. The loam was too wet to be dry-blown, and diggers lit fires to dry the soil, and for days a pall of smoke drifted across the plain. Then the warmth of the sun dried all the soil, and newcomers saw from afar the red dust spiralling from twenty points and the camel teams crossing the plain with water for the parched camp.

Water fluctuated like gold shares, sixpence a gallon one week, 1s. 6d. the next. The water teemed with insects, it turned white damper to yellow, and diggers vowed a bucket of water wanted an ounce of epsom salts to make it drinkable. When stores sold out of salts, a speculator hurried from the coast with a donkey team carrying enough epsom salts to glut the market for months.

The field nearly ran out of food. Men paid reckless prices for the luxury of a tin of treacle or a bottle of Worcestershire sauce. Flour was £10 a ton in the cities, and £120 in Kalgoorlie, and the hungry walked miles along the track to buy a sack of flour from incoming teamsters, opening the bag on the spot, cooking johnny cakes over a hasty fire, and gulping them before they were brown. 'Tucker stolen from three camps yesterday,' Cammilleri notes in his diary on 30 July. 'If the parties are caught it will be made warm for them.'

The alluvial gold lay in shallow soil that was quickly sifted, and soon diggers were hungry for gold as well as flour. The population fell to three hundred. A few with the strength of blacksmiths gouged gold from tough rock. The diggers vanished with curses and the speculators came with praise.

Patrick Hannan, the discoverer, pegged his twenty-acre claim on veins of golden quartz, and as quartz was invariably voted a generous host for gold, he thought he had the richest mine in Kalgoorlie. He ignored the brown ironstone hills a few miles to the south. The early diggers scorned them too, preferring to claim the veins of quartz near Hannan's mine. It was left to Adelaide investors to capture these richest hills in Western Australia.

South Australia was poorest in minerals of the Australian colonies, yet was always eager to speculate in mining shares. That eagerness may have derived more from its adherence to Methodism and the dissenting churches than from its copper mines and Cornishmen. The dissenting chapels denounced gambling on cards and racehorses but were wisely lenient to a form of gambling that was industrially essential. Share gambling thus was one of the few respectable vices in Adelaide. They said share syndicates were sometimes formed in the vestry after church, and it was certainly true that Adelaide stock exchange was strong in mining even before it became a leading market in Broken Hill shares in the 1880s. By 1893, however, the low price of silver and copper choked speculation in Adelaide's favourite shares, so local syndicates sought gold.

One syndicate that sent prospectors to the West was formed by George Brookman, a 43-year-old Adelaide financier. His brother William had worked for gold near Adelaide with Sam Pearce, and the two were appointed as the syndicate's prospectors and sent by steamer to Albany in Western Australia. After walking by their loaded spring cart to Coolgardie, they heard of Hannan's new find, and finally reached Kalgoorlie on the thirteenth day of the rush. Seeing quartz from the Cassidy Hill lease crushed with their own dolly pot in their tent, they reported that they had seen 'some of the richest quartz it is possible to imagine'. They regretted they were too late to peg rich quartz for themselves. Hopefully they went south to the ironstone hills and found small leaders of quartz. Sam Pearce was less dogmatic than many around him and, spending all he could spare on lease rents, he pegged hundreds of acres on the ironstone hills. He sometimes said 'the iron cap covered the golden head', but he was probably wishing rather than believing. His first love was probably gold in quartz, and the first lease his syndicate floated into a company was the Ivanhoe, and the main shaft on that lease actually chased a small quartz reef rather than the massive quartz dolerite greenstone that held the real gold. After all, greenstone had yielded no gold on any other Australian field, and wise old Paddy Hannan visited Cammilleri's claim and argued that it was barren, until he crushed rock and saw the tail of gold in the pan. By then it was too late for Hannan to peg the iron clad hills. The latecomers, Brookman and Pearce, held the best ground on the field.

In Adelaide their Coolgardie Gold Mining & Prospecting Syndicate was jubilant. It floated some of the leases into three small companies within months of the finding of the field, and the members of the syndicate got large packets of free shares in Ivanhoe, Great Boulder, and Lake View. In the following seventy years Great Boulder was to produce just on £100,000,000 of gold, at present gold prices. Many of the other Brookman mines were also rich.

Many men left Hannan's Rush for a new rush seventy-five miles north of Coolgardie, and some did not return. Billy Frost and Bob Bonner had found forty ounces of gold at a place so dry and sunburnt that they warned men of the danger. The distant gold seemed richer for the warning. Teamsters refused to take their horses into the waterless north and men set out on foot, some returning delirious with thirst, others pushing on to that place henceforth known for its misery as Siberia.

From the new field at White Feather, beyond Hannan's, twenty waggons and carts left overnight for Siberia, the first waggon carrying seventy-five swags for men who walked ahead with waterbags in hands. At the end of the mirage was sparse gold and a soak of water so miserable that men waited all day for water to seep, the flies torturing them, their hands prickly with heat. The rush became a retreat, men on foot were in grave distress. Some did not resist the thirst and drank in two days the water they had rationed for four. Some threw off clothes and chased the mirage, silly with thirst. Some strayed to the salt lakes south-west, and were not found. From Coolgardie an alert public servant, Mr Renou, sent two horse teams to place tanks of water along the track in anticipation of the ragged retreat, and went himself with a string of camels and precious water. Perhaps ten men were buried in Siberia's hot sand, the names of some unknown, and there were women in Victoria wondering that summer why sons and husbands ceased to write.

'The first thing prospectors had to do was to discover water,' said Jeremiah McAuliffe, finder of White Feather, and one of those souls of granite who had followed rushes all the way from Charters Towers. He believed in the blackboys as guides and his boy Monkey found the water that was as valuable as the first gold of White Feather. The natives are our 'water prospectors', said John James Brown, young veteran of gold rushes in six Australian colonies, New Guinea and Johannesburg. The wise men of the desert and their blackboys looked for those low rises of smooth granite with rockponds in the cavities and a sward of green grass at the foot, gnamma holes of the blacks, reservoirs of the prospectors. The rush to Siberia taught hundreds to seek granite before gold.

The pace of discovery was swift from 1893, and the nightfires of the prospectors spread like stars in a clearing sky. Pastoralists moving north from the wind-swirled port of Esperance found gold in the Dundas Hills, and gold-seekers moved south from Coolgardie and found Norseman, still rich in gold. Hannan's Kalgoorlie was a stepping stone to I.O.U. and White Feather, names too romantic for the men in collars and ties, who renamed them Bulong and Kanowna. North of Coolgardie the adventurers streamed over claypans and dustbowls to Black Flag and Broad Arrow and Dead Finish, names once known to every sub-editor in Australia, and on to Goongarrie, which had a salt lake, and Niagara, which had a waterfall when there was water. Some in this vanguard were silent men who found much and said nothing, and some were like Leslie Robert Menzies, a dapper camel-rider after whom the goldfields town was named. He wrote late in life the book *A Gold Seeker's Odyssey*. He recalled how in October 1894 he jumped from his camel, crushed his heels into a pile of nuggets, gathered £750,000 of gold in two hours, carried it a hundred miles to Coolgardie with the camels running an even sixteen miles an hour, shouted 4,000 dollars-worth of champagne, and triumphantly wheeled the gold in a barrow from his hotel to the bank. A strong man, a strong barrow, his six tons of gold equalled the recorded yield of the entire colony for 1894!

Beyond Menzies' camp the rushes pushed farther northward into mulga scrub so thick that horses without hobbles and bells could hide for days from their masters. Around Malcolm and Morgans and Leonora and Laverton the prospectors racing north from Coolgardie met those pushing out east from the Murchison, finding Lawlers and Sandstone and Lake Darlot on the way. Some of those Murchison men who penetrated into Coolgardie's sphere of influence were strong men. Booden found the Diorite King, twenty-three miles north-west of Mount Leonora, and on returning to camp with seven ounces of gold, found his provisions stolen by blacks. He was over 250 miles from Cue on the Murchison but he simply walked there, completing the journey in bare feet. He got more supplies and returned to his gold with that bold barrowman Edward Sullivan, known everywhere as Doodah, and Doodah found the first gold at Mount Leonora, and was buried on his lease with his home-made barrow to mark his grave.

Many rough cairns of rock were built far out in the bush to mark the graves of prospectors who died of fever or typhoid. A dry-blower named Sligo wrote in his memoirs the story of three Murphy brothers. They dollied about six hundred ounces of gold from a rich quartz leader at Redcastle, near the Mount Margaret track, and Pat took the gold to the bank in Coolgardie and caught typhoid fever and died. Jim went in to arrange his brothers's affairs and caught the fever on the way back and was buried by the claim. And Dan Murphy became delirious in his tent, called for his dead brothers and a girl in Ireland and babbled of the gold he had found. 'It's too good to be true,' he whispered shortly before his death.

Mapmakers and geologists could not keep up with the prospectors, so quick were their discoveries and so quiet were their journeys. The large area of goldbearing ground in the southern half of Western Australia had been traced to most of its extremities within three years of the discovery of Coolgardie. When at last Western Australia's mapmakers caught up with the gold-finders and plotted all the goldbearing areas they were amazed at their extent. It was clear that men could travel from within a hundred miles of the southern coast of Australia and go by various routes all the way north to the Tropic of Capricorn, travelling through auriferous zones almost the entire way, and finding the longest stretch of goldless country a mere thirty miles.

The defining of the gold belt required many unsuccessful journeys. People believed the goldfields were without limit until some brave party went beyond shade and water to prove them wrong. Their journeys were the more heroic because the goldless country was hotter and drier and less accessible. David Carnegie, son of a Scottish earl, worked for wages at Bayley's Reward and won a few pinheads of gold at Kalgoorlie before he earned enough in 1896 to seek a new route from the Kimberley to Coolgardie, a journey of three thousand miles through sand and spinifex that ranked in merit if not in fame with those of immortal Australian explorers. Billy Frost and James Tregurtha made another amazing journey beyond the gold belt. Both had worked in Croydon in North Queensland and had followed the gold trail around the Indian Ocean to Coolgardie, where Frost was the finder of Siberia, and were thus men of that breed that had opened so many dry goldfields. With money won from gold they went to Oodnadatta in South Australia with a third mate named Schmidt, eight pack camels and a few horses, and went west to the goldfields through some of the most cursed desert in the continent, losing their horses on the way, and completing the journey on foot without an ounce of gold from the thousand miles they had crossed. For them, and for the advance guard of gold-finders, the Western Australian goldfields were losing their glitter. When women and stockbrokers and telegraph operators moved in, the more adventurous prospectors began to move out. Frost and Tregurtha went to the Klondike, Carnegie went to Nigeria to die from a poisoned arrow, and many of the most daring of Western Australian prospectors were in the wildernesses of the world long before the goldfields they had found revealed their magnitude.

Coolgardie in the glorious winter of 1894 caught the two phases in the life of the goldfields side by side. It was still the supply base and the entertainment parlour for the prospectors returning from distant rushes, and they delighted to sleep in its chink-walled hotels and shop in its rows of galvanized iron stores with their frontages of six or seven feet and their French Jews and South Australians and Englishmen behind the counters. A visitor to the town observed clearly the 'hum of the men's voices' after the silence of the bush and saw with surprise the bustle of carts and waggons and camel teams and men on bikes or on foot in the wide main street, reminding him of Adelaide or Gawler on a Saturday afternoon. In the hotel of John de Baun, the paunched publican once famous in Silverton, tattered prospectors with dusty beard and bloodshot eyes drank happily with city men in white shirts and clean pith helmets, whose cheques would soon dominate the goldfields. The rush from the city, with its manners and cheque-books, was following the rush of the prospectors. In July 1894 Coolgardie had its first wedding, sent away its first telegraph and elected as first mayor James Shaw, who had once been mayor of Adelaide. And when the stock exchange of Coolgardie opened in the following month with a banquet on clean linen, and Lord Fingall sitting at the long table as the symbol of British investors, the clink of glasses rang in the second era on the goldfields.

The goldfields for their size were poor in alluvial gold and not much richer in reef stone that could be mined at a profit by small bands of men. The wealth of the field lay in the large lodes that wanted capital to develop them. Isolation and climate and the nature of the lodes made Western Australia hungry for capital, but that capital had not been easy to catch. Many miners thought it would be harder to woo after certain events of 1893.

While Coolgardie was dizzy with rushes the cities across the desert plunged into financial crisis. Diggers sold nearly all their gold to the Commercial Bank of Australia at Southern Cross, and at the start of April 1893 that bank closed its branches throughout Australia. In the following weeks Hannan's Rush heard of countless bank failures, huge insolvencies, and destitution in the cities. Hundreds of its diggers lost savings or the right to withdraw them. Promoters who had gone to Melbourne and Adelaide to float mines saw their old brokers filing insolvency. Hundreds of men reading newspapers by the firelight in the gold camps of the west believed the nation's crisis was a crisis for themselves. In truth, their crisis became a triumph. Tens of thousands of men came west from the impoverished cities to mine for gold, and the trickle of mining capital which in normal times would have flowed from Australian cities came in a river from London.

The London stock exchange was ripening for a mining boom. Financial panic in Australia and the United States and world depression cut the outlets for investment in the new world. By 1894 so many Englishmen were hunting for profitable investment that London banks, saturated with money, paid only 10s. interest on each £100 deposited 'at notice'—only a quarter of the average interest paid from 1884 to 1891. Much of the surplus money went into gold shares. Whereas dividends from most companies had fallen because prices for most commodities were lower in the world-wide trade slump of the early 1890s, the price of gold was unchanged. Therefore gold shares became attractive. English money in the middle of 1894 began to pour into the rising goldfield on the Rand, and by the end of 1894 South African gold shares dominated the London 'change. Never before did Britain invest so much money in foreign mines. No broker could remember a boom to rival that part of the London market known as the Kaffir Circus. Though Western Australia in 1894 mined a mere eighth as much gold as the Rand, its shares too were snatched so eagerly that a 'Westralian boom' began to match the Kaffir Circus.

Albert F. Calvert was the first of a band of exuberant promoters who drummed in the boom. He was a mining engineer and journalist, who had once lived in Western Australia. His first step towards Christian socialism and bankruptcy was to float Mallina Gold Mine to work that reef near Roebourne where the lad had picked up his famous stone to throw at a crow. He floated another seven mines in Glasgow and London, wrote five books in three years, and edited an English newspaper on Western Australian mines to quicken the flow of capital. Promoters fulfil a vital function and Calvert in 1895 was honoured in London at a great banquet in the Imperial Institute with the string band of the Royal Horse Artillery playing music to eminent financiers, and two floral swans in gold as symbol of the colony that was dominating the stock exchange. The floral swans were more golden than some of Calvert's mines, but not even he knew that in 1895.

Sun-tanned prospectors from Coolgardie reached England in every Australian mail steamer to hawk reefs and lodes in London, and Calvert, Horatio Bottomley, Whitaker Wright, and scores of promoters awaited them. The profits from buying untested mines and floating them into companies were so attractive that scores of English and French engineers and financiers went to Australia to buy leases. The market for mining shares was insatiable. Ninety-four Western Australian gold mines were floated in England in 1894, more mines than were floated from Johannesburg and all Africa. The boom had barely begun.

Two amazing fiascos did not halt the rush for Western Australian shares. In the winter of 1894 two Victorian and four New South Wales men went with horse and cart south of Coolgardie, and as the horse got wearier and men's tempers fresher they began to curse the gold they couldn't find. John Mills sat down to rest in the bush under the blue sky and saw mossy stone and looked again. Seeing a flash of yellow, he bent down and cracked the rock, shooting chips streaked with gold. Casually he displayed them by the fire that night, specimens as rich as Bayley's, and his mates began next morning to break into that low mound of quartz. They worked undisturbed for six or seven weeks and the gold became richer as their shaft went down. A lump of quartz they called Big Ben weighed 250 pounds and over a third was gold. On 23 June 1894 they lifted on to the scales of a Coolgardie bank more gold than a giant of a man could carry.

Those who galloped the twelve miles to see the Golden Hole were as surprised at its richness as at the scarcity of gold in the surrounding scrub. Captain Beglehole, mine manager, came to offer £25,000, which was promptly refused. The claim was cut into shares and they were traded in Coolgardie at leaping prices as the gold was cracked like kernels from dazzling white stone.

The Earl of Fingall and his Australian escort, R. G. Casey, cantered out to peer down the golden hole. Breathless at the splendour, Fingall got expert advice and sent cables to his London friend Colonel North, who organized a syndicate. It offered to buy the Golden Hole for the astounding sum of £180,000, plus a sixth of the shares in a new company to be floated in London.

The Earl of Fingall was not familiar with mines but advisers suggested the hole was so rich that the gold would be looted unless it was guarded. Could they have sensed too that he would be more certain of raising the purchase money for the mine in London if all work ceased while the gold still studded the stone in the hole? He ordered that the hole be sealed and cemented, surrounded by an iron fence and guarded by men, while he went to London to resell the mine.

In London Fingall and a covey of lords and the Chilean nitrate king, Colonel J. T. North, agreed to sit on the board of Londonderry Gold Mine Limited. Capital was 700,000 shares of £1 each. Fingall and friends took 233,000 and offered the rest to the public. Their call for a clear £467,000 in cash was amplified by a rosy prospectus noting that 'it is most improbable that the rich quartz of the Londonderry Mine can be surpassed at the present time in the universe'. Investors clamoured to pay that sum for two-thirds of an interest in a sealed hole the size of a grave, and the promoters and discoverers shared £417,000 in cash and assigned a mere £50,000 to develop the mine into the great mine of the universe. Shares jumped to a premium. The company's statutory meeting on 25 January 1895 in the Cannon Street Hotel witnessed a feast of praise. Fitzgerald Moore, hotfoot from the mine, stood with sombrero in hand, one of the most travelled hats in mining, and said such a mine had not been seen before. (It would not be seen again.) Fingall had to speak and said if experts were right, and usually they were, the mine would be worth £300,000 for every twenty feet of sinking: if the reef went down a thousand feet that made £15 million. And now Fingall, in evening dress, ginger-haired and affable, sat at the farewell banquet and was cheered at Charing Cross Station as he left to unlock the mine.

In the bush they took away the fence and broke into the mine. Swagmen called daily for work and wondered why the company did not want them, but only two men could fit into the hole and what they saw as they blasted the rock puzzled them. Fingall climbed down often to hold a candle to the reef and in the thatched hut seemed moody. On April Fool's Day 1895 he called at Coolgardie post office late in the afternoon, and at 5.15 the morse operator tapped out a message in code to Colonel North in London: 'Regret in the extreme have to inform you that rich chutes of ore opened very bad indeed; does not appear to be practically anything important left. . . .'

Once the news spread through London and Paris, 'the Queensberry-Wilde libel case was quite forgotten during the consequent excitement', the editor of London's *Mining Journal* wrote. Mysterious cables from the mine sent shares fluttering up and down. Colonel North raged that someone had stolen the gold. He and three friends who had made clear profit of over £200,000 by buying the mine from the discoverers and selling it to European investors were harried by critics. Lord Fingall, Colonel North, and Mr Myring decided with an honour not common amongst promoters to pay their profits to a new gold exploration company in which all Londonderry investors would share. The fourth promoter, Casey, was reluctant. Perhaps he argued that the discoverers' syndicate should repay the £180,000 they had received; maybe he argued that he had floated the mine in good faith and that sharebuyers had willingly risked their savings. Perhaps he was right. Thousands who flocked to buy Londonderry shares did not really care if the mine were rich. Most bought shares to sell at quick profit, and many got that profit. Investors who suffer most in a mining mania are those who buy late, not early.

Even after the lid had blown from the golden hole Western Australian companies were floated daily in London. Sharebuyers saw no lesson down the golden hole, nor did Colonel North. To the northwest of Coolgardie the brothers Dunn, working for a Perth syndicate, found a dazzling reef from which they knapped specimens as beautiful as jewels. The danger of looters taking gold was so high they summoned police, but at least one lump with sixty ounces vanished while Maltese Charlie and Ragged Jack worked in the trench. This was a mere fraction of the gold that changed hands when the mine was sold to Colonel North. He registered a company called Wealth of Nations in London in July 1895 and offered nearly 200,000 shares of £1 each to the public. The public applied for nearly five times the shares the company owned; 3,600 people were allotted Wealth of Nations scrip and speculation was intense. Those who had faith in Colonel North were again the victors, selling scrip at high profit. Those who bought in hope that shares would keep on soaring, lost. The mine, forgotten in the excitement of the market, again had last say. One of those surface splashes of gold so common near Coolgardie, it died as suddenly as the Londonderry.

Seers said the Wealth of Nations and Londonderry would repel investors from new Western Australian gold mines. Those who remembered the heavy Scottish and English losses in the recent Australian bank crashes agreed that Western Australia was a land of sirens, luring investors to the rocks. Despite the inflow of British money to develop new mines the annual gold output of Western Australia did not reach £1 million until 1896. And where were the dividends from mines that were said to be so rich? These arguments did not frighten British investors who had abundant money but few profitable ways of investing it. The chance of quick profit on the stock exchange spurred them on, and promoters continued to float their companies, paying trusted newspapers to puff shares and hush news.

A new Western Australian company was born in London at the rate of one a day for two years. In one month, April 1896, eighty-one were floated, and the companies floated that month alone collected more money from the public than Western Australian mines had won in gold over the last two years. In two years 690 Western Australian gold companies were floated in London. Minted gold rather than mined gold was transforming the hot sands where men had died from thirst.

Men who left the excitement of the London stock exchange to visit the goldfields found more excitement there. The wooden hotels in Coolgardie's wide street of dust could not accommodate all the visitors who came to speculate and see. The town seemed so certain to become a city that English companies were buying land in the main street and Australian companies were building breweries and hotels. At any hour of the day hundreds of bottles of French champagne chilled in the iceboxes of crowded hotels; a party of prospectors shouted ten cases for their friends at £15 a case the day they sold a Broad Arrow mine to a French doctor, and such sales and sprees occurred almost daily. The predominance of young men and the excitement of money so easily won and spent gave the town a reckless conviviality which the hundreds of deaths from typhoid merely intensified. Every few months fires raced through the flimsy hotels and shops. One fire burned an acre of shops and saloons and the whole acre was quickly rebuilt with grander houses of entertainment and supply. The telegraph line to the coast was clogged with London mining transactions, and one speculator paid £1,150 to send a telegram. Bicycle expresses left Coolgardie daily with cables and letters for visiting mine-buyers at Dundas and the outlying fields, the riders speeding in the sun over soft camel pads.

In March 1896 the Governor of the colony, Sir Gerald Smith, himself an avid sharebuyer and part-owner of a Coolgardie hotel, arrived to open the railway from the coast and five hundred people drank champagne unlimited until after midnight. The railway signified not only cheaper champagne, but the start of vigorous mining, and that soon exposed the shallowness of Coolgardie's hopes and reefs.

17

WHITE FEATHER AND BOULDER CITY

KALGOORLIE GREW FASTER than Coolgardie. Its fountain of champagne shot higher when the railway arrived. Its mines were larger and seemed more permanent. Three years after William Brookman pegged the despised ironstone outcrops south of Hannan's Camp, he was sitting with lords and gentlemen on the boards of twenty-one companies in London. For a fee of £10,000 he revisited the mines at the start of 1897, arriving in a special sleeping car on the express and driving in triumph in a six-in-hand around the field.

Kalgoorlie could not hold its mayors in the face of London's easy money. John Wilson hurried to London to float a mine, returned with mayoral robes, then vanished to the Yukon. Harold Parsons, graduate of Oxford, once on the literary staff of Henley's *National Observer*, pushed a wheelbarrow at Coolgardie for wages, became mayor of Kalgoorlie and successful stockbroker, and went to London with reputedly a six-figure sum before he was thirty.

In the dining-room of the Palace Hotel at Kalgoorlie two hundred sat in style beneath the swirling fans for evening dinner, a rich assembly of men who had inherited the wealth of old English families and men who were about to snatch it. Their coffee sipped and chilled cheeses picked, they strolled under the verandah and into the hot night to assemble again in crowded stock exchanges or the Dutch auction in the wide street. Deland the baker could hear the prices called in the street as he stood at his ovens. 'I don't think there were less than a thousand men in Hannans street last Saturday night when the open-air Mining Exchange was called,' he wrote in September 1895 and eighteen months later the speculation was enormous. Men on the fields knew of rich finds before the news was cabled to London, could buy shares in unknown mines before they were refloated in London, and so brokers came from almost every city in Australia to breathe down the horse's mouth. Some of the small mines floated at the end of the share boom suggest the craze for shares. When the Bank of Ireland Gold Mining Company was floated in February 1897 it spread the new shares amongst stock-brokers in Adelaide, Perth, Kalgoorlie, Coolgardie, Menzies, Bulong, and Kanowna. Two of those places are now ghost towns; from another two even the ghosts have fled.

The greed for information on mines and shares stirred the newspapers. Six of the eight daily papers in the colony in 1897 were printed in the gold towns and ten weeklies as well. Visitors admired the newspapers and respected their young aggressive editors. John Kirwan edited the *Kalgoorlie Miner* and Hugh Mahon the *Menzies Miner;* both were eloquent Irishmen, both were elected to the first Commonwealth parliament, from which outspoken Mahon was expelled in 1919. F. C. B. Vosper, with flat white face and black hair to his shoulders, edited the *Coolgardie Miner* and sparred with his rival Coolgardie editor 'Smiler Hales', brilliant and bellicose, once famed in booming Broken Hill for a galloping pen and stuttering voice. In his newspaper in January 1897 Vosper hinted that Hales had organized a charity race meeting for his own benefit and Hales retaliated by bursting into the office and shouting, 'Vosper, put up your hands, you cur, put your hands up.' After pummelling the seated Vosper he was dragged away and taken to the police court for a £5 fine. Vosper became a politician, dying in his early thirties. Alfred Hales won fame as British correspondent in the Boer War and as author of over sixty books, many of them best sellers in the United States, many of them capturing the raucous and sentimental life of the mining fields. He wrote of promoters with jewelled fingers and of the gold-seeker dying in the desert and murmuring the name of his little daughter in a distant city. 'And with that well-loved name upon his lips,' wrote Smiler, 'he handed back his miner's right to the Great Warden of Creation.'

Scarcity of water did not check the westward migration and the rise of large towns. Salt lakes from which prospectors had drunk with insane thirst became some of the first 'gold mines', yielding fortunes to men who condensed the salt water and sold it around the towns from water carts at a penny to fourpence a gallon. There were hundreds of condensers through the goldfields, with their wood fires below high iron tanks in which water was boiled and the steam condensed. In hot seasons the pipes buckled with the expansion of the metal and the lakes baked dry, forcing condenser men to sink wells in the dry beds. When the rains fell the condenser men cheered as the lakes filled with water and complained because customers had saved enough water in cans and roofs to last them a few weeks. In daylight the condensers were seen from afar, their mounds of salt dazzling in the sun, and at night their woodfires were seen from afar, 'glaring on the salt like blood on ice'. The mines had their own stands of tanks to condense the salt water pumped up the shaft, and it was rare to see a jet of steam escaping from boilers. Great Boulder company spent as much on water for the mill as many Victorian companies spent in the whole milling process: 4s. 7d. for each ton of ore treated. The companies provided each miner with a daily ration of water, usually two gallons of condensed water before he began work, two gallons when he finished, as well as a wash in the water that cooled the engines. Good hotels charged 2s. 6d. for a bath and small hotels didn't even have a bath. The fire sweeping through the crowded streets of Menzies or Coolgardie, the whirlwinds of dust, and the canvas hospitals crammed with feverish miners, were the tolls the parched plains demanded until cheap water could be pumped from the hills near Perth. A weir was built and pipes of 30 inches diameter were laid 351 miles to Kalgoorlie at a cost of £2,700,000, but the first water did not flow from the pipeline until 1903.

Women for long were as scarce as water, being outnumbered ten to one in many gold towns as late as 1897. Most men came west for sudden wealth, and those who found no wealth and remained to mine for wages were relucant to bring wives to their hot, dusty camps. At first their women lived in tent or hessian hut, a stamped earthen floor inside and a camp oven outside, packing cases for chairs and hessian slung between poles for beds. When they moved to houses of corrugated iron and whitewashed the walls they thought they were aristocrats, but when typhoid raged and neighbours were ill they were Samaritans. Sparing the dirty water from the washing, they planted pepper trees and hardy shrubs, nursing them like children. Their life was hard, with few comforts, but most had more money than they had ever owned. One Melbourne woman who visited the great racing carnivals at Boulder City and Kalgoorlie could only envy the style of the miners' wives, their magnificent dresses trailing in the red dust behind the grandstand. After gold stealing was restrained at the mines, it was said that the racing carnivals were not such parades of fashion.

Those goldfields had something noble in their crimson years. Reading through the old letters and writings of those who visited them, reading the brisk goldfields newspapers of that age, and hearing the talk of men who still live to remember, one glimpses in the life of the towns a strain of generosity and hope and zest such as the records of no other new Australian mining field of that decade of the nineteenth century quite evoke. This can neither be measured nor weighed, it is only an impression, but a strong one.

The commercial travellers who had to visit the goldfields towns at the turn of the century to sell explosives or mining machinery were surprised at their extent. To visit the main gold towns from

Norseman in the south to Peak Hill in the north and Laverton in the east demanded a return journey from Perth of at least 1,700 miles by railway, horse coach and bicycle. The travellers were possibly even more surprised at the size and amenities of many of the smaller towns. Broad Arrow had two breweries, and Bulong's Miners' Institute had a grand piano imported from Europe. Menzies had a post office employing thirty people and Coolgardie had as many daily newspapers as Melbourne, and in the opinion of some had a superior shopping centre to Perth. All these towns are now shadows. When electric trams were novelties in Australian cities they ran between the twin outback towns of Leonora and Gwalia and the twin cities of Kalgoorlie and Boulder. For volume of retail trade in an acre of land the Boulder suburb known as Fimiston was possibly the busiest in Australia, and in a small island of ground surrounded by chimneys and poppet heads were six hotels, a brewery, many cramped shops and trades, all so jammed into the 'dirty acre' that the surrounding pavements could not contain the customers.

Melbourne and Adelaide people spent their first night in Kalgoorlie and were elated. Above the plain were the lights in a clear sky. On the plain was the roar of a metropolis and the stench of sweat and warm red soil. Men fresh from London or Johannesburg observed the tram cars swinging round corners, the electric lights overhead, the men spilling over wide pavements and on to roads as wide as boulevards, the broad-brimmed hat of the manager and assayer, the flare of the match struck near the bearded faces of old miners, the aroma of hot pies from the vendors, the scent of spilled beer as the swing-doors of hotels swivelled, horseshoes and cycle wheels crunching the gravel in Hannan Street, the horse manure steaming in the gutter. These sights and sounds filled the senses of men who came not sure what to expect. They paused at sharebrokers and watched the crowd studying the share lists, saw in windows gold in milky quartz from the Golden Link or Hainault, and saw Afghans in turbans and old suit coats, and French and Japanese women from the red lights of Maritana Street. In the distance was the song of the stamp mills and the whistle of locomotives going through the night to the coast, and under the stars the cornets of the Salvation Army played 'Where is my wandering boy tonight'.

The wandering boys and men had come from every large town and farming district in Australia. The goldfields had attracted a flood of migrants such as Western Australia had not known before, and its population in the 1890s leaped from 48,000 to 180,000. The year 1896 marked both the high tide of British money and Australian men, and in that amazing year there were estimated to be over

65,000 people on the Western Australian goldfields. Most were men, most were young, and the great majority had left Victoria and the eastern colonies to escape depression. But for the twin facts of depression in Australia and depression in the British money market the population of the Western Australian goldfields in 1896 or even 1900 may have been only half as much. It was not so much the gold but the English investors who paid the wages. In the two years 1895 and 1896 each miner in the West produced an average of only £46 of gold.

The sickness of the economy helped the goldfields and the debt was repaid. The boom in the West stimulated shipping and trade and manufacturing throughout Australia. Even the money that miners and trades people in Coolgardie and Menzies sent to families in eastern colonies was vital. As most men on the goldfields had left wife or dependants at home and as they earned perhaps £3. 15s. a week, they could send home half their high wage. In 1897 Western Australians mailed money orders worth almost £900,000 to other colonies, and nearly all those orders were for sums of less than £5. In addition, unknown sums were sent as bank drafts, in registered letters, or in gold or coin in the purses of returning miners. It was cautiously estimated that £1 million a year was going to Victoria alone to support families, and if so that sum could support 40,000 or 50,000 people directly and far more after the original money was circulated and spent. On the old Victorian goldfields reports suggest that whole towns such as Clunes and Eaglehawk almost lived on remittances from the West. By a strange spin of the mining wheel the retired colonels in Bournemouth were feeding the grass widows of Ballarat.

Wages paid by English companies provided the loaves for thousands of Victorian homes, and in 1898 a spectacular burst of deep alluvial mining at Kanowna (alias White Feather) channelled more money to Victoria. Twelve miles across the plain from Kalgoorlie the White Feather companies crushed quartz in stamp mills exposed to the sun and a few men fossicked in the dust. Two prospectors, Sims and a young Victorian electrician named Gresson, dug down a few feet in September 1897 to find rich gold in greenish cement and ironstone wash. They found a regular lead of gold, going deeper as it went across country. They won enough gold to set up a billiard saloon for the thousands who rushed to Kanowna and within a year they had got £7,500 of gold.

Others picked up the lead at increasing depths, hauling up the rich cement by windlass and bagging it for delivery to stamp mills. Fifty claims each yielded about £5,000 of gold and the Klondike party won nearer £10,000. Diggers traced the lead to the fence of the cemetery, were halted by the warden, and at last permission to dig under the cemetery was given in December 1897. On the chosen day several thousand men crowded the fence to await the signal from the warden on his white horse, and at the drop of his handkerchief they swarmed in to hammer pegs and dig a grid of trenches to mark their claims. Throughout 1898 the rush continued, the warden computing that at one time twelve thousand men were near the town. The lead was followed to a depth of over a hundred feet, twelve feet wide in places, and there were often five ounces of gold to the ton of cement and sometimes more. In 1898, the one great year, the Kanowna warden estimated the alluvial claims had yielded 132,000 ounces of gold, half of which came from the crushing of the hard cement they sent to private stamp mills and half from the softer ironstone gravels that the diggers themselves could treat. Kanowna's deep leads that year provided about an eighth of the colony's gold and helped it become premier gold colony for the first time.

The diggers of Kanowna could be seen leaving their tents and bough sheds at night to walk through the maze of shafts and trenches to the town where fifteen hotels and saloons were crowded and the fast trains left often for the glamour of Kalgoorlie. Father Long preached in one of the town's corrugated iron churches to Irish diggers who were successful beyond their numbers in the West, and he had the fortune one night to be shown a magnificent lump of alluvial gold. He told others what he had seen and was plagued for news, for in August 1898 hundreds in Kanowna had unpayable claims and were eager for new alluvial fields. At last Father Long promised to announce the news from the balcony of the Criterion Hotel, and on the chosen day the trains from Kalgoorlie and Coolgardie were crowded with gold-seekers, the road busy with horsemen and buggies and the faithful old dry-blowers with swags or barrows. Four to six thousand men stood hushed on the road outside the hotel that afternoon when Father Long walked on to the balcony and announced that the gold had been found near a salt lake on the Kurnalpi road. Before he had ceased speaking cyclists and horsemen on the outskirts of the crowd were away, leading a stampede of men with pegs and shovels to a desolate spot where they dug and cursed. They had been hoaxed, Father Long had been hoaxed, and his 'Sacred Nugget' was, it seems, a piece of iron splashed with gold paint.

Kanowna had deep leads as rich as some of the buried rivers of Victoria, and other fields might have them. A thousand shafts went down in a few weeks at Broad Arrow. Kalgoorlie prospectors sank a hundred feet and found a buried watercourse with patches of gold, and as Kalgoorlie's lodes were infinitely larger than Kanowna's it was not absurd to think that its deep leads could be larger. At the end of 1897 men rushed to the new find and pegged ground all over the mining lease of the Ivanhoe Venture Gold Mining Company. Now Western Australian mining law allowed diggers to prospect for alluvial gold on a company's lease so long as they did not go within fifty feet of the lode, and as it required some imagination to say that the Ivanhoe Venture had a lode, the invading diggers seemed within their rights. The Ivanhoe manager, however, protested that they were breaking the law and the warden ordered the diggers to leave. Most of the diggers continued to sink their shafts.

The Minister for Mines, E. H. Wittenoom, sensed that if diggers could invade leases on which companies had invested heavily, the result would be confusion on the leases and a jolt to the British investors whose money was essential. So he ruled that alluvial miners could not sink shafts on companies' leases to a greater depth than ten feet. His rules were ignored by many diggers on the Ivanhoe Venture lease; they said they were not interfering with the company's mining operations. Their refusal to leave the disputed ground won them heavy support on the goldfields. The incident seemed another instance of the folly of a government which was depriving the goldfields of the public works and the political power their population deserved.

The twin towns of Kalgoorlie and Boulder City were in explosive mood. Thousands supported the diggers on the Ivanhoe Venture lease with torchlight marches and rallies on street corners. When two alluvial miners were sentenced to gaol, a crowd erected an effigy of the Minister for Mines on a lamp post by the Palace Hotel and hanged and burned the effigy with its placard 'the ten feet drop'. More miners were sent to gaol for contempt of court. Sir John Forrest, the Premier, arriving at Kalgoorlie after opening the railway to Menzies, was hooted at the railway station by a crowd estimated at five thousand. When the Supreme Court confirmed the miners' right to dig deep on the Ivanhoe Venture lease the government brought in the law prevailing in other colonies that alluvial miners could not work on another company's lease. In fairness to the government no law could entirely reconcile the conflicting interests of companies seeking lodes and small teams seeking alluvial gold; but in fairness to the miners they surely had the right to seek gold on Crown land that was not being effectively mined.

In 1899 invaders on other leases at Boulder revived the clash with all the former bitterness. A massed meeting on 26 November in the open air at Kalgoorlie resolved, if the diggers' wrongs were not redressed, 'to take immediate steps to shake off once and for all the intolerable tyranny inflicted on goldfields residents by the Perth Parliament'. Parliament's refusal to allow Western Australia to join the approaching Commonwealth of Australia was the final goad to the goldfields, and at the close of 1899 they petitioned the Queen to create the eastern goldfields into a separate state within the coming Federation of Australia. Their agitation virtually made Western Australia join the new Commonwealth as an original state.

In the background to that discontent the alluvial troubles were inflammable. They affected few men, the gold in dispute was trifling, but because the dispute happened on the fields instead of in parliament in distant Perth it fired an outcry far stronger than the particular injustice seemed to merit. There seems too to be something in the nature of deep-alluvial mining to inflame bitterness in the face of unjust laws; the great civil insurrections on the Victorian and Western Australian goldfields were both in deep alluvial mines, and it is probable that at Kalgoorlie, as at Ballarat in 1854, the diggers were determined to fight because they had spent so much in sinking the deep holes in search of gold, and they were able to fight because they were men of more enterprise and courage than the average miner. The man who seeks deep-alluvial gold has to be courageous and enterprising. Ironically, the deep-alluvial gold of Kalgoorlie was not worth fighting for. Except at Kanowna, the colony's deep leads were few and poor; the goldfields essentially depended on lodes and the money of big companies.

18

CHEMISTS AND BUCCANEERS

THE HEART OF THE TOWN of Kalgoorlie was near the deceptive quartz reefs that Hannan liked, but the heart of the mines was in the greenstone three miles to the south, at Boulder City. There the lodes were phenomenally rich. They called this area the Golden Mile. By the middle of 1898 six mines had each produced over a ton of gold, and Kalgoorlie had passed Charters Towers as the nation's leading goldfield. Although no company was mining ore at greater depth than four hundred feet, most engineers believed the ore would live down. Even so, the failure of rich Coolgardie mines at depth worried investors that the gold of Kalgoorlie too might suddenly give out, though its orebodies were quite different. At a banquet in January 1897 the alert American engineer Callahan was suspicious of these freak lodes, and the bland reply of old Captain Oats from Southern Cross that the lodes would live down and 'would yet astonish the world' did not blow away the doubts.

The small size of the leases on the Golden Mile encouraged a quick answer to the question, did the lodes live down? At the vital south end of the field only four companies had more than thirty acres. Rich companies such as Golden Horseshoe and Ivanhoe, Boulder Perseverance and Hannan's Brown Hill, had only about twenty-five acres each, and many had less than twenty. With so many companies standing seven or eight abreast along the line of mineralization, many soon exhausted their shallow ground. As most companies had been floated in England into so many hundreds of thousands of shares, they could only earn sufficient profit to satisfy so many shares by mining on a large scale. Thus they also exhausted shallow ore and had to go deep. All this gave an air of feverish activity to the Golden Mile, and it became crowded with treatment plants, dams and mullock heaps, water condensers and engine houses, railway lines, stacks of timber and managers' houses. Above all, the shafts probed quickly downwards. The Great Boulder passed 1,500 feet in 1902, and still found gold.

The mining of deeper ore created a new crisis. The first millmen at Kalgoorlie had erected the old stamp mill to crush the oxidized ore and had caught the gold by amalgamation. The gold they recovered was so rich they were elated; the gold that escaped in the waste stream was so rich they were puzzled. Endless experiments increased the junk heaps of discarded machinery, but not the extraction of gold. As the shafts went below the shallow oxidized zone the loss of gold in the mills became serious. Many particles of gold were intimately blended into the barren rock, and were as fine as flour, so that the sulphide ore had to be crushed more finely in Krupp ball mills to release them. But fine grinding merely turned the ore into that powdery slime which, since the dawn of gold milling, had been the bugbear of metallurgists. The slime absorbed tons of water, and water was precious. It also made the extraction of the particles of gold by the cyanide process too slow and expensive.

John Sutherland thought about this problem. Aged twenty-seven, he was lanky and scholarly and had studied at Ballarat School of Mines and had worked for the Broken Hill Proprietary before coming west as metallurgist to Lake View Consols. Ambitious to solve a problem that worried hundreds of more experienced men, he adapted in 1897 the filter press that had been used in sugar mills to squeeze juice through cloth, leaving behind the pulp. It worked as well on gold slimes as on sugar, and quickly spread along the Golden Mile and across the seas. It enabled the cyanide to extract the gold more quickly, and it saved precious water.

Arthur Holroyd, son of a Melbourne judge, had a shop in Kalgoorlie and assayed ores sent from small mines. In May 1896 the Block 45 mine sent him a dark calcite stone for assay. He ground it down and panned it without finding much gold, then assayed it in his small furnace and found it astonishingly rich. Not quite sure that he had assayed it accurately, he repeated the test and found the gold was still rich. He thought the sample might contain telluride of gold, so he applied hot sulphuric acid and saw a brilliant carnation colour in the bowl. His announcement that the rare telluride of gold existed on the field was seized upon by the share market, for tellurides enriched the famous Cripple Creek mines in Colorado.

The rich telluride of gold in fact was widespread at Kalgoorlie. It delighted shareholders but did not altogether please metallurgists. Whereas Cripple Creek successfully chlorinated its telluride ores, the lime and magnesia in the Kalgoorlie ores wrecked that treatment. Roasting of the crushed ore in rows of furnaces patented by Edwards of Ballarat turned the gold in the tellurides to metal, which could be recovered in cyanide vats, but the lime in some ores so hardened the mixture in the vats that it had to be quarried out with dynamite. All along the line were pitfalls and problems. The solving of one problem created another.

In an airy laboratory near the Brown Hill mine German chemists with pince-nez and clipped beards sat on high stools testing the ores from the adjacent mine. They were servants of the London and Hamburg Gold Recovery Co. which had taken shares in the Brown Hill mine in return for designing and building a treatment plant; and their leader, Dr Ludwig Dhiel, would pull at his waxed moustache and vow that normal cyanide of potassium was too slow in dissolving the gold from telluride. So he tried bromo-cyanide as a quickener and made it work. The coarser fragments of rock did not release their imprisoned gold to the cyanide, so he decided to crush all his ores into slime. Knowing that cement works had long used a steel barrel of pebbles to grind soft limestone, he adapted their tube mill to gold ores, and the first of thousands of tube mills used throughout the world on gold ores was used on the Golden Mile.

It was a memorable day for Kalgoorlie, Saturday, 14 January 1899, when the Dhiel process was unveiled to the world. Official guests arrived by express train from Perth, and were taken in buckboard and buggy to Dr Dhiel's bungalow, the lofty new plant and the mine, finding everywhere on their escorted tour tables of wines and spirits and cigars to soften the persistent technical chatter of the young Germans. At nine o'clock that evening a hundred guests sat in the Palace Hotel to a menu as long as the toast list and the final reward late in the night, the declaring of a dividend of 7s. 6d. for each one pound share.

The willingness to experiment and adapt in the thousand jobs and reactions that make an efficient mining field was a hallmark of the Golden Mile at the turn of the century. Australia had never had such an alert, invigorating field. In private dining rooms or in the dignity of Hannan's Club, managers with gay bow ties from Colorado, deep-eyed old men from Victorian reefs, young metallurgists from Broken Hill, the Rand, and Freiburg argued and talked. In laboratories and assay offices on the mines young men from Australian schools of mines pottered late at night, eager to work for companies that scorned no new idea.

Kalgoorlie's success in mastering rebellious ores made it probably the world's leading goldfield in metallurgy. Its young managers and its ideas were exported to fields in every continent. Ralph Stokes, mining editor of the *Rand Daily Mail*, often heard that 'the Rand is behind Kalgoorlie', and when he inspected Kalgoorlie in 1907 he thought the comment was true.

Many Western Australian goldfields, however, were not efficient. The sudden gold boom required more skilled managers than the industry could provide. Hence shafts were sunk in the wrong places, rattletrap treatment plants were built, money was dissipated, and promising mines damned by inefficient men who had not before seen a mine. One challenge to inefficiency came from the London consulting firm of Bewick, Moreing & Co. which sent a strong team of engineers to inspect or manage English mining companies from Coolgardie to Cue. It introduced American practices, strove in all its mines for cheap costs, and employed specialists in every phase of mining.

Herbert C. Hoover was one of Bewick, Moreing's men; he reached Coolgardie in May 1897, quiet and ambitious, teetotaller and bachelor, dark hair parted in the middle and a twirled moustache that half disguised his youthfulness. He was only twenty-two and had not long left Stanford University and the mining fields in the Rocky Mountains, and he observed with quiet curiosity the restless haste of the gold towns and their thirst for champagne. For a year he wandered over the goldfields on camel or cart, writing voluminous reports on mines and wildcats for his firm. Even then the Melbourne *Leader* recognized him as one of the colony's best mining men. He knew more geology than most consulting engineers and had the uncommon sense that a mine was not worth a penny unless it had gold or promise of gold. Partly on his advice his firm bought the little Sons of Gwalia mine a hundred miles north of Kalgoorlie, and he went to the mine as manager. For seven months this man who was to live later in the White House in Washington lived in an iron house in the path of the hot winds that blew across the sea of sand and scrub. He insisted that a large mill should not be built at the mine until a dam was full of water and 72,000 tons of ore were ready for mining, a decision sensible and rare amongst the managers of the west, and when he resigned as manager the Sons of Gwalia was employing over one hundred men and beginning one of the longest lives of any Australian gold mine.

The growth of sound mining practice was often slow. London promoters and company directors often wrecked the honest ideas of trained engineers. Directing mines which had little gold, they sometimes boosted share prices by erecting treatment plants; the erection of a plant convinced many investors that the mine was capable of producing gold, and shares would rise and directors would sell their shares. If their mine had rich ore they tore it out quickly, even wrecking the mine for the sake of a high yield of gold that would lift share prices. Financiers who could no longer float new companies with the waning of the share boom now fixed the price of shares and juggled balance sheets. Fleet Street had few watchdogs to protect the small investor, and the mining correspondent of the London *Economist* wrote on 10 July 1898 'that there has been incomparably more dishonesty in the short annals of Westralian mining than ever there was in the Transvaal'.

One man who was probably indignant at such criticism was Whitaker Wright. He was a confidence man with much righteous indignation. Aged fifty-three, portly, semitic in face, mysterious, eminently plausible, he dined with bankers and sat in boardrooms with lords. His name was hardly known in Kalgoorlie but he controlled two of its richest mines, Lake View Consols and Ivanhoe, as well as many smaller mines from the Murchison to the mulga. At his London office, 43 Lothbury, he employed rows of clerks and at his mansion in Park Lane he had galleries of works of art and guards to protect them from the unscrupulous. At week-ends he retired to his estate at Lea Park, near Godalming in Surrey, and indulged his taste for landscaping, hiring hundreds of labourers to level hills obscuring the view and to excavate lakes enhancing the view. He entertained well, playing billiards with his guests in a glass saloon beneath an artificial lake, engaging celebrated artists to amuse them in his private theatre. The estate also had an observatory and on summer nights of 1899 Wright may have studied the stars intently.

Whitaker Wright never visited Western Australia; he had served his apprenticeship in the mines of western United States, going there as an alert young assayer from the north of England at the age of twenty-one. He bought and sold shares, confessing that after he had made the first 10,000 dollars the others were easy. At the age of thirty-one he was a millionaire in dollars, and he won more money at Leadville and New Mexico before going east to Philadelphia where he presided over the mining exchange. He began to lose money, and after he had lost the first half-million the rest was easy.

Wright returned to England and found money hard to gather until the gold boom began in London in 1894. He then floated two large Western Australian companies, which yielded him £238,000 from their manipulations and operations. In 1897 he merged these two companies into the London and Globe Finance Corporation, and he became managing director and held 605,000 of the 2,000,000 shares. His own shares were soon worth £1·2 million in that company alone, for it had large holdings in two rich Kalgoorlie mines, Lake View Consols and Ivanhoe. He floated in 1898 the Standard Exploration Company, which controlled such mines as Mainland Consols and Paddington Consols, and from the holding company Wright drew fees as managing director and a large block of free shares. He also made heavy profits by using the Standard company's funds to speculate in shares of the mines it controlled.

Wright loved to speculate, and in 1899 his Lake View Consols mine gave him a wonderful chance. One of the most productive mines on the Golden Mile in 1899 it ran into one of the best shoots of stone found on the field. Miners named it the Duck Pond; it was full of golden eggs, and month after month the company gathered the eggs and produced a ton of gold a month. The shares jumped from £9 to £28 each in London, and the company paid £625,000 in dividends in the one year. No Australian gold mine except Mount Morgan had paid so much in a year, and shares in Lake View flew in a whirlwind of speculation.

Whitaker Wright sat in an enviable seat. He was the first to receive cables from the mine and he could secrete or release what news he wished. He decided that the Duck Pond was vast and that the mine could maintain its marvellous dividends. Henry Clay Callahan, a crusty American from Cripple Creek, was general manager at the mine and must have been optimistic that he could continue to tear out the riches at the rate of a ton of gold a month; but after five months the Duck Pond was drained. Gold yields slumped, and shares too. The stars told Whitaker Wright that the mine was still rich and he continued to buy shares, manipulating the price in the face of strong selling by hundreds of other shareholders. London and Globe had made huge profits by selling Lake View shares when they were rising and now it lost all the gains by cornering shares when they were falling. By the end of 1900 it had lost £750,000 and Standard Exploration had lost £250,000. On 5 December 1900 Whitaker Wright tried to avert disaster by publishing a spurious balance-sheet showing annual profit of £463,000. But three weeks later his companies' confession of insolvency shocked the stock exchange, ruining many of its members and hundreds of its customers. The official receiver studied the accounts and saw that one company had paid dividends by borrowing money from allied companies and Wright himself. Mr Balfour, leader of the House of Commons, conceded that the public had 'deep and profound indignation' at Wright's frauds but he was not sure that company law would permit a prosecution.

A warrant for Wright's arrest was issued in March 1903, but Wright was in New York and fought extradition by every stratagem of law. Finally he returned to London to face prosecution under the Larceny Act of 1861. His trial aroused intense interest, for it was revealed that the companies he managed had lost £5 million of shareholders' funds and contracted another £3 million of debts. The case against him was overwhelming. On 26 January 1904—Australia Day—he was convicted and sentenced to seven years in prison. He received sentence, went to a waiting room to consult his solicitor, and fell dead. He had taken cyanide, the magic chemical which had won so much of the gold he had squandered. Friends said he had always played for high stakes, and life itself was a counter to be gambled away.

It is dubious if revelations of the corruption much affected the flow of mining capital from London. Wright's gamble on Lake View Consols, however, did have a curious indirect effect on Australia. In his greed for high dividends the company had pillaged the Duck Pond, and when Wright lost control of Lake View Consols it had paid £1,300,000 in dividends but had so exhausted its rich ore that it struggled to make a profit. Francis Govett of London became chairman of the company and sensing that the great mine was finished searched for another. He found it at the south end of Broken Hill, and the money Lake View Consols invested there was to shape the future of Consolidated Zinc and all its international interests.

The year of Whitaker Wright's death was marked by another revelation from the Golden Mile. Not far from the Duck Pond was the rich and admirably efficient Great Boulder Perseverance and the small Boulder Deep Levels Limited. The chairman of these London companies was Frank Gardner, an American who was said to have been a drummer boy in the Civil War. He had black eyes and black moustache, black pearls on finger and shirtfront, he was cold and sullen, a flashy speculator in shares and women. He owned at one time 170,000 of the 280,000 shares in 'Boulder Deeps' and believed he had exclusive right to confidential news from the mine and the right to withhold it from directors and shareholders. He did not realize that others could also load the dice.

Boulder Deeps had paid no dividend but in 1903 its diamond drill found a promising lode. The shaft was deepened and a drive explored the lode for 240 feet in the autumn of 1904. The underground manager, Flynn, chipped samples from the lode and sent them up the shaft in a bag to the assay office of Boulder Perseverance, which Gardner also controlled. Vickers, an assayer, confidentially reported to Gardner that the lode averaged seven ounces to the ton for its entire length. Gardner heard with delight of this unexpected bonanza and coolly spread word that the mine was still a disappointment. Shares were cheap after the company's long struggle and no doubt he intended to buy them. As he allowed no journalist or shareholder to inspect the Boulder Deeps he believed he could hush up the find for a few precious months.

A few men at the mine knew more about the find than Gardner and they bought shares. A Sydney stockbroker received an order to buy and a telegram explaining why: 'Is biggest thing on fields.' A dozen or so Kalgoorlie brokers had orders to buy Boulder Deeps, and as most of the shares were held in England they wired orders to Lionel Robinson & Co., powerful London stockbrokers. With heavy buying, shares leaped from about 6s. to £2 each. Gardner was now obliged to announce the sensational discovery to shareholders, but he did so more than a month after he had heard the news himself.

Gardner employed as general manager of his two Boulder mines a clever New York mining engineer who had graduated from Columbia and had run mines on the Comstock before becoming Gardner's personal adviser in the Westralian mining boom. Ralph Nichols, the engineer, was in London while the lode was being explored and did not see it until 1 June 1904. He sampled the lode and found it averaged less than an ounce to the ton. Where were the other six ounces? Vexed and suspicious, he decided to examine the ore that had come up the shaft to be dumped in the mineyard with an iron fence around it to prevent the stealing of rich specimens. He walked around the dump, saw nothing stealable, and directed that several tons be put through the mill. This test revealed the gold was even poorer, and he realized that someone had salted the mine, or sampled it spuriously, or assayed the samples with intent to deceive. He dismissed the assayer and underground manager; they left willingly, having made £3,050 by speculating in the shares.

Rumours of the fraud made Western Australia appoint a Royal Commission which unfolded the story of corrupt employees deceiving a corrupt chairman who had long been deceiving the shareholders. The year 1904, with the death of Whitaker Wright and the disgrace of black Gardner, perhaps symbolized the end of a swashbuckling era.

Metal mining has perhaps more temptations than any other industry. Therefore it demands a higher standard of ethics if investors and gamblers who give the essential funds are to be protected from company directors and employees. No series of wise laws can adequately protect the investor. The industry itself has to protect him. Nothing was more vital in the long term for Western Australian gold mines than the rise of sound, honest management, for it not only protected the investor but enhanced the value of his mine.

From the end of the boom in British investment at the end of 1896 to the year 1903 the output of gold in Western Australia multiplied by six, even though population on the goldfields declined slowly. British money had provided the employment in the booming West in Australia's depression and now it was gold that provided work and profit. In 1903 the gold output was nearly two million fine ounces worth nearly £8 million. Australia that year was again the world's great goldminer and the West produced half of that treasure.

14 Kalgoorlie stock exchange, about 1895

15 The bough camp at Great Boulder, 1895

16 Speculators at one of Kalgoorlie's share markets, 1890s

17 Sons of Gwalia mine and its American manager, Herbert C. Hoover, 1898

18 The denuded hills of Mount Lyell, scene of another American triumph
(T. Nankivell)

19 The Zeehan-Ringville mountain railway in the 1890s

Some of the promoters' prophecies were coming true at last. Dividends from the West's mines exceeded £2 million in 1899 and again in the three years 1903-4-5. To the end of 1915 they paid £25,500,000 in dividends, of which half came from four wonderful mines—Great Boulder, Ivanhoe, Golden Horseshoe and Great Fingall. The Golden Mile dominated the dividends and ten of its mines, plus the outsiders Sons of Gwalia and Great Fingall, had paid 87 per cent of all the dividends to 1915. Of the several thousand British and Australian companies which had spent money in the state no more than ninety had paid dividends, and some of those paid only one and some had no right to pay even that. Outside the Golden Mile the money invested by the British had rarely been repaid. The investors themselves were perhaps largely to blame for their losses, but the mines took the blame.

The fields entered their long decline. The big mines stayed prosperous but as their shafts went down the ore became poorer and the cost of mining dearer. The prospectors had scoured the sands and hills so intensely that few new reefs and lodes were found, and each year few new companies joined the dividend list. One could travel hundreds of miles through the outback in 1910 and see everywhere the crumbling stone hotels and the neglected graveyards near mines which once were popular names in London. In the sand the traveller saw the pad of abandoned camels and the wreckage of carts and waggons, the piles of tins that had marked a busy boarding house, the gleaming bottles that marked a bush hotel. If he examined his map he saw large towns marked and now he would pass them in train or motor car and wonder why he hadn't seen them. The outer fields were in decay.

The mantle of pessimism settled on the richest fields. Just as the prospectors had hurried to the Klondike and New Guinea and Canada in the 1890s, so a decade later some of the great companies on the Golden Mile decided to search for new deposits before their own mine was exhausted. They were now wary of testing untouched lodes in Western Australia. Like the prospectors they often went to greener fields. Oroya-Brownhill bought a mine in Nicaragua on Herbert Hoover's advice, Hannan's Pty Ltd bought a tin mine in Cornwall, and Associated Northern Blocks turned to Mexico. Great Boulder explored the old Magdala-cum-Moonlight at Stawell in Victoria, and Lake View Consols went to Broken Hill and shaped Australian industrial history. The mining tempo is rarely restrained, and the pessimism of the goldfields by 1910 was as infectious as optimism in the boom.

19

MINES OF THE MIST

FROM MELBOURNE'S RIVERDOCK in the late 1890s a procession of coastal steamers sailed for the Western Australian goldfields, saloon and steerage packed with passengers. As the ships' bells rang and the last mailbags were stowed, men ran from adjacent hotels in Flinders Street, and some who had drunk freely to success in the West ran up the wrong gangway and did not reach the West. For ships left the same wharves for the Tasmanian mining fields that were comparable comets to Kalgoorlie in the tail years of the century.

The island of Tasmania had been settled for two generations before its mining era began. The alluvial gold was not plentiful and its rich deposits of base metals were buried in mountain forests or highlands. Tasmania called for prospectors as skilled as those who made the gold trail in the West, but the Tasmanians fought cold forests instead of hot plains, carried food and blankets on their backs and mining tools in their hands. They mostly carried an axe and rarely a gun, but never a waterbag.

James 'Philosopher' Smith was one of the hardiest Tasmanian searchers. He had dug gold in Victoria in the 1850s, and a decade later his Tasmanian farm near Forth was his supply base for lonely walks into the mountains that soared to the south. Tall and lean, a God-fearing teetotaller, at the age of forty he had the gravity of a man of seventy. 'I cannot remember ever hearing him laugh,' wrote one of his sons, 'but occasionally he would smile at something amusing or pleasing.' His patriarchal beard and manner did not proclaim his skill as a bushman but this he proved in his forty-fifth year. At Mount Bischoff on 4 December 1871 he found the greatest lode of tin then known to man.

Curiously, surveyors had erected a trig station on Mount Bischoff about 1843, and like the surveyors who erected the trig on Broken Hill they missed or ignored the heavy black tinstone. Smith then may not have been the first to see the mineral but he had the ability to recognize it and the courage to risk money in exploring it. Selling property and borrowing money, he raised about £1,500 and paid men to hack a track and cart to the coast some trial loads of ore. He hoped to float a company, for the price of tin was high, the rich New

England tinfield on the mainland was booming, and all kinds of mining shares were high on the stock exchanges. And yet he hawked in vain the most valuable mining property so far offered for sale in Australia until, late in 1873, he sold it to the new Mount Bischoff Tin Mining Company for £1,500 cash and 37 per cent of the company's shares.

The lode's oxidized capping was so rich and soft that it seemed to be a freak deposit of alluvial tin, offering quick profits. Isolation and muddy mountain roads ate those profits. The company spent £7,600 of shareholders' capital and then borrowed £30,000 from a bank. Profits from the sale of tin were invested in large dams and water races and waterwheels and treatment plant, and the company used water power more freely than possibly any other in Australia. A long wooden tramway was laid and horses pulled trams forty-eight miles to the port of Burnie where ketches carried tin concentrate to the company's smelters in Launceston. The company spent £100,000 before it paid its maiden dividend in 1878.

Commercial travellers who rode up in the jolting horse-tram sensed the loneliness and desolation of the new town of Waratah at the foot of the tin mountain. Evergreen myrtle forests swept south and east and west, beautiful in the morning sun, dark when clouds scudded from the Antarctic, and snow slushed the quarry where miners worked in leather leggings and the woollen knee-coats known as 'blueys'. On winter days the wind skimmed the puddles in the main street and piemen hurried to the mine at noon with baskets of pies shrouded in warm flourbags.

The mountain cone of Mount Bischoff seemed to lie on the rim of civilization. Heinrich Wilhelm Ferdinand Kayser, the general manager and stout German god of the town, looked at the rain gauge and swore that his company could not survive without its annual eighty-five inches of rain. Born at the mining centre of Clausthal in the Harz Mountains, the first of the men from German academies who were to have great influence on Tasmanian mining, his was the first Australian industrial plant to be illuminated with hydro-electricity. But the electric light glowing in 1883 from the long buildings in the dark gorge was not the most impressive feature of Mount Bischoff. Its monthly dividends were more famous. The company had paid £1 million by 1889 and went on to celebrate its seventieth anniversary with a dividend. Philosopher Smith's share and his children's share of the spoil would have been about £940,000 if he had held his free shares. In fact he sold them, before they earned a dividend.

The Philosopher's mountain quickened the search for tin. In north-eastern Tasmania farmers were moving to the red rich soil beyond Scottsdale, making access easier for prospectors, and in 1874 the experienced prospector G. R. Bell panned black grains of tin beneath the smooth pebbles of streams. Whereas Philosopher Smith had pegged most of the rich tinland in the north-west, the alluvial tin in the north-east was widespread and could be extracted as simply as gold. So many young Tasmanians rushed to the Blue Tier and the Ringarooma Valley that they found most of the valuable deposits within two years. Goldminers hurried from Victoria and Chinese jogged in single file to the tinlands. For years their joss house, now in Launceston's museum, was sweet with incense, and even in 1900 nearly as many Chinese as Europeans mined tin.

The tin oxide had been washed and eroded over the countless centuries from granite rock and was found in the gravels of rivers or buried beneath hillsides from the grassroots down to depths exceeding one hundred feet. Sapphires and zircons of deep orange-red sometimes mingled with the millions of black fragments of tin oxide, and were prized by Ah Fat and Wong Mee. The tin's high specific gravity enabled it to be washed in the same way as goldbearing gravels, and miners sold concentrates carrying over 70 per cent tin to smelters in Launceston and Sydney. Parties of four or six men built earthen dams in the creeks and hewed or built long flumes that carried water five or twenty miles to their claim. Often they played a giant nozzle like a fireman's hose on a cliff of tinbearing clay and washed the debris down wooden flumes and trapped the heavier grains of tin and allowed the sludge to flow away. In the Ringarooma they used hoses and dynamite and crowbars to remove the layer of basalt that covered deep leads of tin resembling the sub-basalt gold leads of Victoria. Water was the blood of the north-eastern mines, and in dry summers the mines would close and at Boobyalla and St Helens and the small anchorages the ketch sails would be furled, waiting for the rains that would send bullock waggons of tin rumbling down from the uplands. Australia and Straits Settlement were the world's largest tin producers, and from the 1880s Tasmania was the largest producer in Australia.

The small city of Launceston with its Cornish name and bold speculators got much of the wealth and commerce from the distant tinfields, and to cap its triumph the first rich goldfields were opened close to the River Tamar that flowed past the city. In the 1870s, close to the modern aluminium refinery at Bell Bay, gold was won from the shallow reefs of Lefroy. On the other side of the river in 1877 William Dally, a farmer, found the Tasmania Reef that cradled the godly town of Beaconsfield and made it for a time the third largest in Tasmania. And to the Launceston company that had financed that reef went £773,000 in dividends.

Launceston had only twenty thousand people in the late 1880s and there was no poppet head or stamp mill within twenty miles of the city and yet it challenged Bendigo and Ballarat in the dividends its mining companies paid. The great disappointment to its mining community was a mine to the south-east, at the small town of Mathinna. In 1887, after the poor gold mine had been abandoned for about eight years, a fossicker named Loane walked with candle down a muddy tunnel and saw a small band of quartz that earlier miners had ignored. He sank a hole on the floor of the tunnel and saw a gold-spangled reef. Investors in Launceston's rival city of Hobart bought his New Golden Gate mine and for a trifling outlay won one-third of a million in dividends.

4 Mining towns of north-western Tasmania

Investors who drove in phaetons through Launceston's streets or who took their daughters to London for the season did not monopolize the earth's treasure. The government clerk, the grumbling orchardist, the wheat farmer and wharf labourer and railway navvy and shopkeeper gained even more. Tasmania was in the shadow of a depression for twenty years until mining fields were tapped in the seventies. Once found, the mines gained immeasurably from the depression, for they attracted idle men and money. Depression stimulated mining, and tin and gold in turn blew away the depression and lifted Tasmania in the late 1870s into its most prosperous era of the century. When the nation-wide depression of the 1890s drifted like a fog from the mainland, a new halo of fields in Tasmania's western mountains was strong enough to disperse the fog. In 1900 Tasmania had the nation's largest tinfield, largest copperfield, second largest silverfield, and gold as well. And, like Western Australia, more than half of its export wealth was coming from mines.

The mistline drifted above and below Mount Bischoff, revealing on clear days new scars and contours on the slopes where the tinstone had been torn away. The town of Waratah was a foothold in the forest, a supply base and stepping stone to the mineralized mountains stretching south. Those mountains always seemed about to produce great mining fields, but after twenty years they were still tormenting and teasing prospectors. Their large metal deposits, at Mount Lyell and Mount Zeehan, Mount Read and Rosebery, lacked that large oxidized zone whose minerals were rich and easy to treat. Steep terrain, heavy rainfall, and slow geological change had eroded that oxidized zone almost as soon as it was formed. These deposits were cakes without the icing. They offered no sudden fortunes, and were therefore developed much later.

Forest and gorge and rocky coast imprisoned the treasure. The ocean was unbroken by land all the way from South America, and the Roaring Forties smashed the waves against the coast. The only harbour near the mineral region was Macquarie Harbour, where Tasmania's worst convicts had once been exiled on an island protected from the swell of the ocean by Hell's Gates with the shallow sand bar across which tides rose and ebbed like a mill race. The journey overland was often as hazardous as the sea. Bare mountains of white quartzite and pink conglomerate rose above the forest, and swift rivers cut gorges through the highlands. Here and there bleak plains of button grass made travelling easier, as if to harness man's strength for tangled forest where the shoots of bauera twined around treetrunks and the dreaded horizontal scrub stood as solid as a wall. From horizontal branches a thicket of shoots grew up fifteen feet or more, toppled in the wind, sent up more shoots, and so rising and falling created a web of scrub that sometimes prevented explorers from travelling more than a mile in a day. While they could set fire to it in summer the forest for most of the year was soaked and chill, wet branches overhead and sodden decay below. Few explorers were hardier than those few hundred prospectors who axed tracks through these forests.

At first the west coast had small rushes and hollow booms. North of the dark Pieman River was a small gold rush in 1879, and men walked thirty or forty miles south-west of Mount Bischoff with swags weighing up to a hundred pounds on their backs. Others chartered small steamers to the mouth of the Pieman, which even on mild days was lathered with foam. On 5 April 1879 the steamer *Sarah* stuck on the Pieman sand bar for an hour and a half, decks thumped by the sea, helmsmen lashed to railings of the wheel, and passengers and crew clinging to the rigging and looking helplessly on those round green hills that seemed so near. Finally the ship slid off the bar into the calm river, and a week later its passengers were digging for gold in forest as harsh as the sea. The Pieman diggings were rich in alluvial gold, but had no reefs to sustain permanent mining camps. One of the few hotels was sold for four ounces of gold and, before long, shops and tentpoles and scarred hillsides were hidden beneath the spreading blackberries.

Tin searchers won a more southerly foothold on the rocky coast. At Mount Heemskirk tin was found in February 1876 by a government exploring party led by Charles Sprent, and in the following summer Owen Meredith pegged the St Dizier lease before retreating with the onset of winter. Another summer came, and a few more parties hired ketches in Hobart to take them west. They would have been wiser to cut a track overland; one ketch took 53 days and the other 79 days to sail 210 miles. Half summer had therefore passed before the miners reached Heemskirk, but they worked far into the cold months in their eagerness to prove that Heemskirk was another Mount Bischoff. Their proof was flimsy. The Great Western Company spent £2,700 in winning eight tons of tin concentrate, but when the bags of tin were jubilantly landed at Hobart they were seized by the Crown because they had been mined from another company's lease. That dispute was no sooner settled than a dispute began over the nature of the metal. A skilled assayer declared it was mostly titanic iron and sand, and the valueless 'tin' lies on the riverbed at Hobart.

Each summer strings of packhorses and packmen filed from the Pieman River or Macquarie Harbour to Heemskirk, and the rich cargoes of tin oxide they sent to civilization convinced many that the field was rich. But the absence of large oxidized patches on the lodes and the inaccessibility of the region made capital essential, and that capital did not come until the early 1880s when money in Australia was plentiful. Mining booms erupted in most Australian cities, and Heemskirk was a child of the boom. Melbourne and Tasmanian promoters floated about fifty companies to mine its tin, and in the excitement leases were even taken on the beach and far out to sea.

So much machinery was ordered that ships sought a port nearer the mines, and four Scandinavian sailors in their seven-ton cutter *Trial* found a small opening that some called a bight and some called a hole in the rocks. Into that hole they sailed the cutter in 1881, but overnight the winds blew and the surf dumped the little ship on the beach. After the cargo was unloaded the ship was dragged into the water, and it returned again and again to Trial Harbour, the bell-wether of a line of ketches and steamers that nervously supplied the mines whenever the sea was quiet.

Trial Harbour became a busy town and from the decks of the ships one could see the night lights of hotels and shops along the beach and lonely lights high on the granite slopes from mine managers' cottages and battery sheds and shafts where horses worked the whims through the night. In the day one could see long strings of horses hauling stamp mills up the steep mountains to mines that had only tiny patches of rich ore but had to have a stamp mill to assure shareholders that dividends would soon be coming. At least ten companies built mills and not one paid a dividend. At the Orient mine the Cornish manager insisted that all employees hear him preach in the mine store on Sundays, and the small boys would sit on forms in front bellowing out the hymns and giggling at the horses champing in the stables next door, and the manager would call loudly from his pulpit, 'Have silence, boys, have silence in the House of the Lord'. Soon even the thunder of the stamp mills was silenced at Heemskirk.

The mining boom was disastrous for investors but stirring for prospectors. Syndicates in Hobart and Launceston financed small teams of prospectors, and they walked overland or came by ship to find new Heemskirks. Twelve miles east of Trial Harbour members of two Launceston syndicates were searching on 4 December 1882 when Frank Long drove his pick into a vein of gossan and found glittering galena rich in silver not far from Mount Zeehan. Wrapping a specimen in a handkerchief he showed it to his mates at the camp that night and they eagerly pegged leases for their syndicates. Their discovery of a maze of silver-lead lodes was more deserving of a mining boom than Heemskirk's patchy tin, but unfortunately mining booms tend to come when the money market rather than the mines command. Zeehan wanted capital, but the money market was hardening. It was virtually deserted for another five years.

Another rich but disillusioning field was found near Macquarie Harbour during the share boom of 1881. Cornelius Lynch was the discoverer, wiry, red-faced, versed like Long in New Zealand goldfields and the terrain of western Tasmania. His 'Catch 'em by the Wool' claim extracted tin from the tailings of the Mount Bischoff mill, and he got enough money to join a Hobart syndicate and go prospecting with his mate Tom Currie. They went to the King River, near Macquarie Harbour, and found payable gold not far from where they had found unpayable gold on a brave journey three years before.

The gold had to be rich, the men had to be strong, if this isolated gully in the forest was to be tamed. Lynch and Currie and the diggers had to hire rowing boats at the small port of Strahan, row around the bay and up the dark King River to the first rapids, and then jump in the water and tug the boat over twenty rapids until they reached the gorge. 'Only doses of rum made us endure the cold,' wrote Jimmy Elliott, a thin boy of sixteen who dragged a boat up the river and carried a heavy swag seven miles from the gorge to Lynch's Creek. This boy Elliott was eventually employed by Lynch to search for the reef that had shed the coarse gold in the creek, and on a frosty morning in 1883 he found the reef in a trench he was digging in red clay. The quartz was studded with gold, and specimens in Hobart drew crowds to shop windows, and in Melbourne a hundredweight of the stone yielded £830 of gold. But the reef was like all reefs in those razor-backed hills, brilliant in patches, and Cornelius Lynch waited until the Western Australian rushes to win riches that had eluded him.

Lynch's mine was at least useful, for it was virtually to lead to the finding of the famous Mount Lyell field. Diggers came through the forest and fanned out, finding gold by rushing streams and boulder-strewn valleys. Bill and Mick McDonough (alias Cooney), wage-labourers at Mount Bischoff and Heemskirk, and Johannes Karlson, a Russian Finn who had sailed in the celebrated little cutter *Trial*, teamed up and sought gold beyond Lynch's Creek. All in their twenties, raw in the lore of prospecting, they did not scorn deposits that more experienced men might have shunned. Pushing through wet scrub they saw a massive outcrop of iron jutting twenty or thirty feet above the ground, with small trees and shrubs sprouting in crevices on the crest of the outcrop. It seemed as if a large explosion had shivered the rock, spilling boulders of ironstone far down the hill. It was in fact the common iron hat of a pyritic orebody, but for long that knowledge was buried beneath the Iron Blow.

The prospectors found that other men had camped near Mount Lyell. They saw rotting tentpoles, empty brandy bottles and enamel cooking utensils, left behind twenty-one years previously by a large government party led by the geologist Charles Gould. To the credit of the three raw prospectors they sank holes near a creek and found over a wide area the payable gold that had eluded Gould. As they panned alluvial gold they suspected that the gold had eroded from the ironstone outcrop. Karlson, wide-shouldered with fair leonine beard, clambered on to the ironstone and shook into his tin dish some soil from the grass roots. At the nearest waterhole he washed the soil and found fine gold. His mates examined the Iron Blow carefully and found conglomerate boulders with gleaming specks which they thought were gold. They gathered them, hiding them beneath piles of brushwood, until one day they learned that the metal was fool's gold. Fools in knowledge but not in persistence, they carried crowbars and dynamite and screwjack from the coast and began to strip the ironstone. In rain and snow they fought with the boulders, heaving and drilling and blasting, peering for traces of gold in the gaps left by each dislodged boulder. For a time the gold they sluiced in the creek paid for food and supplies, but eventually they had to borrow from the portly little storekeeper, F. O. Henry, at the port of Strahan. The McDonoughs fell too far into debt and sold their shares, and William McDonough a quarter of a century later became a night watchman on the field he had found. Meanwhile, Karlson invited his two burly brothers to join him, but the three also fell into debt and sold their shares.

At last in June 1886 rich gold was found in soft ironstone at the bottom of a water race. The six shareholders were elated. At night the roar of the mountain streams barely drowned the noise of merriment from the tents. New diggers and syndicate men emerged from the forest to peg claims. A white-headed retired colonel from the Indian army offered £12,000 for a one-sixth share in the mine. The government geologist, Gustav Thureau, came to acclaim it as Tasmania's richest gold mine, theorizing that the gold had come from the depths as boiling solution of volcanic mud and predicting that the gold would go deep and be almost inexhaustible. James Crotty, intelligent, religious, balding, a skilled Victorian goldminer who had become leader of the Mount Lyell syndicate, celebrated the news at Strahan with a drunken party that was halted by a magistrate on the third day. 'Begorrah, it's all gold, I tell you. I'll be that rich I'll buy Ireland and make it a present to Parnell,' said Crotty.

The Mount Lyell syndicate was soon troubled. The Karlson brothers believed they had been tricked into selling their shares, and finding a technicality in mining law, they tried to jump the claim. Crotty one night saw from his tent the flickering light of the intruders, so loaded his revolver and warned them not to drive in one peg. 'I warn you before you do it, to say your prayers, if your mother ever taught you any,' said Crotty. He was bound over to keep the peace for a year, but his party won the dispute in the Court of Goldfields and retained the mine.

Many shareholders wondered if the mine was worth retaining. A small Launceston company was formed to manage the mine, and the prospectors and moneyed men who were directors called for another expert. Dr J. R. Robertson, one of the few consulting geologists in Australia, rode to the mine in 1887 and chilled their hopes. He said the deposit resembled the famous Tharsis in Spain, being poor in gold and silver but strong in copper and sulphur. He doubted if it would pay either as a gold or copper mine. The directors ignored his report and decided to mine the oxidized zone in which the copper was poor and the gold relatively abundant. Teams of horses dragged a stamp mill and long boiler from the coast, mules took over when the track became steeper, and then men and ropes when the track narrowed away. The pyrite and barytes in the ore foiled the mill's efficiency in extracting gold. The quarry caved in, the dams dried. Calls of 5s. 11d. on every share were rewarded with a sixpenny dividend that should not have been paid, for the mill did not work at a profit. By 1890 Crotty was mining in the Sydney sewers to earn enough to pay calls on his shares, and next year the valley was silent with only a caretaker living at the mine. That was the penalty for ignoring Dr Robertson's advice that a copper deposit should not be mined for gold.

The long trail worn down the west coast by the boots of men and the shoes of horses led to desolation. That crescent of equally-spaced mining fields from Mount Bischoff in the north to Mount Lyell in the south was like a new moon in shape and promise and permanence. The blackberries spread over the Pieman gold diggings, the waterwheels at Heemskirk were still, and the wind blew the shingles from the roofs of the battery sheds in the valleys near Mount Lyell. Mount Bischoff alone was noisy and alive, its payroll listing the names of many prospectors who returned to work for wages and to refresh their faith that the forests to the south hid treasures just as rich.

It was Broken Hill that shaped and accelerated the growth of the mining fields on the west coast. When the silver boom gripped Australian stock exchanges in 1887 and 1888, promoters sought new mines to feed to hungry investors. Mount Zeehan, twin mount of Heemskirk, was sentry over many idle silver lodes, and attracted speculative money. A few miners dug trenches across the lodes and worked in oilskins down wet shafts, repaying the cheques of the small companies with news that many of the lodes lived down. As late as September 1888 only about thirty men were employed on Zeehan mines but the Tasmanian stock exchanges were raising the money to employ more. The surface ore was barely rich enough to defray the freight over the granite hills to Trial Harbour and by sea to distant smelters, and so the mines could not finance their own growth. Without the rich oxidized ores of Broken Hill the mines depended on the money market.

Gustav Thureau was just the man to entice money. When he reached Broken Hill with his blow pipe he announced that Zeehan's lodes were similar. Promoters from Broken Hill and Silverton, Ballarat and Melbourne, came to see them and blew more bubbles. 'This will be one of the greatest fields in the world,' trumpeted the prospectus of a company which tried to raise £70,000 to build a railway to the silverfield. Unfortunately the silver shares of Broken Hill slumped before Zeehan caught much money, and with that strange logic of the share market Zeehan's prospects crumpled too. Tasmania's exports of silver-lead ore were £7,000 in 1889, and only £26,000 the following year, and not all came from Zeehan.

In 1891 Zeehan again became the mecca of mining men. The government was laying thirty miles of railway from Strahan on Macquarie Harbour, and Zeehan shares were favourites in Melbourne. On mornings when the small steamers called at Trial Harbour upwards of four hundred men would sit at breakfast at the hotel of Gamaliel Webster, a publican who nailed to the floorboards the boots which guests placed outside their rooms for the hotel 'boots' to clean. The road from Trial Harbour to Zeehan was scoured by coaches bringing speculators and the waggons bringing machinery and stores. The crooked main street of the silver town was paved with planks and saplings, and when investors and city men returned at nightfall from inspecting the surrounding mines the auctioneers and brokers would mount platforms and sell shares by the light of kerosene lamps. The official stock exchange had sixty members and seats sold for seventy-five guineas each. Some Melbourne brokers said that price was cheap after Herman Schlapp, metallurgist of Broken Hill Proprietary, said the lodes of Zeehan would live down.

Zeehan had so many narrow lodes that there was scope for hundreds of prospectors and promoters. Sylvester Browne, Fitzgerald Moore, Colonel North the nitrate king, the Earl of Fingall, and men who would be the colourful investors of Coolgardie a year or two hence, vied with Bowes Kelly, Orr, Schlapp and the men who had won wealth at Broken Hill a year or two previously, for the privilege of getting options to buy Silver Hills, Silver Harp, and the mines that reported brilliant discoveries. 'We cannot remember any period in the history of these colonies when discovery after discovery of undoubted value followed each other so quickly,' wrote the *Australian Mining Standard*. In July 1891 at least 159 companies and syndicates held leases on the silverfield, and two companies were building rival railways and smelters to serve the adjacent Dundas mines.

The share boom collapsed at the close of winter 1891, the last excess of the most speculative era of Australian history. The nation's financial system was quivering and one of the shakiest of scores of shaking banks was the Bank of Van Diemen's Land, the busiest bank in Zeehan and intimately linked to its mines. It closed on 3 August 1891, and next day twenty-seven Zeehan mines temporarily ceased work. Nevertheless,the share boom had given the silverfield perhaps a million pounds and that was more than enough to prepare mines for steady production. For long the mines had depended on the investors; and now the investors waited anxiously for the mines to produce.

The two small Zeehan smelters soon closed. The mines now had to ship a concentrate, high in lead and silver, to smelters in Australia and Germany. In 1894, the field exported nearly £300,000 of silver-lead ores, but as the price of silver fell that figure was not again exceeded. For nearly twenty years Zeehan averaged over £200,000 of metals yearly and supported a population that was once nearly eight thousand and often above five thousand, but it did not repay the money invested in its mines. It did not reward the German group, led by the Deutsche Bank, that opened large smelters at Zeehan in 1898. And it did not reward the discoverer of the field, Frank Long, who slept on bags in an engine house when his bustling town had trams and theatres and eighteen hotels.

In cold August 1891 two wealthy promoters from Broken Hill sat by a log fire in a wooden hotel near Mount Lyell. Bowes Kelly was thirty-eight, weighed sixteen stone or more, was ginger-bearded, slow in speech, genial and optimistic, and indeed had reason to be optimistic because he had made a fortune from his early shares in the Broken Hill Proprietary. The other promoter, William Orr, was older and smaller, had auctioned or owned sheep and shares in many places, had promoted companies in the Broken Hill and Zeehan booms but had missed the prizes in silver as conclusively as he was to win them in copper. In Zeehan they heard of the abandoned gold mine at Mount Lyell, and now James Crotty was escorting them on horseback over a track strewn in boggy places with a corduroy of saplings and brushwood. At the Queen River hotel, tough mutton and black steaming tea closed the first day of their journey from the sea, and on the second day they rode around the precipice of Mount Owen and saw the deserted battery and the tumbled boulders of the black Iron Blow.

In driving rain they climbed over the outcrop and in the shelter of short tunnels they chipped samples of rock from the walls and sealed them in sample bags. At Broken Hill their private partner in speculations, Herman Schlapp, assayed the samples and found them suspiciously rich in gold and pleasantly strong in copper. He sensed what Dr Robertson had seen, that if the pyritic orebody was large it might support a large copper mine. On his advice Kelly and Orr parleyed with the small Launceston company, paid a mere £5,000 for 55 per cent of the shares in the mine, and formed a new company in Melbourne with Kelly as chairman and his brother and friends as directors. They had bought a copper mine; Crotty and his Launceston friends still believed they had sold a gold mine, and some evidence suggests they had salted gold in the samples that Schlapp examined with such insight at Broken Hill.

One of the lessons Kelly had learned on the board of the B.H.P. was the wisdom of employing experts. He consulted Schlapp, who was possibly the foremost metallurgist in Australia, and when Schlapp went on holiday to the United States and suggested that another opinion be got, Kelly paid the huge fee of £2,500 to Dr Edward Dyer Peters, a doctor of medicine and North American copper expert. Peters spent some months at the mine in 1893, cursing loneliness, rough food, and rain, working with chemicals on the table on which food was prepared, sampling and mapping the workings, and sitting on moonlit nights at the door of the battery shed to watch the white crags of Mount Owen reflecting a brilliant stream of light 'as from a magic lantern'. Kelly and friends called to hear his verdict, fortifying themselves with the largest cargo of whisky ever landed at the mine, and celebrating for days as Peters explained the magnitude of the deposit. 'In the past twenty years,' he wrote, 'I have never seen a mining and metallurgical proposition that promises so certainly to be a great and enduring property as this.'

After investing £12,000 Kelly and his friends preferred to risk no more. But to build railway and smelters and works which alone could prepare the mine for dividends might cost £400,000. Only the

British investor could provide such money. And so William Knox went to London where in 1887 he had sold for B.H.P. the dearest Australian mine ever sold, the British Broken Hill. This time his offer to sell to British investors 150,000 Mount Lyell shares at £3 each was hopeless when they could buy the same shares in Melbourne for little more than £1. His offer of preference shares was also ignored. His discussions with the Rothschilds and French copper sharks and British companies collapsed. Not even a magician could sell shares on the terms his company sought.

When Knox returned to Melbourne at Christmas 1893 his gloom was dispersed by news from the mine. Earlier that year Dr Peters had observed a seam of mud-like ore in the lowest tunnel and had advised the company to sink an internal shaft and follow the seam. Many deposits of pyrite have enrichments of silver on the footwall and Peters found that samples of this crumbling ore assayed two thousand ounces of silver to the ton. 'The rich ore is only interesting and not important,' he said. The miners found it was not only interesting. They traced it down eighty feet, and one block of ground ten feet square yielded £18,000 of silver. For two years the ore was packed in bags and loaded on to mule teams, sent to the port and then to smelters at Adelaide and Swansea, where 849 tons averaged over a thousand ounces of silver to the ton and over 20 per cent copper and yielded a net profit of £106,000.

If small gamblers had controlled a mine with such a bonanza they would have paid the profits in dividends, played the share market, and maybe ruined the mine. Kelly, Knox and Jamieson were big speculators and preferred to re-invest all profits in the company in the hope of gaining a larger reward. They had a big mine and they would work it as a big mine instead of gouging out the bonanzas for the sake of short and spectacular dividends. Although silver was the company's only real income for two and a half years, it waited patiently until the chance came to raise £195,000 by selling shares and debentures in Australia.

Miners working without jackhammers stripped the waste rock in order to mine the ore in a large open cut. In a valley of forest, axemen cleared the ground for sawmill, limekiln, blacksmith's shop, crushing plant and the long smelter shed with its smokestack 250 feet high. Horse teams arrived daily from the coast with locomotive boilers, furnaces, girders, a million bolts and a thousand sundries, the horses plodding single file with black mud oozing through the waggon wheels. Men with leggings and thick overcoats and swags on their backs walked behind the waggons, anxious to find work in the construction camps.

Robert Carl Sticht, aged thirty-eight and widely versed in smelters of Colorado and Montana since his graduation from the famous mining school at Clausthal in Germany, came along the track with his bride. Sticht was an expert on pyritic smelting, long the dream of metallurgists, for in theory it used the iron and sulphur in the copper ore as fuel in the furnace and so used less coke in smelting the copper. When Sticht reached Mount Lyell he convinced himself that the ore might make pyritic smelting succeed for the first time. He persuaded the directors to abandon their costly plan to roast sulphur from the ore before smelting it, arguing that sulphur was a tonic rather than a poison in the furnace.

The opening of new smelters is always delicate and often calamitous, but Sticht was confident of his huge 150-ton blast furnaces. On the evening of 25 June 1896 rain drummed on the iron roof of the smelters as Sticht lit the furnaces and tipped coke and ore and white silica into the flames and waited to tap the incandescent orange slag. That afternoon in Melbourne the directors got the coded cable, 'Will start to smelt tonight'. They were more anxious than Sticht for news of an event for which they had waited five years.

Sticht's smelters were a milestone in copper metallurgy, and a month later the company completed its mountainous railway from port to smelters. The directors of the Mount Lyell Mining and Railway Company came in a small steamer to Strahan, crossed the harbour by steam ferry and boarded their own train at their river port of Teepookana, admiring reflections of the forest in the dark river and the bustle and order of their railway terminus at a bend of a gorge. They went fourteen miles to Mount Lyell in a small train hauled by the peculiar rack locomotives that Herr Abt had invented for the steep crossing of the Hartz mountains. On four miles of the Mount Lyell railway was a third or middle track consisting of rows of small iron teeth, and on steep pinches the locomotive derived grip from both the normal rails and the central racks of teeth. All the company's works reflected a zeal for new techniques.

The company's railway and smelting town of Queenstown became the third largest town in the island, and high up the hills and in the opposite valleys were the mining towns of Gormanston and Linda Valley and North Lyell, all living on copper. The census takers of 1901 counted 2,900 houses, but 1,800 of them had only one or two rooms: bachelors' huts of palings with iron chimneys, flimsy camps that were tossed in gales or burned in bushfires that swept the hills. In five summers more than 400 huts and houses were burned, for few people stored much water on hills that had 115 inches of rain a year. As Sticht built more furnaces the pall of sulphurous smoke from the chimneys thickened winter fogs in the valleys so that men often walked in daylight to the smelters with a hurricane lamp and some walked all day without finding the smelters. Fire blackened the hills, sulphur killed new growth and rain scoured the topsoil, and those green valleys that the prospectors loved became black as battlefields. Night hid this desolation, and hundreds of lights could be seen weaving slowly down the hills, men picking their way down goat tracks with candle in bottle to guide the way, or lights creeping upward to nightshift at the mines. In the main streets of the towns the hotels were boisterous, shops and theatres and billiard saloons alight, shares auctioned at many call rooms, axemen chopping for wagers, workmen with more money than they had ever owned, towns alive with the exuberance and vigour and easy prosperity that visit most boom mining towns.

A wild share boom began in 1897. The Mount Lyell Mining and Railway Company paid its first dividend in July, and two months later shares reached £16. 10s. to give the mine a market value of more than £4 million. Five or six early investors each held parcels of scrip worth over £100,000 and Bowes Kelly had already recouped about a tenfold profit from the £3,500 he personally invested in the mine and still held shares worth £320,000 at ruling prices. Mount Lyell and the line of adjacent mines dominated Melbourne stock exchange and, as the price of copper was rising from the long slough, the gaudy share certificates of many Lyell companies were soon currency in London.

The Mount Lyell company's mine was a large, isolated deposit but Gustav Thureau, emperor of the optimists, vowed that it was part of a gigantic lode that curved north and south like a sleeping python; and Australian promoters were emphatic that their worthless leases embraced that lode. They raised at least £200,000 from Britain with this spurious claim. Mount Lyell Comstock Copper Co. printed a pretty map with a lode in blue running from the Mount Lyell mine to their own lease, 1½ miles to the north. Companies owning leases in the gorge to the south of the Mount Lyell mine painted the lode stretching their way. If all the copper lodes painted on the prospectuses had been drawn on one map they would have covered the field with colour.

Forty-four companies were floated to search for copper at Mount Lyell by the spring of 1898, and their names were as confusing as their 'lodes'. King Lyell, Queen Lyell, Empress Lyell, Kaiser Lyell, Prince Lyell, Duke Lyell, and Crown Lyell exhausted the royal adjectives but not the points of the compass, and North King Lyell, North Queen Lyell, and many other companies with similar names erected a brass plate in Melbourne and an office at their mine. Chance finds pepped the excitement; roadmakers found a beautiful lode of copper in the North Lyell lease and men levelling ground to pitch tents at Crown Lyell found promising yellow copper. Local investors who wished to predict such discoveries could call on Madame de Rome, clairvoyant and palmist, who advertised from Henry's Hotel that she would advise on speculations, business, and love. Some of the managers of the mushroom companies possibly called on Madame, for they explored the likely ground so thoroughly that they seem to have left little for later generations to find.

The richest deposit on the copperfield was found by accident during the boom. Bowes Kelly almost regretted its discovery, for it was controlled by his enemy, James Crotty. Excitable, clever, and courageous, Crotty had levered boulders from the original Iron Blow, had been so excited by its gold that he promised to buy Ireland for Parnell. When Bowes Kelly and his Barrier friends took over the mine in 1891 Crotty resented their control. He attacked them in court and in shareholders' meetings, unfairly accused Kelly of corruption and stole his confidential clerk. But he could not regain control of the Mount Lyell Company unless he enlisted the votes of those British investors who had quietly bought over 40 per cent of the shares. In 1897, he sailed for London to win those votes and float his own mines.

Crotty was chairman of many small Melbourne companies that owned mines near the Iron Blow and he prepared to refloat some of them in London. His prize lease was North Lyell which had the 'Eastern Orebody', unpayable as a whole, but with patches of rich ore. Crotty's men mined some of the rich copper ore and took it to London where forty tons was found to average 24 per cent copper. He said his mine had £1 million of such ore already in sight, and advertised in London and Paris and Glasgow that his company would make £250,000 a year profit. It was the height of the Lyell boom, and Scottish financiers backed his plan to float the North Mount Lyell Copper Company. He offered 15 per cent of the shares to British investors for £70,000. The shares were rushed.

One month later, on the afternoon of 20 October 1897, men were making a public road along the ridge through the North Lyell lease. An outcrop of white quartzite blocked the way and the roadmen began to blast it. After firing a charge they saw blue and purple mineral permeating the white rock. The mineral was bornite, a rich copper mineral, and the orebody the roadmen had accidentally found was to yield three times as much copper as the famous Mount Lyell mine. The dubious lease became rich, and even Crotty was surprised. The will he had composed before the new discovery made it clear that the only shares he really believed in were the old Mount Lyell.

'In my uphill fight for Mt. Lyell I know there was a supernatural power which assisted me,' said Crotty sincerely. At the end of 1897 he was contriving his own mysterious form of power. Like the buccaneering Western Australian promoters he knew that newspapers could influence investors, and that money could influence certain newspapers. He secretly controlled a Tasmanian paper, the *Mount Lyell Standard*, and he used it to praise his own mines, his own achievements, and to deride the rival Mount Lyell Company and its leader Bowes Kelly. Special editions of his Tasmanian newspaper reached London, and were quoted freely in financial newspapers as the impartial voice of the mines. London journals praised Crotty and attacked Kelly and his company, the *Mining Journal* arguing that Kelly's railway was useless and would soon slide into the river. Such publicity promoted Crotty's plan for his own railway to the copper-field.

During the copper boom railway companies seemed as glamorous as copper companies. Many agreed with Crotty's journal that 'Mount Lyell has been endowed with a golden history without parallel in the annals of the world'. Railway promoters hoped it would yield them even larger fortunes than Broken Hill had given to the small Silverton Tramway Company. Although Mount Lyell already had a railway it was steep, costly, and led to a dangerous harbour with a mill race at the entrance and a sand bar only eleven feet deep at high tide. A harbour expert, Napier Bell, described the cataract of sandy water that passed through Hell's Gates as the most furious ebb tide he had ever seen. Ships rode the swell outside the harbour for hours rather than venture inside, and seven small ships were wrecked there in the 1890s. As ships larger than 600 tons could not safely enter the harbour, shipping and insurance charges were so high that ports on the opposite side of the island could perhaps capture the trade of the west coast by building long overland railways. Two companies fought for the right to build railways 150 miles across the mountains from Hobart to Mount Lyell. Three companies sought the right to build long railways over the mountains from the north coast. One of those companies was the Emu Bay Railway Company and, leasing for ninety-nine years the railway from Burnie to Mount Bischoff, it extracted £150,000 from Australian investors in September 1897 to extend the railway to Mount Lyell. Eventually it shelved its plan to tap Mount Lyell and instead tapped the silverfield of Zeehan. The Emu Bay railway opened valuable mineral country but after sixty-five years it has not paid a dividend.

While Crotty's journals unscrupulously criticized every proposed railway, he quietly planned his own. His North Lyell railway would run from Macquarie Harbour through the gentler country on the east side of the mountain range, linking his North Lyell mine with the harbour of Kelly Basin. His bold answer to the sand bar at Hell's Gates was to commission from Tyneside shipyards a steamer of two thousand tons, drawing barely ten feet of water, and three times as capacious as any ship that had entered the harbour. He had out-generalled Kelly's company; his railway and port and smelters seemed likely to capture most of the custom of the great copperfield.

Crotty did not travel on his luxurious steamer and railway. He died in London in April 1898, at the age of fifty-three. His parcel of shares in many Lyell companies must have been worth nearly half a million at the height of the boom, but with the collapse of shares his Victorian estate was valued at £160,000. Crotty had married an attractive barmaid late in life, and he bequeathed her £100 a year, a legacy of £300, and a provisional legacy of £500 if she agreed to enter a convent: she did not agree. In retaliation she contested the will with the main beneficiary, the Catholic Church, and married her husband's personable lieutenant.

The death of Crotty confronted his company with that dilemma that was to jolt many London companies that owned mines in isolated corners of the world: how to direct mines which they had never seen. The most effective answer was to appoint a general manager of the skill and honesty of Robert Sticht of Mount Lyell or Richard Hamilton of Great Boulder, and follow his advice. North Lyell appointed no such man. Their policy after Crotty's death was summed up by his nephew, J. P. Lonergan, a stationmaster turned company director. 'Money is no object whatever, as long as we develop the property.'

The mine was the company's vital property and the one they did least to develop. It had no rich oxidized zone but its sulphide ore was as rich as the oxidized ore of the old South Australian copper mines. The directors were wont to boast that their mine was producing the richest copper ore in the world, and they could certainly produce ore carrying 15 to 20 per cent copper in tens of thousands of tons. Their mine, however, was so mismanaged that they had no clear idea of their ore reserves. They often struggled to mine the small weekly quota of ore that they had contracted to sell to German smelters. The mine was a sectarian asylum, a rest home for foremen's friends and relations. The general manager one night found seventy men doing work which he said six or eight men could have done. A British shareholder who persuaded the lax London office to pay his fare to the mine reported that the 'whole place was conducted like a free park'.

The company believed the mine was so rich that they lavished money on all their projects. Their port of Kelly Basin had long piers and wharves capable of loading a small fleet of steamers, a large saw-mill and brickworks, and high-roofed stores. The twenty-eight miles of railway that went north to Mount Lyell had powerful green locomotives and Pullman passenger cars fit for a royal train. Unfortunately, the company's mine was on a steep hill and the railway only reached the foot of the hill, so the company built huge earthworks for a branch line to the mine itself; after spending £26,000 the branch line only reached the rival Mount Lyell mine and there it stopped. Finally, the company had to build an inefficient aerial tram to carry the ore from their mine to their railway.

Lamartine Cavaignac Trent, the company's metallurgist and general manager, was decisive, impatient, a striking American with his cloth cap, flat nose, and the trace of Red Indian in his complexion. In his tours of the company's empire he said he found enough crime and corruption to fill Newgate with scoundrels. He was not efficient himself, and despatched special locomotives to fetch his lunch or mail or morning newspaper. He was not skilled in metallurgy and the expensive smelters he built at Crotty, halfway along the railway from the port, were superbly fitted for everything but the smelting of his company's ore.

North Lyell's ore was harder to treat than the Mount Lyell ore, but curiously both ores were made for marriage. The North Lyell ore was strong in silica and very weak in iron; the other ore was strong in iron and very weak in silica. Both companies would have saved enormously by smelting their ores in the same furnaces, but the North Lyell Company were so ambitious for their own mining empire that they refused Mount Lyell's offer of a merger. Their blunt refusal probably cost them about £750,000, too much to pay for a vendetta.

At the close of 1901 the Crotty smelters were veiled in secrecy. They had failed, and the company refused to admit failure until feuds within the company made every shareholder aware of the failure. The advisory Melbourne board employed their newspaper, the *Mount Lyell Standard*, to ridicule Trent, and Trent retaliated by refusing to carry bundles of the newspaper on his railway. As the mine manager was a friend of the Melbourne board, Trent secretly imported two new managers from Broken Hill in February 1902 and tried to install them at the mine. The miners refused to allow them to enter the tunnels, so Trent cut the telephone to the mine, returned to Crotty in his special train, and sent hysterical telegrams to newspapers denouncing all at the mine. The mine and the smelters were at war, the company almost ceased to operate. Two directors arrived from London and dismissed Trent, who sailed away with dramatic adieu: 'Farewell Australia! I leave you to the working man and the servant girl!' Back in the United States that year Trent made another unfortunate decision; he neglected a chance to buy cheaply the Utah mine at Bingham, which eventually became the largest copper mine in the country.

J. S. MacArthur, inventor of the famous cyanide process, was a director of North Lyell and the company's metallurgical adviser. After he had dismissed Trent he rebuilt three of the furnaces and stood day and night in the hot building, watching the hot slag tapped, supervising charges and puzzling at the inability of the furnaces to smelt swiftly and efficiently. He had won fame in wet processes and knew little of fire processes, but was naturally reluctant to admit his ignorance or employ a good metallurgist. For weeks he fretted and experimented at Crotty, his reputation in peril, his melancholy deepening on that misty plateau shadowed by the towering mass of Mount Jukes. He decided to scrap the furnaces and return to London.

North Lyell now copied Sticht's pyritic smelting and brought over four small old blast-furnaces from Dry Creek near Adelaide. They were lit in September 1902 and smelted sluggishly. Whereas Sticht could make a profit on 2 per cent copper ore, North Lyell's 'coffee-pots' only made a profit on 7 per cent ore. And when on the last Sunday morning of 1902 a storm broke on Crotty and lightning toppled the high chimney stack, it seemed an omen that the company would topple.

The two large mines, Mount Lyell and North Lyell, stood on the same ridge less than a mile apart, but a spine of mountains separated their rival smelters and railways and ports. A mountain also separated them in efficiency. Robert Sticht, the brilliant American metallurgist, had both a flair for efficiency and the ability to attract skilful lieutenants. Visiting engineers applauded his small mining kingdom. The special mining commissioner of *The Economist* of London named it the 'best managed mine in Australasia', and few would have dissented from his verdict. The company employed over two thousand men in its quarries, railways, smelters, foundry and brickworks, and its large coke ovens at Port Kembla, New South Wales. By 1897 its mine was yielding, as morsels, more gold than any mine in Victoria and more silver than any mine in Zeehan. Two years later, when its annual output of copper reached 9,500 tons, it was the largest copper mine in the British Empire. Total profit passed a million only five years after the first copper had been smelted, yet profit was the sole criterion by which the company did not measure well. But for the steady rise in copper prices after the company had planned its enterprise and computed its costs, the dividends from the Iron Blow would have barely repaid the capital invested.

The Mount Lyell mine had been disappointing. Dr Peters' predictions had been overthrown. In four years the average grade of the ore fell from 6 per cent to 2½ per cent copper, and the company soon had to mine and smelt a hundred tons of ore to win the copper it had once got from forty tons. Above all, the deposit was narrowing and was finally to cut out like the bottom of a basin. The company tried to merge with North Lyell, tried to buy smaller mines in order to prolong its life. Bowes Kelly lost faith and sold half his shares at the end of the century. And the tumble of copper prices in 1902 threatened to extinguish the company.

On 13 November 1902 Sticht gave his company new hope by fulfilling the ancient dream of metallurgists. He smelted ore without coke or coal. Blowing cold air instead of hot air into the furnace, and increasing the blast of air, he defied the precepts of his craft and succeeded in smelting copper ore without the aid of coke, and smelting it so quickly that the capacity of the furnace was dramatically increased. There was now no more famous metallurgist in the world than this short, quietly-spoken, kindly man from Brooklyn, who spent his evenings with his music and Rembrandt etchings and Dürer woodcuts in his mansion on the hill, and who spent his afternoons watching the molten rivers and shaking flames in the lofty, sulphur-tanged smelting shed. Such hopes were fixed on pyritic smelting because it was so cheap and seemed relevant to so many deposits of copper or gold that Sticht's advance seemed of international importance. In fact, he was the perfectionist of ancient fire techniques that were about to be deposed at the height of their usefulness by a chemical technique. That technique was largely contrived at Broken Hill, and eventually was to supplant pyritic smelting in Sticht's own treatment plant. But that lay far ahead in 1902, and to his own company Sticht's success was more an urgent source of profit than a triumph in metallurgy.

Sticht was master in the fight for supremacy. His company had the asset of efficient smelters that could make poor ore profitable. His company had everything except a good mine. North Lyell had nothing but a good mine, and through neglect in exploring its mine it did not even appreciate its one asset. After blast furnaces had yielded scant profit at Crotty, the North Lyell had another outbreak of dissent. MacArthur and his Scottish friends resigned, and new directors explored the chance of a merger. The private rail motors dashed through the forest between Queenstown and the northern port of Burnie, taking representatives of the two companies for discussions. Each company sent men to inspect the rival mines, but neither could compute how much ore remained. Each general manager sensed that if he could hold on a touch longer he might outwit his rival. Sticht expected the collapse of the Crotty smelters. North Lyell hoped for the exhaustion of the Mount Lyell mine, and that was nearer than they knew.

People in the affected towns snatched every rumour. If the companies merged, one railway and one port and one smelting town would be redundant. In Queenstown the man who had to tar his roof thought seriously whether he should waste the tar. In Crotty the publicans wondered whether to continue building hotels in a town that could suddenly vanish.

In May 1903, the companies decided to merge after six years of open warfare. For the shareholders it was an equal merger, each company taking half the shares in a new Mount Lyell Mining & Railway Company. Sticht was general manager at £5,000 a year, all his staff remained, and the Melbourne board led by Bowes Kelly controlled the new company. Henceforth the rich North Lyell ore went to Sticht's smelters in Queenstown, and for another generation was the main source of the field's wealth.

It was a cheerless Saturday evening along the east side of the mountain range when the people of the North Lyell towns knew that their jobs had gone and their properties were valueless. The smelters of Crotty closed; in the main street of white road metal, carts carried furniture to the railway. A hotel of thirty rooms had been completed that month, and it closed without selling a drop of beer. A town of nine hundred people was quickly deserted except by the watchmen and caretakers who guarded the machinery before it was sold to mining fields throughout Australia. Along the railway the quarrying town of Darwin employed eighty men, and its shops and hotel and workmen's club vanished so completely that not a sign of it remains. At the port of Kelly Basin, most lavish of Australia's minor ports, converter vessels had just been unloaded at the pier on their way to the smelters. The port was idle, the railway useless, and today Kelly Basin is uninhabited.

James Crotty, the prospector from County Clare, has two

memorials. One is the ruins, the concrete foundations, the piles of yellow Scottish bricks, and the grassy main street of Crotty, heart of a mining empire on which £1,200,000 was lost. The other is a tablet in one of the noblest cathedrals of the nineteenth-century Gothic revival, St Patrick's in Melbourne.

20

MOUNT MORGAN

NEAR THE QUEENSLAND COAST, a few miles south of the tropic of Capricorn, boulders of black ironstone lay on a timbered hill. For more than twenty years men chipped fragments from the boulders or washed the surrounding soil without gaining more than a hunch that a deposit of gold lay below. This deposit was Mount Morgan; its gold shaped the life of thousands of Australians and millions of Medes and Persians and the powers of the world.

For a quarter-century after the first settlers came the gold of Mount Morgan was untouched. Only twenty-four miles across the steep ranges the town of Rockhampton spread along a wide brown river, shipping away the wool and hides and minerals of a rich hinterland. It shipped gold down that river long before Mount Morgan had a name. The river bore the tide of gold-seekers to the fiasco of a gold rush at Canoona—the Port Curtis rush of 1859. Seven years later yellow diggers and white diggers rushed to Crocodile Creek and dug and fought and rioted within ten miles of Mount Morgan, then fanned into the hills and washed alluvial gold at the foot of the mount itself. Cattle by then fed on the grassy flats of the river that meandered past the ironstone mountain, and stockmen watched the cattle and the hungry diggers watched them too. One of the stockmen was William Mackinlay, and as he lived in a bark hut near the ironstone mountain he had time to prospect. On his rides he found the outcrops of the Adolphus William copper mine and the Crow's Nest gold mine and, almost certainly, Mount Morgan. But he made only casual efforts to test the gold in the iron, and those efforts soon had to be surreptitious because a settler entered the valley and his square mile block covered not only the grassy valley but the sides of the ironstone mountain.

John Gordon, the elderly Scottish settler, and his sons fenced their land and dug holes for the posts by the massive lode without seeing gold. They ran their cattle for six years over the richest little farm in Australia before they chanced to learn from a daughter of Mackinlay the stockman that their farm held gold. The Gordons examined the ironstone but saw no riches. They could not even see wealth in cattle, and by 1882 the family had scattered and the property was idle.

Why did so many men who valued gold and knew how to mine it spurn this ironstone mountain? The gold was in the ironstone, and many goldminers had a prejudice against iron deposits. It was almost a dogma from Castlemaine to Charters Towers that quartz was the mother of gold. Moreover, the gold on the surface at the ironstone mountain was hard to recognize both because it was often coated with oxide of iron and because it was extremely fine. 'The ironstone,' wrote Dr Jack the geologist in 1884, 'contains gold of extraordinary fineness, which, however, after a little practice can be detected in almost every fresh fracture.' The first men who examined the deposit had little practice. Few Australian gold prospectors of the 1870s could recognize the gossan or iron cap of a mineral deposit. The deposit therefore awaited men with flexible minds, men with the drive which the Scottish cattlemen in the Dee valley lacked.

The Morgan brothers were such men. Frederick Morgan had dug gold at Bathurst and had worked a rich lode of tin at Stanthorpe in the cold south of Queensland in the early 1870s. At the nearby town of Warwick he owned hotel and butchery and racing stables, a sporting speculator who gambled freely on untried mines and untried horses. When he became the publican of the Criterion at Rockhampton in 1879 he continued to invest, reviving the faded hopes of the district by winning £7,000 of gold from the Galawa at Mount Wheeler. His brothers ran that mine and they too were not bound by geological dogma, for they had mined tin and gold on many fields, and it was their open mind, their interest in all minerals, and a long coincidence that took them to the ironstone mountain.

In a hut by the Galawa gold mine lived one of the Gordon boys. He drank hard at night and worked slowly in the day, and was dismissed. Now Sandy Gordon's wife had been a Miss Mackinlay, daughter of the old stockman from the ironstone mountain, and she pleaded with the Morgans to give her husband another chance. With womanly persuasion she promised that her husband would not only drink less but lead them to a silver lode. Edwin and Thomas Morgan agreed, and Gordon soberly led them south through Rockhampton and over the steep ranges into the Dee valley. They did not find the silver lode but instead found the gold on ironstone mountain. How they found it will not be accurately known; none of the men described the journey until after the mine had proved rich, and the Morgans were then determined to show they had found gold without Gordon's aid while Gordon and the intermarried families were just as determined to prove that Gordon had found the gold. Whatever happened one fact stands: the Morgans had the experience to value what the valley cattlemen ignored.

On the hill they took samples of the ironstone, crushed them on a shovel, and washed them in a pan of water at Gordon's old hut by the creek. 'We were still doubtful what it was,' said Ned Morgan. 'In panning off there was apparently more gold than stone—if gold indeed it was.' For three days the Morgans prospected the ironstone mountain, then sent their guide to the distant Galawa mine with samples of stone for assay. The feckless guide called at the Dairy Inn and stayed.

The Morgans suspected the deposit might be rich. They got supplies and equipment in Rockhampton and returned to dig and sample. The senior Morgan, Fred the publican, now wanted to see the mine and he hired horses and buggy in Rockhampton and drove to the land-locked valley. A surveyor went with him and marked more than thirty small gold claims on the boundary fences of Gordon's farm. But the richest deposit lay inside the fence, and under Queensland mining law the owner of freehold land owned the gold. Anxiously the Morgans negotiated with Donald Gordon, who was overseer on a sheep run far west. They offered him a pound an acre, said nothing of the gold, and he gladly sold the land. Folklore says he was robbed and scores of writers have upheld the lore. The Morgans in fact paid £640 for land that was poor in grasses and uncertain in gold.

The Morgans now had 640 acres of freehold and a row of mineral claims covering the entire gold deposit. No rich goldfield in Australia had ever before been dominated by one syndicate. Here was a goldfield where a few men could gather vast riches. This proved of vital importance not only for Australia but for Britain.

The Morgans were eager to give their name to the mountain but reluctant to give their money. They were still hesitant about its richness, so Fred Morgan invited his partner in the Galawa gold mine, Thomas Skarret Hall, to buy a half interest in the new mine. As manager of the Queensland National Bank on the waterfront at Rockhampton, T. S. Hall eschewed the rule that bankers should not gamble in mining shares. He lent his own and the bank's money for mining flutters, for he liked gambling and gamblers, and handicapped the horses for the local turf club. And so he eagerly read the letter that Morgan wrote on 5 August 1882 and no doubt laughed within himself at the old salesman's bluff: 'In conclusion we are not anxious to take any person into this Reef as we are in a position to carry it on ourselves.' The Morgans were not anxious to take partners into their reef but neither were they anxious to spend £1,200 on machinery at the reef.

Hall had six days in which to consider the offer. He called on friends in the town and showed them the letter, and possibly with temporary aid from his bank he got the £1,200. That was virtually the last capital spent on the mine. Thereafter it paid for itself.

The mine could not release its wealth until a crushing battery was hauled across the mountains. Bullock teams hauled it from Rockhampton to the foot of the hill known as Razorback and could go no further. The teamsters tied a chain to the heavy boiler and tried to haul it up the hill with block and tackle, but halfway up the hill the chain snapped and the boiler thundered to the plain. An old sailor called for rope, and a carrier came from the port with rope as thick as his arm. Then, fastened to tree-trunks, the rope pulled the boiler up that hill of oaths. A crab winch, with four men turning the handle, hauled the lighter machinery. In all a month was spent in tug-o-war before the machinery was piled high on the crest of the range, a feat long toasted by teamsters on their St Crispin's Day.

While the ten-head battery was being assembled on the creek, miners were busy on the hill. In the autumn of 1883 they drove hammers in the quarries, pulling rags from the belt to wipe sweat from face and hands. Drays hauled the broken ore down steep slopes to a spur, where they tipped the ore into a wooden chute so makeshift that much rich stone jumped away and rolled to the creek. Edwin Morgan devised this poor method of cartage, and he was a better boxer than engineer. In front of the men he fought Red Harry the blacksmith with bare fists, and so rough was the fight that one spectator fell dead.

The gold was rich. No records were kept but the ore must have averaged at least ten ounces to the ton. But what the ore contained and what the battery extracted did not tally. The German manager decided that the mercury on the battery plates was not attracting the gold that was coated with a thin film of iron oxide. Careful assays suggested that four fifths of the gold was lost. Experiments by millmen at Gympie could find no remedy, but then Gympie had the kindliest ore in Australia and was not used to problems. As the syndicate did not have the enterprise to employ a skilled metallurgist from Victoria or abroad, they muddled on.

Rockhampton did not know that gold was quietly coming from Mount Morgan in bulging saddle bags. Over seven thousand people lived in the town, and few had heard the name, Mount Morgan. Even those who lounged about Morgan's hotel did not know why he was so often absent. The town had two daily newspapers but the journalists wrote and yawned. Not much seemed to happen at the Morgans' 'discovery of gold and copper at the Dee Range', wrote one newspaper late in 1882. The Morgans' battery was working for nearly a year before the *Morning Bulletin* became curious. On 19 March

1884 a reporter went out behind a pair of ponies to find a hundred men at work and the battery crushing night and day. Astonished at the bustle he became suspicious when the manager refused to divulge the gold returns. Back in Rockhampton the customs house revealed that exports of gold were soaring, and the press knew at once the source of the gold.

Mount Morgan seemed born to deceive. Its richness had deceived the old valley families of Mackinlay and Gordon and now it was the Morgans' turn. Edwin Morgan, the manager, thought the deposit plunged straight into the earth, and so he drove a tunnel to explore the deposit at a depth of four hundred feet below the outcrop. This tunnel would answer the vital question: was the gold rich at depth? Unfortunately, Morgan did not realize that the deposit was dipping to the east, and accordingly his tunnel cut the poorer fringes. After driving through hard rock for nine hundred feet the Morgans concluded that their mine was like so many promising Australian mines—a jeweller's shop on top and a blacksmith's shop below. They argued that 'the bottom was out of it' and tried to sell their shares. Offered £88,000 for half the shares in the mine, they took the money. Fred Morgan was so delighted with his shrewdness that he shouted champagne to all comers in Rockhampton.

A red-headed Rockhampton solicitor became the big shareholder and he was soon to be Australia's richest man. William Knox D'Arcy was only thirty-two when his bank manager offered him a large interest in Mount Morgan in 1882, and as he was not thrifty he had to borrow most of the purchase money. His years at Westminster School in England seem to have schooled him less in books than in sport, and in Rockhampton he was better known as a sportsman rowing on the river or gambling on the turf than a partner in his father's solicitor's office. His gambling sense made him plunge on Mount Morgan, and he spent his dividends in buying the Morgans' interest until by 1886 he owned 36 per cent of the shares in the mine. When three years later he left Australia to live in London those shares were worth over £6,100,000, and the huge dividends he had already received bought him a house in Grosvenor Square and a country estate and reputedly the only enclosure at Ascot outside the Royal Box. Mount Morgan was his purse, and to the thousands of fashionable Englishmen who shared his lavish entertaining that purse seemed as bottomless as the mine. For, while Fred Morgan had argued that the gold deposit had a shallow bottom, D'Arcy believed it was bottomless.

The Halls were the other chief shareholders in the mine. T. S. Hall retired from his bank and went to Melbourne in luxury, leaving his brother Walter to supervise the family share in the mine. As a partner since the 1850s in the famous stage-coach firm of Cobb & Co., Walter Hall knew how to run a large business even if his time as a Victorian gold digger had taught him little of mining. At the invitation of his banker brother he had bought a share in Mount Morgan, and it was so rewarding that he retired from Cobb & Co. Often he visited the mine, and he would sit proudly near the pay window and watch the miners collect their pay, or walk around the open cut to see if the men worked hard. On one occasion he was puzzled to see a crowd of 'Mountain Top' men rushing down the hill.

'Who are those men?' he sternly asked. He was told they were the day-shift going home.

'Well,' he snapped, 'if they have been working hard all day, how is it they can run like that now?'

Walter Hall was said to be a little mean but on his death in 1911 his wife gave a million to charity. One foundation of that bequest was the Walter and Eliza Hall Institute of Research where the Nobel prizeman, Sir Macfarlane Burnet, has done his celebrated research.

This was far ahead in 1886, when the little syndicate decided to form their Mount Morgan Gold Mining Company. So valuable was the mine that they created a million one-pound shares, and as they needed no more capital to develop the mine they took all the shares themselves. D'Arcy took 358,000 shares, the Halls 283,000, William Pattison took 125,000—a lot of shares for a man who had recently worn the butcher's apron in Rockhampton. A local contractor and politician named John Ferguson took 117,000 shares, and though all shares were in his name some probably were in trust for local speculators. These five men held about 88 per cent of the shares; Melbourne and Sydney graziers owned the rest. Eight men held all the shares when the company was registered in Queensland in October 1886, and though some shares were soon fluttering about the stock exchanges the lucky eight had faith in the mine and rarely sold. They had no need to sell shares in a mine which, working on a humble scale, had yielded perhaps £350,000 of gold by the end of 1886.

Their reluctance to sell made shares scarce—and therefore dear—in a mine in which thousands were eager to speculate. Journalists spurred on the share buyers. W. H. Dick edited the penny *Daily Northern Argus* and returning from the mine at the close of 1886 he told Rockhampton that 'it is beyond the power of any man or woman born to compute the capacities of the mine'. That didn't restrain him from computing that shareholders would soon get profits of £1,500,000 a year. And it did not restrain Rockhampton from believing their mine would make their city great. The price of street frontages and river frontages soared, and the dear land was covered with ornate buildings that stand today as shabby witness to Rockhampton's boom.

Even attempts to rob the company of some of its rich ground seemed to inflate the shares. While the company's freehold land was safe, the claims on the sidelines were held under mining law, and that law could be challenged. The company had envisaged the danger, and on the advice of a Gympie specialist in mining law, Horace Tozer, they converted their claims into leases. To their dismay Tozer later advised other clients that the legality of the leases was dubious, and the clients jumped the ground. In Rockhampton one director, Pattison, wept when he heard the news. 'If I have to stand behind a butcher's block again I'll fight them to the last.' The jumpers were as determined to fight and they floated companies in which subscribers were offered shares in the disputed ground should the lawsuit succeed. They engaged the smartest silk of Melbourne and Sydney, pursued the case from court to court, and two of the suits reached the Privy Council before they were finally dismissed.

In all there were nine attempts to jump the company's ground, and forty or fifty men with sharpened poles once invaded the ground and scuffled with company's miners. The company spent much on defending its ground, but it gained more than it spent. The jumpers and invaders so emphasized the richness of the ground in dispute that the price of Mount Morgan shares increased.

The big company had so much ground at Mount Morgan that intruders barely had space to swing a wildcat, but there was space enough for shrewd promoters who held a lease far to the west. Reaching London just when the Charters Towers mines boomed on the London market, they easily floated Mount Morgan (West) Gold Mine in October 1886. If Mount Morgan was a hill of gold the West Mine was a hill of deceit. The prospectus in the *Pall Mall Gazette* claimed their deposit of gold was identical with Mount Morgan and was part of the fabulous lode. Investors flocked to the fraudulent banner. The Queensland government, fearing the name of the colony would be impaired, wired London that the new mine was probably a wildcat. A judge in Chancery less than six months later agreed. 'Clearly a most audacious fraud has been committed,' said he. Another eighteen companies still sat like crows on the fence of the Mount Morgan company's acres, and one by one they flew away. The Mount had only one gold deposit and soon only one company.

Even the fumes and the dissolving chemicals in the company's laboratories seemed to the directors a sign that their mine was unique. The gold in their ore was so fine that it was rarely seen with the naked eye. That explains why the first settlers in the valley ignored it and why the first stamp mill failed to save it. The company tried so many processes that the valley became a junk yard of machinery. They tried all the old grinding and amalgamating methods that had failed to save fine gold at Bendigo and Charters Towers, and at the junk museum at the Mount students of metallurgy could see everything from Chilian mills to arrastras.

Perhaps chlorination was the answer; chlorine had dissolved fine gold at many fields since Plattner used it in Silesia in 1848. Two Gympie experts therefore contracted with Mount Morgan to extract the gold from the rich tailings by the Plattner method, but their chlorination works failed. Two more such works were erected and they also failed to save sufficient gold. After the failure of the third chlorination plant Pattison the chairman vindictively told shareholders on 29 June 1887: 'I have now the pleasure—and it is a pleasure—of telling you that the whole of that plant, with the exception of one barrel, has been abandoned as entirely worthless.' Finally, through trial and error, they succeeded in chlorinating their gold ores, and now they enlarged their plants and began to produce gold in enviable quantity.

Everything conspired to boom the shares. As the mine was directed from Rockhampton the townsmen who had paid dearly for their shares pestered the directors to pay commensurate dividends, and directors drove in their carriages to the mine and pestered the manager to raise the output. With small plants the company paid £275,000 in dividends in 1888, and in the next year the new works extracted so much gold that the company's dividend was £100,000 a month. The company strained to keep up this astonishing sum, tearing out the richest ore and paying all profits in dividends. The wonderful returns inspired a booming money market to inflate the shares until the mine had a market value of £18 million, and that made even the Broken Hill Proprietary seem small.

The general manager, Wesley Hall, went away to marry in July 1889, and his return to the Mount suggested that even the workmen rejoiced in the success of a mine from which they got seven shillings a day. Sixty or seventy horsemen awaited the return of the bridal pair at the foot of the Razorback to read an address of welcome and escort them to their house, to be serenaded by a brass band on the lawns and a thousand people at the gates. By November the company had paid £1·1 million for the eleven months, more than any other Australian gold mine had earned in its life. In December, alas, there was no dividend. The fabulous year closed with the richest ore gone and the shares tumbling.

Mount Morgan was still a remarkable mine, its dividend cheques handled with reverence by bank clerks in hundreds of towns, its size and richness extolled by journalists. They went by train along the grassy flats, walked by the coach as it crawled up the steepest pinch of Razorback, reboarding it for the run down the red road to the valley. They found four thousand people at the Mount in 1889, all depending on the rich mine. Below the vast quarry and the chimney stacks and the long iron roofs of the works were tents and iron huts and the weatherboard stores that stood on stilts capped with tin through which no white ant could chew. Visitors applauded a company that copied Edison, for electric lamps illuminated the new treatment works, and the general manager had telephones to save the messengers who dawdled up the mountain on hot days. Visitors in the town saw the duller lights of hotels and shooting gallery and sly-rum shops, and observed an Indian hawker with his basket of oysters and pepper and vinegar, and a towel to wipe fastidious hands. Here stood a bush camp at the foot of a mine that was winning enough gold to sustain a city of 40,000 people in jobs and trades. But the gold at Mount Morgan was so abundant that only a small proportion went in wages; instead it went in dividends that built mansions and villas and paid valets and footmen in distant cities.

Ten years after the finding of Mount Morgan the return on the actual capital invested was over 200,000 per cent and dividends had exceeded £3 million. But the company was troubled. As the workings deepened the gold became poorer, falling from the fantastic figure of 4·3 ounces in 1889 to one ounce in 1898. This was in part caused by and in part compensated for by enlarging the mine. In the last five years of the century the company trebled its annual output of ore to 239,000 tons, and the army of new men increased the town's population to ten thousand. A railway at last came from the coast to the foot of the quarry with its silica and kaolin dazzling white in the sun. Captain Richard—a skilled and cunning chlorinator who insisted on being called captain because he had led volunteer troops against striking shearers—built new plants to treat more ore and poorer ore. Vigorous expansion depended on the mine, and by the year 1903 it was for the first time yielding most of its ore from the sulphide zone. That sulphide ore carried less than eleven pennyweights of gold to the ton, and only seven of the eleven could be extracted in Captain Richard's Mundic Works. The company's dividends reflected the crisis: £300,000 a year in the late 1890s and a mere half that in 1903.

William Knox D'Arcy, the sporting solicitor who had left Rockhampton to become a London millionaire, sat on the English board of the company and was not pleased with the failing mine. A social giant in London, he spent his Mount Morgan dividends and more. His second wife was a Boucicault, and at 42 Grosvenor Square, at Stanmore Hall in Middlesex and Bylaugh Park in Norfolk, she was one of the fashionable hostesses of the seasons. Melba and Caruso are said to have sung to her guests on the same night, and they sang for more than their supper. D'Arcy himself was now over fifty, a portly man in immaculate suit and heavy watch chain, made no doubt of Mount Morgan gold. In social habits he copied the Prince of Wales, attending the same watering place and the same doctor. The English magazine *The King* in June 1904 noted that he had just entertained a hundred guests in his private enclosure at Epsom races, but significantly he was entertaining less in his mansion in Grosvenor Square.

It is not clear how much money D'Arcy had, but he had far less than he should have had. He had sold half of his Mount Morgan shares at high prices and he had already received a clear million in dividends on those he had retained, and yet by 1900 he felt the pinch. He had spent too much, entertained too many Persian dukes and Russian generals. Mount Morgan was his capital and his income, and when about 1900 he met the company's general manager, Captain Richard, on world tour and was told that the company's rich days were over and that the £1 shares were worth barely half that price he was disturbed. And, just as he was willing to take deep risks to win a gold mine twenty years previously, now he was willing to mortgage his own future in search of another. In 1901 he was approached to invest in the search for oil in Persia and he listened carefully. He himself had no wish to visit Persia but his emissary went to Teheran and signed in May 1901 a concession to search for oil over 480,000 square miles. D'Arcy financed the search and on the border of Turkey they drilled the first holes. Everything went wrong. The boiler that provided steam for the drill broke down. A sledge-hammer was dropped down a drill hole, and the tanks of drinking water nearly ran dry in burning heat. With these delays the men in the field wanted money, and D'Arcy paid his thousands monthly.

The future of Mount Morgan remained gloomy, but the gloom was more in the mind than the mine. Its manager had long been in the shackles of geological theory, and those shackles were loosening. A clever Scottish geologist, Robert Logan Jack, had shaped the thinking on the nature of the Mount Morgan deposit, and from his first visit in 1884 he had preached that it was exceptional, even unique. This idea was wrong but it had deep effects. So long as the company believed its deposit was a freak they misconstrued evidence in the rocks and folds. They did not suspect that Mount Morgan was like many famous pyritic ore bodies and so they did not watch for increasing signs of copper as the workings went down.

Jack was thirty-nine when he first walked the slopes of Mount Morgan. A scar on his shoulder from the spear of a black testified to his adventures as a geologist in northern Queensland; the theory he expounded on the origin of the Mount Morgan deposit testified to his adventurous mind. He argued that the gold had flowed as a boiling liquid from the depths of the earth. He argued that Morgan had once been a thermal spring such as still flowed in New Zealand and Iceland and America's Yellowstone National Park. Though T. A. Rickard described Jack's theory as 'the last resort of a perplexed geologist' Jack was understandably perplexed in 1884. Prospectors in Australia had then found few of the pyritic ore bodies that were to prove rich in the next decade, and Jack had possibly seen none in his years in Scotland and Queensland. Moreover, in 1884 few holes had been dug at Mount Morgan and evidence was therefore sparse. Nevertheless, his strange theory became a dogma. The company did not believe it but they did not supplant it, for they were happy to think that if their mine had been a golden geyser then the gold would descend to immeasurable depth. Visiting geologists and engineers might not support it but as they spent few days examining the mine they rarely found enough evidence to frame a contrary theory.

The richness of the ore proclaimed the mine was unique, the stock exchanges affirmed it. Dr Liebius of the Sydney Mint told the Royal Society in that city in 1884 that Mount Morgan gold got the exceptional price of £4. 4s. an ounce because it was virtually free of silver and other metals. His suggestion that it was the richest native gold ever found enhanced the idea that nature had performed in mysterious ways along the River Dee. J. Macdonald Cameron, member of the British House of Commons and self-styled 'mining expert', called at the mine for a fee of 500 guineas in 1887 and told the eager directors after four days that he could see no end to the gold, and only assumed there was an end because 'there is a limit to all things'. A young American engineer, T. A. Rickard, was at the mine for two rainy days in February 1890, and in the first of many famous articles he was to write for the American Institute of Mining

Engineers he came close to seeing that Morgan was a decomposed pyritic deposit. But not close enough, and he added another strange theory. Karl Schmeisser, a German engineer, spent the customary few days and called it 'the mightiest deposit of gold that has been discovered in the world up to the present time'. The twin ideas of unique richness and unique origin were married, and events were slow to divorce them.

Signs of a divorce came only when the mine ceased to be rich. The underground workings were eight hundred feet below the outcrop, and they gave more evidence than earlier visitors had seen. The Mount Lyell copper mine in Tasmania had become famous and it resembled Morgan—a pyritic deposit strong in gold at the surface and strong in copper at depth. Dr Jack wrote his fourth report on Mount Morgan in 1898 and abandoned his geyser theory. Edward J. Dunn, a Victorian geologist who had made one of the first prophecies that the Transvaal would be rich in gold, inspected Mount Morgan at the end of the century and predicted that it would eventually be worked for its copper. Members of the Royal Society of Victoria listened with respect, but the directors of Mount Morgan nodded sceptically.

For more than a decade the copper had cried for recognition. Metallurgists saw it upsetting their leaching process so they told the mining engineers to dump patches of golden ore rich in copper. When mines throughout the world turned to cyaniding, Mount Morgan tried and rejected the new process; the copper in their gold ores gobbled the cyanide before it could dissolve the gold. It was not until the penurious times of 1901 that the company exploited the copper, and the balance sheet of November 1901 notes ominously that a tiny precipitation plant had saved copper worth £57. 7s. 10d. That sum paid the wages of one man for half a year; copper was still small change to D'Arcy.

The metallurgists led the company and their works were highly efficient, but the fact that they depended on a mine was often forgotten. The increasing presence of copper in the mundic ore and the new tide of geological thinking should have persuaded them to drill for copper. The waning profit should have made them test the idea. As it was they ignored the copper until it stared at them. Early in 1903 a drive at the 750-foot level passed through an area of payable copper, and only then did the company act. It built a small furnace to extract the copper and scavenge gold, but such was the pessimism on the size of the copper deposit that the company looked to the small copper mines in the hills to keep the furnace going. By November 1904, however, they found that the copper covered a larger area than they had anticipated. They computed that possibly two million tons of copper-gold ore lay underground, enough to justify a new copper smelter.

In London D'Arcy must have been exhilarated by news of copper in the mine which nourished his life of luxury and his search for oil. Moreover, it came in the year when his dividend from Mount Morgan was halved. The company's shares were slipping. By the end of 1903 he had spent some £150,000 in drilling for oil in Persia, and had mortgaged tens of thousands of his shares to secure an overdraft from London banks. The falling price of Mount Morgan shares thus endangered his ability to borrow. And he had to borrow if he was to keep those drills churning in the Persian desert. He tried in vain to borrow from the Rothschilds in Paris and from English financiers. 'Every purse has its limit, and I can see the limit of my own,' he said. The limit on his purse was the market value of his gold shares, and during 1903 Mount Morgan shares fell from £4. 4s. to £2. 10s. 6d.

In January 1904 D'Arcy's men struck oil. As they drilled deeper the oil gushed over the derrick. 'Glorious news from Persia,' shouted D'Arcy, and his bankers agreed. After a few months oil ceased to flow and D'Arcy was borrowing again. In November came serious worry. The directors of Mount Morgan decided to raise £220,000 for a copper smelter, and D'Arcy knew the company could not afford it.

The Mount Morgan Gold Mining Company had no reserve fund, having paid all profits to the shareholders. To finance the smelter they had to take steps that would indirectly lower the price of shares. As all shares were fully paid to £1 the company could not issue new shares without reconstructing—and that would lower the shares. If the company financed the plant from profits and cut its dividend, shares would fall. If the company issued debentures, that would, in the opinion of at least one director, encourage a bear raid on the shares.

'During this stage D'Arcy was nearly frantic with anxiety,' recalled Captain Richard. He was leaning on credit and a sharp fall in Mount Morgan shares would ruin his credit and even his solvency. He sent his son to investigate the mine and its books; he instructed A. J. Callan, who sat on the board with his power of attorney, to oppose any plan that would affect Mount Morgan shares. Callan could do nothing. Then the company announced that it could finance the new copper smelter from profits. The announcement was bluff or optimism, but it buoyed up the shares. D'Arcy continued to drill for oil, his overdraft secure.

The company was slow to realize or slow to announce that it could not finance its copper smelters. On 30 December 1905 the company announced that it had borrowed a mere £15,000 to complete them. Six months later the new smelters were producing copper and the company could safely announce that it had a temporary loan of £75,000. The company had emerged from its crisis without damaging its name, but it was only saved by the giant American combine, the Amalgamated Copper Company, which cornered the world market for copper in 1906-7 and lifted the price. Mount Morgan gained from high copper prices, repaid its loans, and kept its monthly dividend at three shillings a share.

While an American trust unknowingly saved Mount Morgan, who was it that lent the money to prop Mount Morgan shares? Walter Hall, a director, lent £75,000 at bank interest. As a large shareholder he wished to sustain the value of shares. Moreover, it is likely that the executors of his brother Thomas, the retired banker who died in London in 1903, were selling his shares and were therefore anxious to avert a slump. One more motive has been attributed to Hall by Captain Richard, who as general manager knew the tortuous problem of financing the copper smelters. Richard claimed that Hall wanted to help D'Arcy because, though disliking him as a man, he believed that Britain should acquire a reservoir of oil. Whether this was Hall's motive is dubious. However, it is clear that Mount Morgan's shrewd effort to sustain the shares saved D'Arcy in his crisis, enabled him to drill for another six months or more, and thereby gave Britain one of its richest, most strategic assets. For D'Arcy spent £225,000 in his quest for oil and in May 1905 he had resolved to abandon the search. As he could get no capital in Britain he began to negotiate with the French Rothschilds for the sale of his Persian concession. The British government at that time was studying the merit of coal and oil as fuel for the navy and decided that oil might supersede coal. As the British Empire produced scant oil the Admiralty heard of D'Arcy's interests and approached him with only days to spare. He agreed not to sell to the French, and on 20 May 1905 he signed with the Burmah Oil Company to finance the search.

Burmah Oil continued to drill for oil and after three years they had spent enough. At four in the morning of 26 May 1908, the drill tapped oil, which spurted above the derrick and bathed the drillers and crew. They had found the largest oilfield then known and on that field began the gigantic firm known successively as Anglo-Persian, Anglo-Iranian and British Petroleum. It was John Strachey who wrote that 'for sheer wealth there has never been anything in the history of Imperialism like Middle Eastern oil', and it was the wealth of Mount

Morgan and the fact that it was held in few hands that gave Britain a prize which even in 1956 was providing nearly half of its income from overseas investments.

The year 1908 was a triumphant one for D'Arcy, but sad for the town of Mount Morgan. Underground they were mining copper in large chambers, with heavy rock drills each worked by two miners. There was little timber to support the roof of these chambers on which millions of tons of rock pressed from above. In some places dykes or lines of weakness crossed the massive deposit, and heavy bulwarks of brigalow logs were built there from floor to roof as precaution. The resourceful mine manager, H. P. Seale, had devised this system of flat-back chambers in 1904 because the old system of square-set timbering was costly and in the year of D'Arcy's plea for thrift all costs were pruned. The cheap method of mining had been introduced carefully. Engineers and miners were convinced of its safety, so they extended it to all the new copper workings, saving tens of thousands of pounds and helping the company switch from gold to copper in the crisis.

Fourteen weeks after D'Arcy struck oil he heard bad news from the mine. On the afternoon shift of 5 September, in one of the vast chambers 850 feet from the surface, a fragment of rock fell from the roof and crushed seven men to death. The company's engineers were puzzled at the collapse of a way of mining that had seemed so safe, and they examined the chambers and increased their safety. Seale himself supervised the changes, then fell sick with peritonitis and died. On the afternoon of 4 November 1908, two months after the accident, miners at the 750-foot level were building heavy pig-styes to support the roof in a suspected place of weakness, when the ground began to talk and rumble, a warning that rock was stretching and settling. Just as they were preparing to leave the chamber a slab of rock weighing 13,000 tons fell from the roof, crashed through the timber and killed five men. Two others had a miraculous escape. They stood directly below a winze or shaft that rose one hundred feet to the level above; it was as if they stood in the lift well of a building that suddenly dropped into the earth and the falling rock surrounded them but did not touch them. From a hundred feet above, a courageous miner came down a rope and hauled them to safety. Although a government board of enquiry could not decide the cause of the accidents, the company agreed that the old method of heavy and regular timbering was the safest.

In London D'Arcy's courage in seeking oil was rewarded. He was repaid for all he had spent in the search and got oil shares worth

£900,000 at the time and worth more later. His interests in Mount Morgan too were enhanced, for it was both a steady gold mine and one of Australia's largest copper mines. He was inclined to be sentimental about the mine that had shaped his life and in 1913 the chance came to make another bold investment in the mine. Walter Hall had died and the family's holding of 350,000 shares was privately for sale, so D'Arcy and the London broking house of Robinson, Clark & Co., formed a syndicate and reputedly paid about £1,250,000 for this block of more than one-third of the shares in the company. D'Arcy had made his first spectacular gamble in the faith that Mount Morgan was a great gold mine, and now thirty years later he invested in the belief that it was a great copper mine.

In the English spring of 1917 D'Arcy proudly observed that the oil of Abadan and the copper of Mount Morgan were weapons against the Germans. A stout clubman with his cigars and his memories, he died that year, not knowing that his third big speculation was to fail. The ironstone mountain that had paid 413,000 per cent in thirty years was soon misted in storms.

21

REVIVAL

AT THE START of the twentieth century thousands of poppet heads stood like watchtowers on the plains. Australia's five big inland cities, Ballarat, Bendigo, Broken Hill, Kalgoorlie, Charters Towers, had all been built on metals. There was scarcely a town that did not get remittances from fathers and sons on distant fields. One in each four members of the first Commonwealth parliament came from districts where the mining vote was strong. Mining and smelting towns employed about one-eighth of Australia's male breadwinners. At least half a million people set their tables with money from the mining and smelting towns. This was the rich age of Australian mining, surpassing even the 1850s, which old men croaked could never be surpassed.

Metal mining stirred from the long slump. Output was worth barely £6 million in 1886, nearly £11 million in 1896 and close to £25 million in 1907. Australia was the world's largest producer of gold in 1903, and for ten years in a row Australia mined more gold than in its own record year of the 1850s. Output of silver and the base metals soared with the gold, increasing from about £1·7 million at the start of the rise to a brief peak of £11 million in 1907. There were sharp swings in production. Gold had provided four-fifths of Australia's metallic wealth in the nineteenth century and now its lead was challenged. Victoria had provided half of Australia's metallic wealth, and now its eminence was eclipsed.

The mining revival was strongest in the 1890s and was decisive in arresting depression and spurring recovery. Australia had suffered one of its recurring crises in the balance of payments, and the solution was to expand exports. While exports from soil and pasture were stagnant in the 1890s, the export of metals soared. Metals exceeded wool as leading export in the first years of the twentieth century. In the domestic economy the mines were only half as productive as agriculture in the early 1880s, but were more important at the end of the century. Bare statistics to us, they were not to the hundreds of thousands of people whose poverty and hopelessness was relieved by the money that went coastwards from Coolgardie, Charters Towers, Canbelego and a hundred busy towns.

The causes of the mining revival puzzled thoughtful men of those days and historians of ours. The first cause was obvious—new mining fields. More than half of the wealth mined in 1900 came from deposits that had not been touched twenty years previously. This Indian summer of prospecting was so productive because it broke into the Pre-Cambrian zone. Pre-Cambrian is the oldest of the five main eras in the earth's history and covers a fifth of the world's surface. Its rocks were formed at least 500 million years ago, in an age before plants and creatures. Perhaps because Pre-Cambrian rocks were formed over a long period of time they had more opportunities to be mineralized, and in fact the Pre-Cambrian rocks throughout the world have so far proved richest in minerals. Australia's three largest known metal deposits are all Pre-Cambrian—Broken Hill, Kalgoorlie and Mount Isa—and none of these deposits had been found before 1880.

Virtually all the Australian fields found before 1880 lay in the younger Palaeozoic rocks that form the eastern highlands from Tasmania to North Queensland. Most of the highlands had ample rainfall, good soils, a temperate climate, proximity to harbours, and these advantages spurred agriculture and grazing and, of course, mining. These highlands were also rich in alluvial gold and tin, and could be easily mined because they had been slowly concentrated in Nature's mills. So deposits of the Palaeozoic or Cainozoic era dominated the first era of Australian mining.

A line drawn between the ports of Normanton on the Gulf of Carpentaria and Warrnambool in Western Victoria roughly divides Australia's two main mining regions. Virtually all important deposits east of that line are in more recent rocks. All important mines west of that line are in Pre-Cambrian rocks. In the vast dryness to the west only the South Australian copper mines and Northern Territory gold mines had been found by 1880, being the only deposits that lay both close to the coast and in areas of adequate rainfall. All the other great deposits of the Pre-Cambrian zone in Western Australia, western New South Wales and western Queensland were remote from the coast. These were regions of scant rain and high evaporation, and were therefore sparsely peopled. Hence their minerals were found late and developed slowly. Their growth was a major cause of the mining revival.

The other mining fields found after 1880 were in the Palaeozoic rocks of the east, and their belated birth had a similar explanation. Some, like Mount Boppy and Croydon, were in parched areas that were sparsely settled. Other deposits such as Mount Morgan and the fields of western Tasmania lay in broken country, well timbered and weak in alluvial metals. Hence few people paused by such deposits and they had small incentive to develop them. In short, most new mining fields found after 1880 were inaccessible places with extreme climates. They provided a firm cause for the revival but the pace of the revival came more from the flood of money which isolated fields needed.

Why did money flow so freely down mines from the 1880s? It was attracted less by new fields than by changes in the money market. Before 1880 Australian mining investors rarely bought shares in mines outside their own district or colony. In the 1880s the scope for investing in new mines within two hundred miles of Melbourne, Adelaide or Sydney was narrowing, so speculators had to invest in real estate in the cities or in new mines in remote regions. The Broken Hill boom of 1887-8 was the first inter-colonial mining boom. The willingness of investors to look beyond their own colony increased the available money for any new field and possibly strengthened the market for mining shares. Thus in the late 1890s Adelaide, capital of a colony that was poorest in gold, speculated so heavily in gold shares that it was probably the busiest stock exchange in Australia.

The mainspring of the mining revival, however, was probably British money. In the world recession of the early 1890s Britain no longer lent so much money to foreign governments, railway companies, foreign banks and financial houses; and as mining was one of the few alternative investments, it attracted London and Glasgow money. More and more Australian mines were quoted or floated in London. Dr John McCarty of Sydney University estimates Australia got about one-quarter of the capital which Britain invested in new mines overseas in the 1890s. British investors moved from Western Australian gold shares to gold mines in eastern Australia and then bid excitedly for shares in copper and lead and tin mines. By 1900 they probably had more money at stake in Australian mines than the Australians themselves. From 1886 to 1907 British investors must have paid Australians from £40 million to £50 million cash for shares in mines and holes in the ground. Without prolonging a nightmare of statistics, this investment was a disastrous loss, less because so many mines were worthless than because Britain paid too much for both the rich mines and the poor. In the history of British investment in many lands this was one of the leanest chapters.

British money had strong impact. It stimulated so much prospecting and the sinking of shafts that many concealed deposits were found decades before they might otherwise have been found. It reopened old mines that would have otherwise been shunned. It equipped many mines so efficiently that they could work poorer ore and make higher profits. Although British companies worked scores of gold and copper mines that should probably not have been touched, such mines increased the output if not the profits of Australian mining.

The rise of gold output in the 1890s owed much to the workmen. But contrary to economic theory, the wages of skilled miners on the old fields do not seem to have fallen during the depression. Nor is the tradition correct that the tens of thousands of men who dug or fossicked for alluvial gold in the old eastern Australian fields found most of the new gold. The reefs and deep alluvial mines in eastern Australia largely lifted the gold output and they gained in the depression from the willingness of men to work harder to keep their jobs and to work their own mines for small reward. The Western Australian goldfields gained from the depression because so many men were unemployed in towns and cities that they rushed west to work. Even the copper and lead fields gained likewise from the influx of workmen but unlike the gold mines they lost most of the advantage by having to sell their metals at depressed prices. But in the last years of the century the price of tin, silver-lead, and above all copper, began to rise, and hence these mines expanded just before gold output began to taper away.

There was one quiet, unmeasurable, change during the revival: a technical awakening. The engineer and chemist did nearly as much as the prospector and investor to revive mining.

The scope for technical change in the 1880s was wide. Most managers were miners who had won promotion through their physical strength and their skill in handling men. Overseas visitors often commented on the skill with which they mined and the clumsiness with which they milled. They received only slightly more pay than skilled miners, and spent much of their time doing manual work underground. The thumb was their ruler. They did not know how to sample an orebody accurately, could not compute the amount of water their machinery had to pump from their mine. Their uncanny skill in dangerous ground and their capacity for improvising did not always compensate for ignorance of scientific principles.

'There is probably no other industry that depends upon so great a variety of other arts and that involves so many branches of science as does mining, nor one that includes such a large complexity of operations,' wrote Professor Henry Louis in 1927. The mining fraternity would have disagreed. When the Victorian parliament created a board in 1889 to decide if all mine managers should hold certificates of competency the evidence was strong against the idea.

'I hold,' said John Cock, a Chiltern manager, 'that a man might not be able to answer any question intelligibly and yet be one of the best men to manage a mine.' Another popular argument against examining managers was that it would penalize men 'who may have reached an age at which it is very difficult to learn'. Many managers had reached that age.

Cornishmen managed most of the large Australian mines and by 1880 they tended to be old and to romanticize the old ways. Most were fine practical miners but ignorant of those scientific principles that lead often to efficiency. They had left Cornwall before there was mining education outside the mines. Above all, the Cornishmen who managed gold mines tended to ignore metallurgy because Cornwall had plenty of cheap female and juvenile labour to assist in the ore dressing and because it sent its concentrates to Wales to be smelted. The ease with which gold was extracted in the mills of Bendigo, Ballarat, Gympie, and even Charters Towers, confirmed their idea that metallurgy was unimportant. The number of Chinese who made a living on the gold that escaped in tailings from the stamp mill seemed to many managers proof that their mill was efficient. They argued that so little gold escaped that Europeans could not live on the scavengings. The small size of the typical mine discouraged the employing of specialists. On the other hand the few companies such as Port Phillip at Clunes and Moonta in South Australia which had large leases and large-scale mines and well-paid managers, were noted for their efficiency and sympathy for new ideas.

One of the few clear milestones in the strong upsurge of technical efficiency was the decision of the Broken Hill Proprietary to import experts from the mining fields of the Rocky Mountains in 1886. That decision linked Australia to a new powerhouse of skills and attitudes, and many brilliant American engineers and metallurgists crossed to Australia to manage new mines. The inflow of British capital at the end of the century brought more skilled engineers. Most big mines opened after 1886 were managed by imported engineers.

Australia was again in the mainstream of world mining. Australian engineers moved from Stawell to Johannesburg, from Coolgardie to the Ivory Coast. The Ballarat School of Mines graduate went to Siberia and the Munich doctor of philosophy came to Kalgoorlie. All inhaled what Toynbee called 'the stimulus of new ground'.

The discovery of unusual orebodies challenged the mining engineers to find new ways of mining them. Unusual ores made the chemist experiment in treating them. The increasing area of ground allowed to each company, and the massive size of so many of the new orebodies, promoted large mines which engaged experts and paid dearly for their advice. The salary and social status of the mining engineer leaped forward. Mining was so prosperous and expanding that it attracted talented young men. The fields were so short of experienced managers that young men had high positions in which they could try new ideas. So many new companies were floated monthly and so many directors were new to mining that they tended to be open-minded; side by side with their bursts of incompetence was a bold sponsoring of new processes. Even the optimism and the excitement of mining booms stirred the desire for innovation. The contrast in the mental attitude of Kalgoorlie in 1898 and 1928, one a year of boom and the other a year of decline, was sharp.

New institutions were rising to educate assayers, foremen, shift bosses and managers. The opening of schools of mines at Ballarat and Bendigo at the start of the 1870s, less than a decade after the early United States mining schools, gave premature promise of a vigorous era of mining education, but by the last years of the century their graduates managed mines throughout the continent, and new schools from Charters Towers to Zeehan taught at night the men who worked by day in mines and mills. The universities of Melbourne and Sydney began to educate engineers and scientists for the mines. Two young Melbourne graduates, E. J. Horwood and Lindesay Clark, ran the two largest open cuts in Australia—B.H.P. and Mount Lyell—but even those two companies that owed much of their success to trained engineers were reluctant to employ the graduate fresh from university. Robert Sticht, of Mount Lyell, refusing R. M. Murray's plea for a job, argued that a big mine risked its efficiency by employing inexperienced graduates. Nevertheless, hundreds of trained engineers were entering the industry, and A. J. Bensusan, old mining engineer, confessed to the New South Wales Chamber of Mines in February 1901 : 'I shall be glad to see the day when most of the men at present managing our mines all over this country retire in favour of trained engineers.'

In 1893, the first national professional body of mining men was born, the Australasian Institute of Mining Engineers, modelled on the American Institute founded in Pennsylvania in 1871. The founder of the Australasian institute was Uriah Dudley of the Umberumberka mine at Silverton, and the hub of the Institute was Broken Hill, but soon it stirred the interchange of ideas on geology and mining and mineralogy and metallurgy throughout the land. 'Speaking from personal experience,' wrote John Howell as general manager of B.H.P. on the eve of the Institute's inaugural meeting, 'I am free to state that the want of practical and scientific knowledge in the management of mines, and the treatment of ores, has caused more failures, and surrounded mining enterprise with more doubt and distrust, than all other causes put together.' At the Institute's first conference at Adelaide in April 1893 the delegates chattered much on the problem of treating sulphide ores at Broken Hill ('the question of the hour,' said Zebina Lane), and in solving that problem Australian mining revealed its quality.

The new attitude to change blew down hundreds of shafts. John Craze of Zeehan's Montana developed his mine far ahead of its immediate needs—a practice rare outside Moonta and Broken Hill. Hamilton of Great Boulder was one of many managers who believed in tearing out the ore in bulk instead of picking the eyes; such men carried the economies of mass production to the cramped space of the underground. Broken Hill imported and adapted square-set mining from the Comstock, and the method spread to massive lodes such as Mount Lyell and Mount Morgan. Broken Hill Proprietary and Mount Lyell imitated some of the big Spanish open cuts. At Beaconsfield in Tasmania and in the Loddon deep leads in Victoria electricity pumped mines as wet as the sea, coughing up seven or eight million gallons a day. The diamond drill and rock drill and high explosives, all used sparingly in Victorian mines in the 1870s, were freely used on large fields at the end of the century. These methods made poor mines payable and good mines rich.

Gold boats were floated on hundreds of artificial lakes in the mountain spine of eastern Australia. Inventors from New Zealand to California had long tried to mine the millions of grains of gold that had eluded the crude appliances and burrowing of the vanished legion of diggers. Some tried to dredge gold just as ships dredged the silt of harbours. In Victoria by 1891 five suction dredges were floating on their own ponds in the Beechworth district and pumping up the golden gravels to their treatment plant on deck. Then New Zealand perfected the bucket-dredge which excavated the gravels with an endless chain of buckets; 236 dredges worked in New Zealand in 1900 and some of their dredgemasters crossed to Australia where new gold-dredging companies were booming. In ranges of north-eastern Victoria and central New South Wales travellers on country roads soon saw cumbersome steamboats floating on dams in green paddocks, the grating clank of the buckets making din in sleepy valleys. Pearson Tewksbury, a frail and dour young Victorian, started companies that won over a million of gold from dredging, and his gold financed a fleet of yellow cabs in the cities. Dredging was so cheap that it paid in valleys which even Chinese fossickers had abandoned. From 1900 to 1914 over £5 million of gold was dredged in Victoria and New South Wales, and tin dredges were heard through the night from Queensland to Tasmania.

The awakening was clear in mining methods and even clearer in metallurgy. An astonishingly large part of the increased output of gold and other metals was extracted with new techniques. Broken Hill was enriched by Colorado smelting methods, by the new Huntingdon-Heberlein process for roasting lead ores, by the similar roasting process devised by the Australians Carmichael and Bradford, by the American gravity mills and old Captain Hancock's improved jig, and by the wonderful flotation process that was born and nurtured at Broken Hill itself. Copper mining was revolutionized by the Bessemer converter and by the pyritic smelting that Robert Sticht of Mount Lyell perfected. More gold was squeezed from rock by the filter press and tube mill that Kalgoorlie men adapted. Mount Morgan developed chlorination to its highest efficiency. Every goldfield gained from the Scottish cyanide process which Australians first used at Ravenswood in north Queensland in 1891 and which Kalgoorlie men perfected. (The peak year of Charters Towers was 1899, and about one-third of its gold that year came from cyaniding the discarded dumps.) In innovation and adaptation no mining country matched Australia during this triumphant era of world metallurgy.

Other ideas in metallurgy were not spectacular but nevertheless made wealth. Most mining companies now employed their own assayer with his small furnace and shelves of chemicals: a man from a school of mines who provided that accurate measuring without which no industry is a science. And in every colony in the 1890s large custom mills and smelters were built by companies that usually owned no mines. They experimented in difficult ores, served mines that could not afford a plant, and saved the export of ores to foreign works. The Rothschilds invested in the large Queensland Smelting Company near Maryborough and the Deutsche Bank and German metal trusts built the large Tasmanian Smelting Company at Zeehan. On the coast of New South Wales large smelters at Dapto and Cockle Creek treated gold, silver, lead, and copper ores shipped from distant ports. Ballarat and Bendigo works extracted gold from defiant pyritic ores, and Fremantle and Northam had large mills and smelters to treat the complex gold ores from Western Australian mines. Only one of these major industries survived beyond World War I and few made much profit, but hundreds of mines gained from their existence.

From 1850 to 1880 ran the era of the small mine. Of the thousands of companies floated in that era only one gold company, the Long Tunnel in Victoria, paid over a million in dividends, and the only base-metal companies to exceed the million were Wallaroo & Moonta, and Mount Bischoff. In the second era the large lease and the large company were dominant. Of the companies floated between 1880 and 1900 at least sixteen paid over a million in dividends: Mount Morgan, Mount Lyell, five Broken Hill companies, seven Kalgoorlie companies, and two gold mines on the Murchison in Western Australia.

The big sixteen had strong influence on mining. Their massive lodes aided mass production and new and efficient techniques. Employing enterprising engineers and amassing huge profits, they had the skill and the wealth to branch beyond mining. Some of these companies financed and directed much of the industrialization of the nation, becoming leaders in iron and steel, shipbuilding, heavy chemicals, fertilizers, paper, paints, aircraft production, aluminium, the refining and fabricating of metals. Their technical mastery in mining was the prelude to industrial mastery. Nevertheless, their huge impact on Australian heavy industry would have been even greater if the nine great Western Australian mines had been directed from Melbourne instead of London. Most of their profits returned to England; their gold created no new Australian industries. Although British venture-money was vital in reviving Australian mining, it exacted its price.

PART IV

Age of Giants

22

FROTH AND BUBBLE

BROKEN HILL was becoming the Athens of metallurgy but its arts were cultivated in laboratories, and did not adorn the streets. Sand from the plains blew across the city in dirty red clouds. Sand drifted against iron fences and piled high until it pressed them down. The treatment plants devoured rock from the mines and spewed out waste sands on to pale hillocks that seemed like pyramids to visitors crossing the plains. Winds scythed the tops of these hills, drove gritty sand into keyholes of doors and mouths of playing children, angered washerwomen and vexed mayors. But when rain crisped the air and washed the leaves of pepper trees in the streets and fruit trees in backyards and cleaned the stone of noble buildings in Argent Street, the city had an air of provincial solidarity.

Broken Hill in 1900 also had a medieval air—it was like a walled city. Every company stacked valuable tailings near their mine, and the huge dumps ran for miles. From the main streets of the city the Proprietary mine was obscured by a man-made ridge of tailings half a mile long. The dumps held about £30 million of potentially recoverable metals in 1903 and every day they grew larger.

Broken Hill extracted only a fraction of the wealth that its miners were sending up the shafts. From the ore they got less than half of the silver, two-thirds of the lead, and virtually none of the zinc. During half of the year 1903 the Block 10 mine, for example, treated ore containing £380,000 of metals, but managed to extract only £90,000 worth of metals. That mine's losses were not exceptional. Only the richness of the ore and the large scale of mining gave Broken Hill prosperity in the face of this fantastic waste of metals.

Even since the shafts had penetrated into the deeper sulphide zone of the lode the metallurgists had faced grave problems. The first dilemma was the toughness of the rock which contained the minerals. The garnet was particularly hard, harder than the steel of the Cornish rolls that crushed it. If the ore was not crushed enough in the mills, the minerals could not be physically liberated or separated from the barren particles of rock. The ore therefore had to be finely crushed, but much of the mineral was thereby pounded into a fine powder or slime that itself defied extraction.

The metallurgists' second dilemma sprang from the density of the various minerals that made up the ore. The concentration mills of Broken Hill—and for that matter of every major base-metal field in the world—worked on the principle that most minerals had a different specific gravity. Now, of the constituents of the ore at Broken Hill, galena or silver-lead had the highest specific gravity, 7·5. Accordingly, when the crushed ore was shaken on the rattling jigs or tables most of the particles of galena fell to the bottom and the lighter minerals escaped in a flow of water. Unfortunately that dirty water not only contained some of the galena that had been crushed too finely, but also the valuable zinc. Nature had played a trick on Broken Hill. The zinc had a specific gravity of 4, virtually the same as the garnet and rhodonite in the ore, and where they went the zinc went too.

Companies lost money and inventors lost reputations in attempting to recover this fortune. The Germans suggested that strong magnets could separate the minerals, for minerals had varying qualities of magnetism just as they had varying specific gravity. An Anglo-German partnership known as the Australian Metal Company came to Broken Hill in 1898 and erected forty-five magnetic machines by the line of lode at Railway Town. They passed the tailings between two magnets and jubilantly collected the highly magnetic garnet and rhodonite that had previously been inseparable from the zinc. Then they produced a concentrate carrying about 40 per cent zinc and sent it to German smelters. Ocean freights and smelting charges ate the profits. The company lost money and the employees lost health; fifty-nine of sixty-three cases of lead poisoning at Broken Hill in 1900 came from these dusty works. 'This is a horrid process,' reported *The Economist* in July 1905, and even shareholders of the Australian Metal Company agreed. Yet such was the prize at stake and such was the promise of the magnetic processes that five other Broken Hill companies built magnetic works. Not until 1911 was the last magnetic plant scrapped.

Neither magnetic nor gravity methods could tame these ores. Expensive fuel ruled out furnaces. Roasting did not pay. Crackpots came with strange methods. One inventor argued that if the zinc refused to sink in a bowl of water it could perhaps be taught to float.

This flotation process resembled bubbles rising in a glass of beer, and in fact brewers were amongst its architects. The strength of the rising bubble had been observed as early as 1789 by Richard Watson, Welsh bishop and Fellow of the Royal Society, who put powdered lead ore into an ale-glass of water and added nitric acid. He observed that the acid generated air bubbles, and the bubbles attracted the mineral and lifted it to the surface. This observation was often made and sometimes patented in the next century but never practised commercially.

One man who independently noticed that certain minerals would float was Charles Vincent Potter, a Melbourne consulting brewer and chemist. The breweries were struggling in the 1890s so Potter turned to other fields. In his Balaclava laboratory he invented an improved bottle and stopper, a fluid that killed germs in butter, and 'an improved nose bag' for horses. He experimented also on methods of treating sulphide ores and it is probable that he tested zinc tailings from Broken Hill, the problems of which were common talk in Melbourne.

About the end of 1900 Potter discovered a process which his paid publicist vowed would place him 'with the great Edison, Bessemer, and other world-renowned inventors'. He applied for a patent on 5 January 1901 and after more research he applied for three more patents. His process resembled Bishop Watson's forgotten process but Potter preferred sulphuric to nitric acid, and moreover he made his process work on a useful scale. Potter interested Block 14 company in his work. That company had closed its mine because of low metal prices and treatment worries, and eagerly it built an experimental mill to test the power of the bubble. The translation of the process from backyard laboratory to a working mill was delicate and frustrating, for this was a new branch of physics and even today its theory is imperfectly understood. Block 14's mill worked from 1903 to 1905 and recovered upwards of 60 per cent of the zinc as well as much of the lead and silver in the ore, but the process was dear.

Guillaume Delprat, sixth general manager of Broken Hill Proprietary, discovered a floating process. He found it by a different chain of experiments, and he found it without any knowledge of Potter's patent which was then being secretly studied in laboratories less than half a mile away. Delprat had walked past his tailing dumps and seen that part of the dump was hot from slow combustion of the iron and sulphur. At once he remembered that Spanish mines recovered copper from dumps of poor ore by slow combustion, and he realized that if he kept hosing the dumps with water the slow fires might produce sulphate of zinc in payable amounts. But water was dear at Broken Hill. Perhaps a mixture of salt and tailings would draw water from the air. Delprat discussed it with Carmichael, his metallurgist, who tried salt and then nitrate of soda on small samples of tailings, without success. Delprat urged him to persist, suggesting that he try the salt-cake of which the company had large heaps. 'That's no good,' said Carmichael. 'Never mind; try it,' said Delprat. 'Boil the stuff and see if it will go into solution.'

Carmichael boiled the stuff and saw bubbles of gas lifting a scum of tailings to the surface. He told Delprat that the experiment had failed; the mixture would not dissolve. Delprat examined the beaker and tried to submerge the scum, but it continued to float. Suddenly he realized the significance of the experiment. He had found a quicker and cheaper way of separating the particles of zinc and lead from the barren particles of rock. He had found it in the year the discovery was most needed, for B.H.P.'s dividend had fallen for the first time below £100,000. After more research, Delprat disclosed his secret to the company's secretary in Melbourne on 17 November 1902 and urged him to apply for patents. 'Our German friends,' he confided, 'are rather fond of stealing a march on anyone.'

In Melbourne, Potter soon complained that B.H.P. was stealing a march on him. Except for Delprat's liking for salt-cake, the processes seemed identical. Potter took B.H.P. to court, and while the litigation slowly weaved through many courts, B.H.P. erected a rival flotation plant and produced zinc concentrate early in 1904 on an increasing scale. It owned the largest dumps along the line of lode and in 1904 they held about £11 million of metals which the process could reasonably be expected to extract.

B.H.P. and Potter fought for four years until the Supreme Court of Victoria ruled in 1907 that Potter's patent was invalid for want of utility. The Court consoled Potter that if his patent had been valid, Delprat's would have infringed it. Potter was not consoled, for legal expenses in one court alone had nearly reached £100,000. Even for B.H.P. the victory was pyrrhic, for after spending a fortune on the law it chose to use the sulphuric acid that had been the pith of Potter's patent. Henceforth the process was known as the Potter-Delprat.

Most mines still favoured the magnetic process, and even Block 14 eventually abandoned flotation for magnetic treatment. Another company that trusted in magnets was the Sulphide Corporation, a London company which owned the Central mine and the second largest dump of tailings on the field. In the will to experiment it led the field at the turn of the century. It had erected a large treatment plant at Cockle Creek near Newcastle to roast and leach and electrolyse concentrates by the novel Ashcroft process. This was the first electrolytic zinc plant in the world, but it was not successful. At Broken Hill the company also made the magnetic process pay, though the margin of profit was delicate. Not content with magnets it tried the granulation process invented by the Englishman Cattermole, who filled vats with crushed ore and acid and oil and stirred the mixture with a revolving agitator. The oil coated the metallic minerals and clustered them into heavy granules, and the valuable granules were easily collected on a shaking Wilfley table. Fortunately the Cattermole plant at Broken Hill did not always work efficiently. In 1905 metallurgists noticed that when insufficient oil was added to the vat the granules of metal ceased to form. Instead the violent agitation created bubbles that floated the metals to the surface in a thick froth that could be skimmed off. They decided that the froth was better than the granules. They had found a major principle of flotation—the use of a mechanical stirrer to create more bubbles on which the minerals could ride to the surface.

In a London laboratory at the same time the same discovery was made. A London syndicate led by the talented metallurgists Sulman and Picard had bought Cattermole's patent which Sulphide Corporation was using at Broken Hill. In 1903 the syndicate became Minerals Separation Limited and widened its scope by paying £225 for the flotation process which the Italian engineer Froment had invented soon after Potter's discovery in Melbourne. Minerals Separation experimented with both Cattermole's and Froment's processes in a small way in its laboratory, and during those tests they found the value of vigorously agitating the mixture of crushed ore to create a froth.

The Sulphide Corporation and the British company worked secretly together at Broken Hill on the new process. In July 1905 the mining journalist James Curle, himself a director of Minerals Separation, wrote in *The Economist* that Broken Hill was on the eve of metallurgical triumph. He predicted that it would soon produce zinc commercially in huge quantities. He did not mention that B.H.P. was already producing it. The secrecy on that company's plant was thick, and even today the mining world does not realize how far ahead it was in this technical revolution. When in June 1906 the new Central zinc mill had produced 8,000 tons of zinc concentrate, B.H.P. had produced 150,000 tons and was far and away the largest flotation plant in the world.

Yet another flotation process was launched by a Melbourne brewer. Auguste de Bavay was a dapper, dignified chemist in his late forties. He had migrated from his native Belgium by way of a Ceylon plantation to Melbourne where he became head brewer at Foster's in Collingwood. Increasing deafness seemed to aid his concentration and his enquiring mind found many problems to study. He corresponded with Pasteur in France and distinguished German chemists, and in his laboratory with the aroma of cigar smoke about him he studied the influenza germ and the making of paper pulp from Australian timber. He was an expert on bacteriology and discovered that fire hydrants in Melbourne streets conveyed sewage into the water mains and endangered the health of the city. It was not surprising that he should be called as a witness by Delprat in his long litigation with Potter, for many principles of flotation were as mysterious to metallurgists as to judges.

Soon de Bavay was immersed in the process and in July 1904 he patented a novel version that became known as skin or film flotation. It differed from all other methods tried at Broken Hill in the early years. Instead of the particles rising with the bubbles from the bottom of the treatment vat, oiled particles of ore were dropped straight on to the surface of the water where the minerals floated and the barren particles sank. The minerals formed a skin on the surface only a particle deep, whereas in the orthodox plants the minerals formed a froth several inches thick. That was the grave defect of de Bavay's method, for he required a huge surface of water to ensure a large output.

Like all the flotation processes de Bavay's was not translated easily from laboratory to mill. Efficiency came only after the five years of costly trial and error at Broken Hill. These trials were made by a small company, De Bavay Treatment Company, in which the brewer united with two Melbourne financiers, Montague Cohen, a director of Foster's brewery, and W. L. Baillieu, a Melbourne auctioneer and financier. He was fortunate in his backers. They were confident that the process would work, and they risked money buying huge piles of zinc tailings. It was then easy to buy tailings at the line of lode, for many large mines distrusted the new process and were also short of money. Baillieu thus signed contracts with the North Broken Hill, South, and Block 10 mines, to buy hundreds of thousands of tons of tailings at a few shillings a ton. With raw material assured he engaged a young chemical engineer, Herbert W. Gepp, who had managed the explosives factory near Melbourne and had built a plant at Broken Hill to feed sulphuric acid to a Potter flotation plant. Gepp set up a trial plant by the North mine and by 1907 it had spent £65,000 and was floating zinc concentrates with intermittent success. By 1909 the syndicate was at last certain of success and, refloating itself as Amalgamated Zinc (De Bavay's) Limited, it went to the public for nearly a quarter of a million pounds and began to build a rambling plant at Broken Hill—the world's largest flotation plant in size and output. That plant not only amply rewarded the shareholders, but financed the two greatest Tasmanian companies launched between the world wars.

Meantime the Zinc Corporation, now one of Australia's largest companies, was nearly wrecked on the wayward bubble. Like De Bavay's, it was created to exploit the flotation process. Herbert C. Hoover of the London managerial firm of Bewick Moreing was in Australia in 1905, examining the mines his company managed in Kalgoorlie and the buried rivers of Victoria, and he heard many whispers and a few facts about the inventions then being tried at Broken Hill. He knew that the price of zinc had risen sharply over the previous four years, and was told that the campaign to crack the zinc problem would almost certainly succeed. In the winter of 1905 he hurriedly visited Broken Hill, possibly on the advice of W. L. Baillieu and of Lionel Robinson, a shrewd mining operator on the London stock exchange. He walked over the dumps, learned what he could about the hushed experiments, and finally arranged to buy several million tons of tailings on time payment. In September 1905 his small syndicate floated the Zinc Corporation to treat them.

The company had to pay large sums for the dumps in the next few years, so quick production of zinc was essential to raise revenue. The company chose Potter's process, believing it could succeed where Block 14 mine had failed. It was mistaken, and after spending over £100,000 it abandoned Potter's technique. Instead of testing the other rival processes Hoover decided now that the Central mine and the Minerals Separation Company had evolved the best method and therefore sent out a New York engineer named Queneau to install a second plant.

Zinc Corporation was slow to learn the costly lesson that even a proved process had to be manipulated through a long merry-go-round of trial and error before it attained efficiency. In March 1907 it abandoned Queneau's plant and turned to the Elmore vacuum process, an English invention of 1904 that had not been tried at Broken Hill. But to erect a new flotation plant they wanted money, and wanted it desperately. They were deep in debt, and even if they did manage to finance the new plant they might slip into liquidation if they could not pay for the tailings dumps to which they were committed as their only source of raw material. 'Believe me,' said F. A. Govett, the chairman in July 1907, 'I am overwhelmed with the sense of the shareholders' misfortunes and their loss.' Shareholders could well believe this sigh of dismay from a man who with Hoover held nearly £100,000 of shares in a tottering company.

Govett pleaded for the shareholders' patience while he erected half an Elmore plant. If that failed, the company would liquidate. He hoped to finance the Elmore plant by selling preference shares on highly favourable terms, but only three of every five preference shares were sold. A loan from a bank was the last hope and that was only secured by the guarantee of the rich Kalgoorlie gold company, Lake View Consols, of which Govett was chairman. Thus the company raised another £180,000 and by the winter of 1908 the rows of Elmore machines standing like little space ships in their long hangar were producing zinc at slender profit. At the same time the Minerals Separation process which the Zinc Corporation had recently forsaken was improving so dramatically that the company regretted its change of policy. Nevertheless, it clung entirely to the Elmore process for two years, and the skill of its metallurgists made the best of its plant until the company in 1910 could finance the costly reversion to the Minerals Separation process. This was the company's fourth plant in five years, and it was very successful.

The Zinc Corporation had survived its ordeal through the grit of its financiers and the steady march of techniques, and in 1912 the ordinary shareholders got the first dividend. By then the company had got itself a mine, realizing that the flotation process had developed so much that few companies would sell their mill tailings except at high prices. The buying of the Broken Hill South Blocks mine through the good graces of Govett's Lake View Consols gave Zinc Corporation the entree to the unsuspected riches of the south end of the line of lode, which today produces more than half the wealth of the field.

In the mills and laboratories of Broken Hill the chemists continued to experiment. The whole line of lode was a laboratory. A froth of secrecy hid the experiments just as a froth of mineral hid the process that went magically on day and night in the long mill sheds. The metallurgists who met after work in the quiet comfort of the Broken Hill Club rarely discussed their work. The directors of each company commanded secrecy, knowing the royalties a successful patent could yield. Little was written about a process that was to change metallurgy possibly more than any process since fire and water. The one tangible sign of its merit was the powdery zinc concentrate sent to Germany and Belgium; half a million tons were shipped from Port Pirie in 1910.

The uses of the new process were far from exhausted. It did not yet recover much zinc and lead from ore that was unavoidably crushed to a fine slime in the old gravity mills; and while it could separate the valuable metals from the barren rock it could not separate the metals from one another. That was a grave weakness. The field in effect produced two concentrates, one rich in silver and lead, and the other rich in zinc, but each concentrate carried a large amount of metal that was lost in the fire processes that followed. The silver-lead concentrate that went to furnaces at Port Pirie or Europe carried much zinc that was not recovered in the furnace. Likewise European zinc smelters paid very little for the smaller portions of silver and lead in the zinc concentrate, and furthermore they did not pay much for the zinc contents. A few German cities thus indirectly made more than Broken Hill from the process they had neither invented nor seen.

E. J. Horwood was B.H.P.'s resident manager in 1908, a civil engineer, born on a Victorian gold reef, untrained in metallurgy. Seeking a way of floating each metal separately he observed in 1908 that particles of galena oxidized more readily than zinc. Accordingly he took the usual mixture of both from a flotation plant and roasted it in a small furnace until the surface of the particles of galena were slightly oxidized. He returned the mixture into a flotation tank; the particles of lead stayed on the bottom of the tank and the zinc floated. He had separated them by the simple trick of altering the surface of the particles of lead. Whereas all the earlier flotation processes were blind processes that floated all the metals, his process could select the metal it wanted to float.

A German company nearly bought his patent for £200,000 and the Zinc Corporation tried it on a large scale in 1913. Inventing a process, making it work in a small laboratory, and making it pay in a large plant, are three separate achievements, and Horwood's process was one that collapsed dramatically at the third and crucial stage. No company, in Australia or overseas, really made it work.

Other men at Broken Hill sought a simpler way of separating zinc from silver and lead. One method caught their eye as they walked about their plants. Sulphide Corporation men observed as early as 1901 in their gravity mill that ore would form a froth or foam whenever it cascaded into the water. Examining the floating froth they saw that it contained many small particles of galena. Accordingly, they set up a small plant that cultivated froth, and small boys were employed to skim it off. It was slow and costly and the froth only carried about 30 per cent lead, so they abandoned the plant. Even when flotation began to be practised they did not realize the importance of their find. They did not realize that somehow the old gravity mill occasionally practised flotation, and above all that it floated only the lead, whereas the proper flotation plant indiscriminately floated both the lead and the zinc.

It remained for a man of no scientific training to make the idea work. F. J. Lyster was a carpenter before he became foreman of the small gravity mill of the South Blocks company. In 1912, soon after his company was absorbed by the Zinc Corporation, he saw fragments of lead and silver forming in the froth and wondered why. Conducting experiments, he seems to have created a stronger froth than other mills had formed, and his froth was surprisingly low in zinc and strong in the two other metals. In May 1912 Lyster's employer, the Zinc Corporation, applied for a patent.

Chemists were long puzzled how and why Lyster's process managed to select the various particles of metal. One can only conclude that he had an unusual chain of luck and the alertness to realize his luck and to follow it through. Fortunately he had used ore that had come straight from the mine instead of lying on the dumps, where it would change chemically with exposure to the air. Also, the individual particles of the ore he used were easy to float, because the ore had been finely crushed. In short, he used a different ore from that which most mills were using. Finally, Lyster had one other fact on his side. In most flotation plants the small amount of copper sulphate in ore tended to activate the zinc and make it rise with the lead and silver in the froth. Lyster's mill, however, chanced to be using water pumped from the mine. That water carried many minerals and when tested with methyl orange was found to be definitely alkaline. That water prevented the copper sulphate from activating the zinc and making it rise with the bubbles, according to a modern authority on flotation, Dr Ian Wark. If that is so, then the mine water was an essential part of Lyster's magic formula. As Lyster himself was not a scientist, and as the scientists who examined his process could not immediately find the distinctive tenets of his process, his patent was woolly. But it worked.

In September 1912 the Zinc Corporation worked the process of selective flotation on a commercial scale, and before his process was perfected T. M. Owen of the Junction North mine and Leslie Bradford of B.H.P. had invented distinct processes to do the same thing. Once again the break through was not the achievement of just one man but the work of many who worked secretly and competitively. By the end of World War I nine companies at Broken Hill had selective flotation plants in their mills.

Slowly the magic process was understood. By blowing or sucking air into the tank of crushed ore and liquids, a myriad of fine bubbles was created. By adding certain chemicals to the mixture, all the particles of one particular mineral were coated with chemical instead of water; they alone attached themselves to bubbles of air and rose to the surface. To prevent the bubbles from breaking, other chemicals were added to preserve the froth. Thus the bubble was cultivated and tamed, taught how to capture some minerals and reject others.

As the process required so many separate advances there was wide scope for chemists in the United States or Germany to build on the early advances of Broken Hill. As it was soon clear that the process might be relevant to every metal mine in the world, there was incentive for men in other countries to experiment. It is therefore surprising that Broken Hill men or Melbourne men should make nearly every major advance in the process. It is more surprising that between 1902 and 1915 at least forty men on Broken Hill advanced the process in a manner which a historian of metallurgy would think worth recording. Those men worked for eleven different companies, and even in one small struggling company, the Junction North mine, six members of its staff made important innovations or advances.

Why was Broken Hill so successful in developing the flotation process? While many scientists contend that co-operation best promotes science, the inner history of this process suggests that secrecy can also further scientific discovery. The men who heated flasks beneath sweltering iron roofs of laboratories, or who experimented in the draughty mill houses, were tight-lipped. They had to be. By its very nature the process made for legal dispute. Discoverers such as Potter knew that their process worked, but so novel were the problems in physics involved that they could not define the essence of their patent, define how or why it worked. Their patents were therefore vulnerable to litigation. Likewise some companies had patented processes that were valid in a law court but useless in a metallurgical plant, and they were able to sue companies that in fact had performed the far more important and meritorious service of making that patent work. In the first decade of the twentieth century hundreds of thousands of pounds were spent in legal fees; in some years a score of wearying cases were before courts in many lands. At Broken Hill, in the mills, secrecy was essential. A company did not dare reveal the essential features of its operations for fear that it might be sued for breach of patent. Even when they got a patent they were not secure, and so secrecy grew.

Men in white coats worked only a hundred yards away from men in a rival mill, worked on the same problems, made the same mistakes, and said nothing. One company may already have spent £10,000 proving that one technique was worthless but in the distance of three miles perhaps five other companies were busy repeating the same experiments. The waste of effort and thought and resources seemed enormous, but because flotation was a delicate process, because it behaved differently on two lots of ore from the same level of the same mine, much of the work was not really wasted. The same experiment could be performed a hundred times by ten teams of men, could appear hopeless, and then suddenly it worked. A chemical in the mine waters, a chemical they added to the tanks (de Bavay's castor oil or Lavers' eucalyptus), a chemical slightly more prominent in the ore from some part of a mine, all could dramatically change an experiment. 'It is to manipulation, learned empirically in the laboratory and mill, that the flotation process owes its metallurgic success,' wrote T. A. Rickard; and Broken Hill had a legion of manipulators seeking the combination that unlocked the vault of wealth.

In attacking the problem Broken Hill was fortunate that it had so many large mines instead of one or two massive ones, thus diverting more energy and excitement to the search. It was fortunate that at first it had the huge dump of zinc tailings which were cheap to experiment with, being already mined; it was fortunate that the particles of mineral in the ore could be separated easily by crushing the ore, for the particles had to be separated if the bubbles were to work effectively. Fortunately too, Broken Hill had many men trained in sciences, and was blessed with many directors who believed in science. All these advantages enabled Broken Hill to master in one decade a problem that had defied a solution in hundreds of mining fields for hundreds of years.

The United States of America mined more lead and zinc than any other country, but was slow to try the formulas from Broken Hill. American inventors independently found some of the techniques. Arthur Macquisten tried his own de Bavay brand of process in Nevada. The Elmore bulk-oil process, not a true flotation process, was tried in Utah. T. A. Rickard, the most famous journalist-scientist of his day, called these ventures 'mere ripples on the calm surface of American apathy'. Not until 1911 did the United States start its first successful froth flotation plant, James Hyde's plant at Butte, Montana, and by that year Broken Hill had already floated about eight million tons of crushed rock and was producing one-fifth of the world's zinc concentrates. When after long litigation, Hyde's flotation plant succeeded on a large scale, Broken Hill was racing further ahead with the discovery of selective flotation. It is therefore strange that most American histories, textbooks, and encyclopedias hail the process as a fruit of American ingenuity.

The bubble was important for lead and zinc ores; it was vital for copper. Copper had a low specific gravity, so did not readily separate itself from barren rock in gravity mills. Moreover, the specific gravity of copper was similar to that of iron, and the two minerals lay side by side in most copper mines. Wallaroo and Moonta was an efficient company but its mills could not extract more than about seven-tenths of the copper in its ores. Throughout the world copper companies at the turn of the century were unable to extract enough copper from ore that in theory seemed perfectly payable. Pyritic smelting was one answer to this problem, but only a few mines could employ this technique. The copper industry in fact had more incentive to develop the flotation process than the lead-zinc industry.

Many copper mines called for bubble experts as soon as the process worked at Broken Hill. The flotation experts arrived with their portmanteaus of plans and chemicals at copper mines from Norway to Cornwall, from South Africa to Queensland. By 1912 small flotation plants were working spasmodically at Chillagoe and Great Fitzroy copper mines in Queensland, Great Cobar and Kyloe in New South Wales, and Moonta in South Australia. However, each ore required research in the laboratory and manipulation in the working mill, and the new flotation plants developed slowly.

Not until Broken Hill had developed the idea of selective flotation did the copper mills gain the full benefit of the bubble, and then the details of the process were applied to copper ores most quickly and effectively in the huge American copper mines during the copper boom of World War I. Anaconda in the United States increased its recovery of copper from 79 per cent to 95 per cent simply by changing from gravity mill to flotation mill in 1915, and scores of other companies copied it. Many companies were now able to produce copper so cheaply that flotation in effect depressed the price of copper. The triumph of Broken Hill in a true sense was a calamity for many of the small Australian copper mines, and the metallurgists who lost their jobs in the early 1920s as copper mines closed along the ranges and plains of eastern Australia could well point a finger to their gifted colleagues in the Barrier Ranges. Some of those copper men, however, were to gain jobs at two new great leadfields, Rosebery and Mount Isa, which might never have been developed but for the flotation process.

The flotation process continued to improve. It had revolutionized lead and zinc mining, and had vastly increased the world's store of mineable copper. It spread to gold, tin, scheelite, zircon, mica, clays, limestone, manganese, replacing fire and water as the dominant method of extracting the world's minerals. In the last thousand years in metallurgy it stands with the cyanide process, and the Bessemer process, as one of the three greatest advances.

23

THE BOOMERANG LODE

On the eve of World War I Broken Hill had over 35,000 people. Man for man it was Australia's most productive city. It had earned as much wealth as Bendigo and more than any other mining town. It had sold metals for nearly £80 million—more than the money all the savings banks of Australia held in 1913—and from these metals smelting towns in Australia and Europe had earned tens of millions more.

Port Pirie was one town thriving on Broken Hill's wealth. Second largest town in South Australia, it smelted lead concentrate from the Broken Hill Proprietary and several smaller mines. The comet tail of black or white smoke from its chimneys trailed far down Spencer Gulf. At night sailors at sea saw the sudden glare on land as horse-boys tipped their ladles of red-hot slag. In the port carved from mangrove swamp the wharves were stacked with oregon timber for the mines and bars of lead for the East, and the strongroom of the refinery glittered like a bank vault with neat piles of silver bricks.

Port Pirie was a town of strong men who drove draught horses and stoked boiler fires and loaded ships and fed furnaces and fought at night at the smelter gates. It was a town so muddy in winter that often men could not cycle in the main street, with soil so salty that few plants grew and sulphurous air that rusted roofs and killed green blades of wheat that sprouted in the pier cracks.

Port Pirie was on the eastern shores of the narrow Spencer Gulf, and on the opposite shore bare hills gleamed like a skeleton in the morning sun. That Middleback Range had bold outcrops of red-grey hematite containing over 60 per cent iron and, so prospectors hoped, precious or base metals. The Mount Minden mining company found no valuable metals, and forfeited its claims at the end of 1896. Broken Hill Proprietary was then transferring all its smelting operations to Port Pirie, and as its furnaces used ironstone as flux the company took the idle iron mines. It built a tramway from Iron Knob mine to the beach at Whyalla, and in 1901 the barges were carrying the ironstone to the smelters' wharves.

This was the most massive deposit of rich ironstone so far found in Australia, but for a time that did not matter. Ironstone filled a negative function for the B.H.P.; it was merely a purgative in the silver-lead furnaces, entering as iron and emerging as hot barren slag. All the ironstone mined at Iron Knob finally came to rest on the black slag heap at Port Pirie. Not until 1907 did the company tip ironstone in a blast furnace to produce a few tons of pig iron. The ironstone made excellent iron, and five tons were sold for £25.

The seven directors of the Proprietary were elderly. Duncan McBryde, Bowes Kelly, and Harvey Patterson were pastoralists who had bought shares and joined the board when the mine was a gamble. William Knox had been the company's first secretary and William Jamieson the first general manager. The other directors, John Darling and H. C. E. Muecke, had joined the board in 1892 as representatives of the Adelaide shareholders who then held the most shares. In the minds of the seven the idea of making iron and steel had for a decade been only a fancy, or an absurdity. They had possibly the largest silver-lead-zinc mine in the world, employing 1,900 men underground and nearly as many above ground. They had responsibility enough, directing a big mine.

The long lode at Broken Hill was like a coat-hanger or boomerang, and B.H.P. owned the apex. They had most of the oxidized ore—the ore near the surface—and that was fantastically rich. Thus they gained great profits and in Broken Hill's first thirty years the Proprietary won half the wealth and paid over half the dividends. The gigantic company, however, was energetic and mined its most accessible ore long before other companies. Through selling some of its leases to the British Broken Hill, Block 10, and Block 14 companies in the 1880s it shortened its own expectation of life. Nature too had shortened its life. On the B.H.P. leases the lode had originally protruded hundreds of feet above the present surface of the hill, and that part of the lode had eroded away over millions of years. The silver had slowly migrated downwards to enrich the underlying ore, and thus the B.H.P.'s section of the lode was unusually rich but shallow. The company mined most of its ore within 500 feet of the surface. The ore went down another thousand vertical feet but much of it was in a small udder that hung from the main run of lode. By 1911 the company had mined more than 80 per cent of the ore it was eventually to win from the mine, and it was clear to the directors that the years of the mine were limited. Would the Broken Hill Proprietary survive? That question must have perplexed many of its officers and directors about 1910.

The general manager of the Proprietary was Guillaume Delprat. Of Dutch birth, he had managed large mines in Spain and Canada, and at Broken Hill he had conceived one of the flotation processes. He had an impressive face, strong nose, some skill at ju-jitsu and a salary of £4,000 a year which would be jeopardized if the Proprietary mine began to fail. He knew that the company itself would survive, for it owned the great lead smelters at Port Pirie and could buy ores from the other mines at Broken Hill and so exist profitably. He also knew that his company could possibly turn to another asset, its huge hill of rich ironstone at Iron Knob, and manufacture iron and steel.

Even if the Proprietary's mine had not been waning, Delprat would probably have seen strong reasons why his company should make steel. Australian's imports of iron and steel were rising. Hoskins had recently begun to make steel at Lithgow in New South Wales, and were smelting an iron ore that was inferior in grade to the Proprietary's ore at Iron Knob. Therefore Delprat and some of the company's directors argued that it was an opportune time for the Proprietary to venture into steel.

On 2 June 1911 the B.H.P. directors met at Port Pirie, and before the meeting Delprat urged Darling to discuss the steel venture that day. 'Told him Broken Hill mine would give out and Company would finish,' wrote Delprat in his diary. Possibly that was an argument to press his point rather than a compelling motive; because even if the mine closed B.H.P. could still buy and smelt ores from other Broken Hill mines.

At the meeting on that crisp winter day the board agreed that Delprat should go to America and Europe to collect information. Mr Muecke, who was German consul in Adelaide, was the only dissenter. 'I don't need any information,' he snapped. The information Delprat gathered abroad suggested that Muecke was wrong and that steelworks would pay.

The directors were not young and yet were personally willing to shoulder the worry of expanding into iron and steel. They knew that initially they could lose heavily. Moreover, the Proprietary, like all rich Australian mining companies, had paid nearly all its profits in dividends, and from £10 million had saved only a small reserve fund. The company would have to borrow to build the steelworks, and that might not be easy. Three of every four shares in the company were held by British investors who might not think steelworks in Australia could succeed. The company's adventure in steel, therefore, was cooler and bolder than tradition allows, for it was neither forced by the decline of its Broken Hill mine nor financed by its profit.

Guillaume Delprat wisely chose the United States rather than Germany for advice and skills. David Baker, a Philadelphia consulting engineer, came to Australia, applauded the project and designed it skilfully. The company had the wisdom to endorse his plan for a comparatively large steelworks, although that entailed an urgent search for capital. Making its first call for capital since it collected £18,000 in 1885, the company raised more than a million in 1913-14 by issuing new shares and debentures.

Baker advised the company to take the iron ore to the coalfields, and on the old site of the Wallaroo Company's copper smelters at Newcastle in New South Wales the B.H.P. built its iron and steel works. The second day of June 1915 was momentous for Newcastle with official guests and governors and industrialists and ladies arriving by special train, old Duncan McBryde's hearty Scottish voice lost in the din of the steelworks as he read his long speech, and a red-hot ingot of steel rolling from the mill as the Governor-General declared the steelworks open.

The company shaped Australia's secondary industry. It moved deeper into steel by buying in 1935 its only rival, Hoskins' Australian Iron and Steel, which had moved from inland Lithgow to coastal Port Kembla. The Proprietary made not only iron and steel but ships and cement and chemicals and alloys, until today it employs 43,000 people and has 86,000 shareholders and ranks about twelfth in the world's large steel companies.

The silver-lead mine that shaped this is now an empty crater, surrounded by a few pepper trees, concrete foundations and broken bricks. The company's only possession at Broken Hill overlooks the city, a black mountain of smelter slag, rich in zinc that may some day be recovered. Nevertheless in the public's mind B.H.P. is still linked with the broken hill from which it hewed its first wealth.

In the Proprietary's ornate offices in the Equitable Building in Melbourne's Collins Street it was the habit of Bowes Kelly and Jamieson and some of the older directors to play two-up with sovereigns after the board meetings. Their love of speculation was easy to understand. All had chanced to buy shares cheaply in a magnificent mine, and that mine at times had been the largest silver-lead mine in the world. All had plunged their money into the despised Mount Lyell mine and it had become the greatest copper company in Australia. And then, when unknown to them their Proprietary mine was failing, the most valuable deposit of iron in Australia fell into their hands.

There was one rich prize that these brave speculators missed, and that prize was in their own Broken Hill. They courted the rich sisters in the middle part of the lode and spurned the cinderellas at the distant ends of the lode. In their own lifetime they would have become wealthier if they had followed the lode to north and south instead of moving into steel.

The long boomerang of a lode dipped deeper into the earth to the south and north of the Proprietary leases, and the companies at the remote ends had to sink deep before they found massive ore. They lacked oxidized ore—the surface gilt of the massive lode. Without easy profits they could not explore their leases quickly. Thus in some years their shares could be bought for pence at auctions.

Broken Hill South was one of these weak companies. It tried five general managers in five years. It tried the patience of shareholders for twelve years before it paid a maiden dividend. The mine closed and opened like a valve. The company's shrewdest venture was to sell one lease for £25,000, then buy it back again at public auction for a mere £700. It was handy for damming rainwater, they said. Later the Zinc Corporation acquired the lease and found it so handy for mining metals that it became a great mine. That still left the Broken Hill South Silver Mining Company with nearly 200 acres of ground. In 1905 its dividends began to soar, and by 1962 the grand total of dividends from the once-despised mine exceeded £25,000,000.

Pessimism also ruled the north end of the field. The huge lode there plunged deep in the same unpredictable manner and for a time the North Broken Hill Silver Mining Company thought it had lost the lode. The company ate the icing on the lode, then became hungry. Five managers failed to make the mine pay. Wilson and Jamieson and the B.H.P. magnates moved on to the board then slid off. The company went into voluntary liquidation in 1895 and the mine was sold by public auction for £1,750 to Halliburton Sheppard, one of the innumerable bank clerks who had turned to sharebroking in Melbourne. A new company was floated, and fourteen years after the mine was first developed it paid a dividend. Then the mine closed again for three years because prices of silver and lead were low and mill losses were high. The north end of the field even in 1904 was so despised that the North Broken Hill was able to buy the adjacent Victoria Cross mine for £9,400 in shares. That lease was to prove as valuable as a mint.

At this time a band of speculators quietly visited the field. Often acting independently and sometimes together, often the closest of friends but not always so, they formed a loose alliance that was to have deep effect on the world mining industry. If they had a leader it was William L. Baillieu. Handsome, ascetic, assured but not dogmatic, he had a strain of magnetism that won loyalty. Son of a Victorian publican he was a young bank clerk at the start of the 1880s, Australia's most successful auctioneer at the age of twenty-nine, and virtually bankrupt at thirty-four. He sorted out the confusion, and early in this century ventured into mining and finance. He backed the inventor de Bavay in his flotation experiments in Melbourne and bought large dumps of zinc tailings at Broken Hill as grist for de Bavay's mill. He raised a loan for North Broken Hill in 1905 and became an influential director just when it was beginning its long run of dividends. At the other end of the field he was in the syndicate which cheaply bought half the shares in the mine that eventually became the Zinc Corporation's mine. Thus in two years he had pinned his faith in the neglected mines at the further ends of the lode and in two companies that were testing the flotation process. All his geese were swans.

Another speculator who came by special train to Broken Hill with Baillieu was Lionel G. Robinson. Son of the financial editor of the Melbourne *Age* in the days when that title signified mining editor, Lionel was first a humble clerk for the government in Melbourne before he followed his father into the share halls. He became a sharebroker in the 1890s when brokers were mobile. He moved from Melbourne to Adelaide and then to London, a specialist in Western Australian gold shares who could read investors' minds so easily that in his mid-thirties he led one of the most vigorous sectors of the London exchange. On the decline of gold shares Lionel Robinson turned to base metals and was a promoter of the Zinc Corporation and partner with Baillieu in raising capital for North Broken Hill and the Broken Hill South Blocks. He had the same courage and vision as Baillieu, the same scent for a good mining risk, but living in London and rarely visiting Australia his influence and name are now virtually forgotten.

Robinson was one of several strong speculators who moved on Broken Hill about 1905. He was joined by his partner, William Clark, a London stockbroker who had once been on Adelaide's exchange. He was joined too by Francis Govett and Herbert C. Hoover, London directors of rich Kalgoorlie companies. And in Melbourne W. L. Baillieu was joined in many of his plunges by such men as Montague Cohen, Alex Campbell, H. J. Daly and John Wharton. This small group, loosely organized, gained wealth from Broken Hill and seats on the boards of its rising companies. For long they seemed interlopers to the famous men who controlled the rich mines in the centre of the lode. No one envisaged that the intruders would eventually control the field.

Baillieu and the Australian investors were strongest at the north end of the Broken Hill field. Robinson and Govett and Hoover and the Londoners were strong at the south end. Their belief in the mines at the far ends of the lode and in the flotation process was sound, and they made much money.

Late in 1905 they turned to another problem. They owned no smelting works, so they had to sell their lead concentrate to German or Australian smelters at low prices. Why not buy a smelter to produce bullion from the concentrate of the friendly mines? Large smelting works at Dapto, fifty-six miles south of Sydney, were idle, so they formed the Australian Smelting Corporation to buy the idle works and rebuild them at Port Kembla. However, metal prices slumped and they did not complete their smelters, and lost dearly.

The Melbourne-London alliance did not dissolve, though many members had sore heads. They turned from lead to copper. They splashed money into the Great Fitzroy mine and the brilliant Mount Morgan in central Queensland and unless they recouped their outlay by share dealing they must have lost. They invested boldly in two good copper mines at Hampden and Mount Elliott in the hot hills of western Queensland, but lost many of their profits in their Laloki copper mine in Papua.

The members of the alliance did not own a majority of the shares in these companies, they did not always make policy, but when they did they were creative and efficient. Their power seems to have depended as much on their initiative as on their wealth, and when World War I began in 1914 their initiative pushed them into visible leadership of Australian mining and made their Melbourne headquarters, known as Collins House, the Australian synonym for industrial power.

There had been anxious faces in the corridors of Collins House when the newsboys shouted war in 1914. Three tenants, North Broken Hill, Broken Hill South, and Zinc Corporation, had become three of the big six companies at Broken Hill but they sold most of their lead concentrate to German smelters. They were now cut off from the final stage of the treatment process. Australia's three lead smelters at Port Pirie (S.A.), Zeehan (Tas.) and Cockle Creek (N.S.W.) were capable of smelting into metallic lead only half the output from the mines. The British Empire was weak in metallurgical works, and even if ships had been available the British lead smelters would not have been large enough to smelt all the concentrates which Australia was accumulating, for Australia was the largest lead exporter in the world.

The men of Collins House tried to buy a share of the Proprietary's lead smelters at Port Pirie before the war, and in 1915 in the pressure of crisis they succeeded. A decade later control was converted to complete ownership. Three Broken Hill companies—North, South, and

Zinc Corporation—poured money and skill into the smelters. Known as the Broken Hill Associated Smelters they became the world's largest, producing one-tenth of the world's lead and much of its silver. They devised a new blast furnace and probed new processes, and at Port Pirie in 1932 Dr G. K. Williams created the world's first continuous process for separating silver, arsenic, antimony, gold and zinc from lead.

For Australian mines the smelting of zinc was even more difficult than lead in 1914, and was not solved until long after the war. The zinc produced in Broken Hill's flotation works was in the form of a moist powdery concentrate containing about 40 to 50 per cent zinc as well as some lead, silver, sulphur, and barren particles of rock. Before World War I this zinc concentrate went by train to Port Pirie and by steamer to zinc smelters in Belgium and Germany. Each ton of zinc concentrate usually contained at least £12 worth of recoverable metals but for each ton the foreign zinc companies paid an average of only £3. To 1914 the Broken Hill mines earned about £10 million for the sale of zinc concentrates that were worth over £40 million when finally treated abroad. This dependence on Europe was costly in peace, costlier in war. The bluish-white zinc, being the enemy of corrosion, was essential for munitions. Australia in 1914 was the world's second largest producer of zinc concentrate but local and British plants could turn only a fraction of this zinc into pure metal.

The production of metallic zinc was costly and delicate. It needed workmen who would sweat and labour for low wages, it needed cheap coal and high skill in supervision. Belgium and Germany had these advantages and bought zinc ores and concentrates from all over the world and roasted and distilled them into metallic zinc. The Broken Hill Proprietary at Port Pirie was Australia's only producer of metallic zinc in 1914, but its roasting furnaces and long ovens of clay cylinders were capable of treating only a fraction of Broken Hill's output of zinc concentrates. At the end of that year huge dumps of zinc concentrate were piled at the mines, virtually unsaleable because of scarcity of ships and scarcity of neutral or Allied zinc smelters.

Australia's prime minister, W. M. Hughes, eventually persuaded Britain to buy in 1917 the large windblown dumps of zinc concentrate at Broken Hill and the field's future output until in effect 1930. Meantime, in the absence of buyers, there was strong incentive for the main Australian zinc producers to build large works to produce metallic zinc. Moreover, the imminent success of the new electrolytic process at the Anaconda Copper Mining Company's zinc works in U.S.A. in 1914 suggested that Australia could copy the process and produce metallic zinc from local ores. The process was cleaner, easier for workmen, and required fewer of them since it involved no stoking of thousands of small clay cylinders filled with zinc concentrate.

Herbert Gepp was manager of Amalgamated Zinc, which made high profits at Broken Hill by floating zinc the de Bavay way. He visited the Anaconda plant and was cautious but impressed with the prospects of the new process. Robert Carl Sticht of Mount Lyell, which now owned the deceptive Mount Read and Rosebery zinc mines in the mountains of west Tasmania, visited Great Falls too and was delighted when it efficiently made metallic zinc from a trial sample of his mines' ores. He engaged an American metallurgist named Crutcher who produced metallic zinc in a small laboratory at Queenstown, Tasmania, before his company sold its Rosebery zinc mines to the Collins House group.

The rising companies that straddled the far ends of the Broken Hill lode—North, South, and Zinc Corporation—had united to buy control of the Port Pirie lead smelters and they united again in 1916 to erect zinc works in Tasmania. In the Tasmanian venture, however, the leadership and the largest slice of the capital came from Amalgamated Zinc (de Bavay's), which had a huge flotation mill but no mine at Broken Hill. Its chairman, Baillieu, went to Hobart, got a mile of waterfront at Risdon for the works, and bought hydro-electric power from the Tasmanian government. The Collins House group formed the Electrolytic Zinc Company of Australasia and financed the long and baffling experiments that have to precede the mating of a new ore to a new metallurgical process. At war's end they produced a few bars of metallic zinc.

Much money was necessary for research and plant. A public issue of £1,100,000 preference shares virtually failed in 1920 and a debenture issue in 1922 nearly failed, but W. L. Baillieu and his friends rescued the company. The cobalt impurities in the zinc concentrates created serious trouble until a young tobacco-chewing American, Royale Hillman Stevens, rescued the company. The first unit of the works opened in 1921, and produced ingots of zinc on a commercial scale. Today the Electrolytic Zinc Company has the third largest zinc works in the world and uses one-third of all the electricity generated in Tasmania.

The loose 'Collins House' alliance of companies had risen with the success of the flotation process and the farther mines of the Broken Hill lode. Their huge lead smelters in South Australia and zinc works in Tasmania strengthened their alliance. For a time they became far more important than Broken Hill Proprietary in both the Australian and the world economy. Their output of cheap refined metals and sulphuric acid gave new Australian industries cheap raw materials and so stimulated their growth. They increased the value of Australia's exports. They gave more security and profit to the Broken Hill mines.

The alliance had bold and skilled leaders: W. L. Baillieu, his son Clive Baillieu, W. S. Robinson, Francis Govett, F. C. Howard and Sir Walter Massy Greene. On the technical side were gifted men such as Sir Colin Fraser, Sir Herbert Gepp, Sir Alexander Stewart, Royale Hillman Stevens, W. E. Wainwright, O. H. Woodward, Harry Somerset, and many more. This formidable row of talent guided the alliance into new industries. They manufactured white and red lead and BALM paints in Sydney. At Port Kembla they bought the copper refinery and smelters from the Mount Morgan company, the largest copper refinery in Australia. They bought control of Metal Manufacturers and Austral Bronze and eventually made brass and copper sheets and rods, lead and plastic and rubber cables for power and telephone lines. They provided a third of the share capital for the Commonwealth Aircraft Corporation which employed upwards of eight thousand people in World War II in the making of aircraft and engines. They provided half the capital for the great pulp and paper mills at the Tasmanian port of Burnie which made the first fine papers in Australia. They dipped into light and heavy engineering, fertilizers, gold mining, oil searching, timber milling.

The source of their wealth was the long lode at Broken Hill. By the 1920s the North and South mines at Broken Hill had become the largest and most profitable in Australia, and the Zinc Corporation was not far behind. Their profits financed much of the expansion into other industries, and their silver and lead and zinc fed the metallurgical works at Risdon and Port Pirie. These mines, the pygmies at the start of the century, became the giants. And from 1939 they controlled the entire field.

Collins House was the home of the most influential group of financiers in Australia between the wars, and that period was its hey-day. Never a tight alliance, it became looser from the late 1930s. The Zinc Corporation slowly began to move away. It sold its shares in the zinc works at Risdon and so severed one bond with its old partners. Whereas the North and South companies remained Melbourne companies the Zinc Corporation increased its contacts with London, merging in 1949 with Imperial Smelting Corporation under the name of Consolidated Zinc, which in turn merged with Rio Tinto in 1962 under the name of Conzinc Riotinto. The new company was one of the strongest mining and smelting groups in the world with mines and plants in America, Europe and Australia. Its office was no longer

Collins House, but Melbourne's tallest skyscraper; and, curiously, its competitor in the huge new Australian aluminium industry is the reformed Collins House group.

The alliance, while it lasted, was an impressive generator of industrial growth, but it had one grave weakness. It neglected the risks of mining, concentrating instead on safer secondary industry. In the 1920s it controlled mining and metallurgical companies that produced about half of Australia's metallic wealth but it reinvested only a pittance of their profits in the search for new mines. It had the chance to take the new field at Mount Isa, but rejected it. Had it used the same argument at the start of the century in Broken Hill it would never have succeeded there, and accordingly would not have become one of the world's great metal groups.

24

DESERTED TOWNS

THE GREAT REVIVAL left its trail of optimism years after mining ceased growing. The alchemists who taught minerals how to float passed on their formula to struggling companies. Silver and lead had never been so glamorous, zinc lost its terrors, copper furnaces blazed and roared from the Gulf of Carpentaria to the Great Australian Bight. The Commonwealth statistician carefully forecast in 1910 that the future of mining would probably be even more remarkable than the past. It was, but not in the way he intended.

In the saturated heat of Queensland, in the north-western ranges, most people in the Cloncurry copper belt counted 1918 a wonderful year. Erle Huntley drove his noisy automobile through Kuridala on pay night at his mine to show visitors the prodigal crowds in the main street of the largest town in half a continent. William Corbould sat in his bungalow at Mount Elliott and heard through open windows the din of blowers at his smelters and the laughter of miners walking from the afternoon shift and the rattling ricochet from the marshalling yards. At Cloncurry, Ballara and Mount Cuthbert, storekeepers wrote away for record lots of Christmas toys. Seven thousand people lived in the long copper belt and mined £1,500,000 in 1918. It was the largest copperfield in Australia.

Much of the copper they had smelted was buried in the mud of battlefields in France and Flanders, and the end of the war exploded the copper boom. The field's four smelters were idle two years later. The purest copper slid from £136 a ton in 1917 to £101 in 1919 and £75 in 1921 and did not recover. The locomotive that had once hauled the long copper trains six hundred miles to Townsville now carried away the people, the luggage vans jammed with crudely packed possessions. Buyers of scrap metal bid for the beautiful machinery in the smelting works. Wind banged the open doors of hundreds of deserted cottages, sheets of loose iron creaked on the roof. The goats multiplied, sheltered from the heat in their masters' houses, foraged in the deserted market gardens, met in the main streets, and scattered from the railway line when the trains came to collect machinery and iron worthy of salvage. In the fast planes that now fly from Sydney to Darwin passengers peering down on the bleached red hills sometimes see the iron chimneys of the Kuridala smelters and wonder what they are.

Far to the south-east, on steep hills near the Queensland coast, the telegram boys walked along the cool verandah of the office of Mount Morgan Gold Mining Company and handed in messages of falling copper prices. The company now depended more on copper than gold, so it tried to cut wages; the men protested. For a generation of nights the scattered lights of the works had shone on the steep hills like stars, and now the sky was black. For long, at even intervals in the night, the smelter slag had been poured down a hill, a yellow splash, a sudden waterfall of molten orange that changed slowly to red and faded as if frozen in its fall. No slag was poured while the company and government and men argued on who should make those sacrifices without which the mine was doomed. After a year the works re-opened and earned no profits. The company blamed the men and they blamed the company. The mine caught fire in 1925 and all argued while Mount Morgan burned. An American engineer derided the company's bold plan to lower costs by working the deposit in a large open cut. When at last in 1927 the company called for the liquidators it had produced more gold than any other Australian mine; it had sold £11 million of copper and £22 million of gold, and that was more revenue than any except the Broken Hill Proprietary had earned. It had paid a total of nearly £11 million in dividends, and that was over 500,000 per cent on the original outlay. These statistics did not add up for the women of Mount Morgan who remained in the houses on the grassy hills while husbands went to distant port or sugar mills and children went south for jobs which the town could not give.

At the busy town of Cobar far west of Sydney, copper mines were never busier than during World War I. Hundreds of men went to the war and in trenches that had the dank smell of familiar mines they read letters assuring them that jobs awaited their return. Early in 1919 they were returning to the town, swallowed on the railway platform by brass band and bunting. Some arrived that week in March 1919 when the big mine, the Great Cobar, finally closed. Out at the C.S.A. (short for Cornishman, Scotchman and Australian) the smelters still treated copper from the local mine and from mines far across the plains until the Saturday afternoon in March 1920, at hotel-closing hour, when smoke was seen rising from a mine shaft. Underground a forest of oregon timbering was ablaze. Men sealed all entrances to the mine to smother the fire but the fire burned. The mine was not reopened and the smelters closed, and neighbouring mines had nowhere to smelt their ore. The discovery of copper half a century before had quickened the prospecting that had led to the finding of Broken Hill, and now Cobar's smelters were dismantled and sent as scrap iron to feed the furnaces of the Broken Hill Proprietary at Newcastle. Half the houses and rambling two-storeyed hotels of Cobar were worth even less than scrap iron.

In South Australia, fabled Moonta and Wallaroo hardly celebrated their sixtieth year. They had united in the copper slump of the 1880s and had branched out, making safety fuse for miners and bluestone for orchard sprays, building lavish custom smelters for lead ores from Broken Hill and gold ores from Kalgoorlie, making sulphuric acid and fertilizer for the farmers, and still employing as many as 2,700 men in mining and milling copper; a bold and enterprising giant, unable to get away from dependence on copper. Once the peace was signed in Europe and copper fell, nothing could save its mines.

H. Lipson Hancock, Cap'n Hancock the Second, began to dismiss the men at successive pay days. On Sundays, calling for order with a swing of his handbell in the vast Sunday school at the mines, he found order easy to maintain as roll-call fell from 800 to 600 to 400. Oswald Pryor drew his celebrated cartoons for the Sydney *Bulletin* and sharp readers saw that the Moonta chapels which he often sketched had empty pews. Old farmers driving buggies in the dark across the plains saw no light shining from the tall stone engine houses. In town halls and church halls there were ceaseless farewells to old families, much singing of 'God be with you till we meet again'. South Australia produced in 1922 less copper than in any of the previous seventy-eight years.

One large copperfield remained. Robert Carl Sticht sat at night in his book-walled study on the lonely spur at Mount Lyell and worried. He was sick with cancer, worried by his own heavy losses in a copper mine at Balfour, worried by the money Mount Lyell was losing, worried that his famous pyritic smelting might vanish from its last stronghold. A familiar world was fading. When he was driven in his Daimler rail car along 130 miles of railway to Burnie to board the steamer for Melbourne and talks with his directors, he saw all along the railway the old mining towns decaying, tracks into the forest overgrown, black rivers now untainted by coloured sludge from mines. He died in 1922, a gracious scholar, a name 'familiar throughout the world', wrote one New York journal in panegyric.

They saved his Mount Lyell. Russell Murray floated the copper ore instead of tipping it in the furnace for pyritic smelting; he closed the old Mount Lyell open cut, relied solely on the underground North Lyell mine, and spread frugal efficiency. A portfolio of safe investments backed the flimsy mining profits of the 1920s and financed an electrolytic refinery and the most ambitious tunnel any Australian mine had driven. The company's foresight before World War I in harnessing hydro-electricity and storing profits and treating its men as men gave it strength to survive. More than any copper mine it deserved to survive.

Salesmen of explosives and machinery who regularly visited the mining towns hoped in vain that other metals would atone for the collapse of the copper mines. Charters Towers, Gympie and Croydon had slumped as goldfields before World War I and continued to slump. Queensland in 1930 mined a bare £30,000 of gold and only three Chinese carried the long-handled shovel where once were thousands. Queensland engineers who had entered a booming industry when they were twenty saw it die before they were forty. In New South Wales a maze of small mines and some big mines closed: no footprints in the morning frost in New England tin towns, no whistle to wake the miners of Mount Boppy and Mount Hope, no hot steam rising from change-houses at the Junction and the British and the older mines at Broken Hill. In Victoria the Rose, Thistle and Shamrock mine at Harrietville was the only gold mine to pay a dividend in 1930. In Tasmania the snow lay unswept on the paths of empty houses at Mount Bischoff, the iron roofs of Zeehan were brown with rust, and the rich gold mine at Beaconsfield was flooded. Western Australia now mined eight of every nine ounces of gold produced in Australia, but even the Golden Mile was sick and Richard Hamilton's annual report at the Chamber of Mines was culled from Lamentations.

Many mines received a London cable of one word, 'Pinpoints', and the manager turned to page 574 of his black Bedford McNeill book of mining code, and read the translation: 'You had better shut down the mine'. The blacksmith left his bellows and anvils and tongs. The storeman issued his last candles to miners who went underground to retrieve picks and pinchbars and shovels and flat sheets of iron and rock drills and ladders. And the manager sat in the office and wrote his inventory of things to be auctioned: one safety cage with eccentric grips, two sinking kibbles of 60 gallons each, '1 only monkey wrench, quantity of bolts, iron, etc.', two jacksaws and a handsaw, gelignite, fuses, firewood, wire rope, grindstone, and so on through shackles and crab winch, until he came to the set of poppet legs and signed away the life of the mine. These last wills and testaments jammed the auction notices of the metropolitan newspapers in some months of the 1920s.

The stunning of the copperfields was sharp and unheralded; their directors accused Versailles and cheap aluminium, also the Industrial Workers of the World and the flotation process that the big American mines adapted more quickly than the smaller Australian mines. But some tin and lead fields had fallen when the price of those metals was exceptionally high, and the gold industry, which usually wanes in times of national prosperity, had not merely waned but crumpled. Gold mines employed 75,000 men in 1900 and barely 6,000 in 1929. They mined 24 per cent of the world's gold in 1903 and 2 per cent in 1929. Gold had vied with wool as the strongest export at the turn of the century, and now it was passed in turn by wheat, flour, meat, butter, and even hides and skins. In most years of the 1920s Australia's metals earned barely one-third of the sum they had earned in the triennium 1906-8.

Most mining fields had three stages of life: many small rich mines, a few large producers, and then many little mines in the decaying years. The tragedy of the third stage was that the little mines were worked by gougers and tributers who lived on courage rather than ore. At Charters Towers 207 goldminers worked at the end of the 1930 but for the whole year they averaged less than £30 of gold. Zeehan had hundreds of men tearing out tiny veins of silver and earning just enough for bread and beef. Everywhere old miners who had lost jobs when large mines closed now scavenged for gold under rusting stamp mills, re-treated old tailings dumps, sank narrow shafts in search of mysterious reefs that folklore said had eluded earlier miners. In mountains or plains in the 1920s travellers came across small parties camped in vast millhouses or deserted stone hotels, unwatering old shafts with vibrating oil pumps or excavating long water races to treat a patch of tin or gold that persistent burrowing had found. A geologist inspecting an old shaft that had shut in 1914 would find yellow newspapers dated 1919 or 1920 or 1923, the calendars of men who came in buggies and on pushbikes and camped a few weeks while they searched for machinery or ore to salvage.

Visitors wondered how old mining towns survived until they sensed the excitement at the post office on pension day. Many mining regions retained their member of parliament long after their electorate should have been abolished, and the member invoked the state to keep the town alive. The train time-table was often unchanged for years, giving work to station staff and engine-drivers and fettlers though the railway had few passengers. The post office employed four men when it had business for two, and one small Victorian post office remained after the town had vanished because the postmaster used to collect at the Masonic Lodge in a distant town all letters which members wished to post. The school wanted painting, the roads needed repairs, and so the state paid and money fitfully flowed. State governments themselves took over the Yelta mine in South Australia, the Chillagoe mines and smelters in Queensland, and a mine at Zeehan in Tasmania. The state sent geologists and drilling teams to worked-out fields to search for ore, lent or gave money to prospecting parties and small mines, crushed gold ores for gougers in State Batteries, subsidized the deepening of shafts in vain effort to prolong the life of big companies that were about to die.

The mining industry, compared to other industries, has been highly taxed in good times but has got comparatively little aid from governments when that aid is essential. The state steps in, not when a new field is born and urgently wants railways and schools and telegraph and water, but when the field has faded away beyond even the power of governments to resurrect it. Perhaps that is because mining regions usually have more political pull when they are old and declining than when they are new and vital to the country.

Most mining towns vanished from the map. Some like Cloncurry and Cobar survived as pastoral towns, Charters Towers had boarding schools and old men's homes. Daylesford had mineral springs and cheap houses, and purified its shabbiness with spa. Ballarat, Bendigo and many Victorian gold towns had become commercial capitals before gold gave out. Derby in Tasmania had rich red soil and Moonta stood in the wheatlands. The old gold towns of Creswick and Maldon and Beaconsfield had an asset in the cheapest cost of living in Australia during the depression. Even those fortunate towns suffered with the collapse of the mines.

More than a hundred thousand people were forced to leave the waning mining towns, abandoning neat gardens or dearly tended pepper trees, selling houses and furniture for absurd prices, leaving the churches and public halls they had built. Thousands in North Queensland went to the sugar farms and small ports, and thousands left the Western Australian goldfields for the wheat belt but most went to capital cities and wore their sentiment on a brassplate on the front verandah of a suburban house—Walhalla or Cobar or Kanowna —and met in reunion once a year in the summer shade of parks and botanical gardens for the swapping of sepia photos and thick sandwiches.

Often the old expatriates of the mining fields sat on the park benches, their tieless shirts buttoned to the neck, sleeves rolled up, arguing why their field had collapsed. If only the company had bought new pumps, if they'd extended that drive at the 1,600 foot level another seventy feet, if they'd sacked that mean manager. Newspaper compositors set type for millions of words of post mortem. Some blamed the young who preferred cities and motor bikes to prospecting. The young prospectors said that few new deposits were left to be found and there were no promoters to buy their leases. The promoters, those who had not vanished into real estate and manufacturing and finance, said the investor had lost his gambling sense . . . The investors said government royalties and taxes and the extortionate demands of the miners had killed the goose . . . miners said the companies were inefficient, would not . . . These complaints had sufficient truth to spin the merry-go-round of blame.

One clear cause of the decline of mining was the price of metals. Comparing metal prices in the decade before the Kaiser's War and the decade after the war, the prices of tin and lead and zinc were higher, and copper and gold the same. As wages and all costs had risen, gold and copper therefore suffered severely. The river of mining capital dried, and that was both cause and effect of the decline. As more and more mines closed and dividends became smaller, many investors became weary of losses and turned from the financial to the sporting page of their newspaper. Mining shares ceased to rise and fall with bewildering speed, and so mining was no longer an effective gambling system. Governments prowling for easy revenue encouraged public lotteries that fed like parasites on a gambling instinct that had once flowed to a socially useful form of gambling. Workingmen and shopkeepers in the mining towns found two-up and horse-racing and lotteries safer and more exciting. There was no long wait for the result of their gamble. The dice could not be over loaded against the investor. When the small Queenstown stock exchange in 1901 decided to close and invest its few remaining pounds in Tattersall's Melbourne Cup Sweep and their ticket won £13,000 that was an omen and lesson.

Mining investment is usually heaviest when other investment outlets are weak. For most of this period the Australian investor had ample scope for investment, and for the first time industrial shares superseded mining shares as leaders on the stock exchanges. The growth of factories not only drew away capital that might have gone into mining but heavily burdened the cost of mining. Australian factories grew behind the tariff wall which so increased prices and wages in Australia that a committee of economists estimated in 1929 that the tariff added about 10 per cent to the cost of production of the mines and farms and sheep runs that sold most of their products on the overseas markets.

This burden of costs turned many profitable farms into struggling farms, but the farmer could meet the challenge by working harder and living more frugally and borrowing from banks. For many mines this was not a solution; whereas the farmer worked harder without additional pay because he worked for himself, the miner on wages could not be expected to work harder without more pay. Undoubtedly many Australian gold mines would have survived the 1920s but for Australia's policy of protecting urban industry at the expense of primary industry.

Britain lost interest in overseas mines even before it ceased to be the world's great financier. When George Meudell went to London in 1908 to float a company to mine brown coal in Victoria he found disillusionment with mining. London experts who claimed that British investors had lost £20 million in Australian and £50 million in African mines may have been close to truth. Moreover, even fewer good mining properties were being offered to London investors. Britain ceased to hold the majority of shares in the main Broken Hill companies and Mount Lyell and Mount Morgan, selling them back to individual Australian investors between 1910 and 1930. Thus thousands of Australians who might otherwise have bought shares in new Australian mining companies invested instead in old. And the boards of the old and rich mines increasingly spent their profits not in seeking or equipping new mines but in manufacturing. In twenty years from 1908 the Broken Hill companies earned such huge profits from mining and smelting that they could pay about £22 million in dividends, add heavily to reserve funds, and finance some of the largest secondary industries in Australia. From that bulging sum they spent a pittance in the risky search for new mines.

Australians were dazzled by distant fields. Malaya lured young engineers, promoters, and investors. Back in 1892 Brisbane had floated the Raub Australian Gold Mining Company to mine gold at Pahang in the Malay Peninsula and Raub proved a wonderful mine. Into the jungle trekked more Australians to seek new Raubs or to manage little mines for Chinese owners; my grandfather was one of many who left Bendigo to manage Malayan mines. Captain E. T. Miles, Tasmanian politician, promoter and shipowner, got valuable tin leases in Siam and floated in 1906 the Tongkah Harbor Tin Dredging Company in Hobart, and its profitable fleet of dredges enticed more Australian companies to the Orient. The Prattens of Sydney and the Miles of Tasmania, and Sir Walter Massy Greene and J. Malcolm Newman became Malayan magnates in the tin boom of the mid-1920s. Australia had once been, as Malaya now was, the world's great tin producer, but in the 1920s Australian investors rushed Malayan tin shares and ignored their own tin lands.

Australians turned to New Guinea gold, and in 1926 the old Adelaide drums were beating again, summoning the most speculative city in the land to invest in gold deposits high in the mountains of New Guinea. Australian money also went a few years later into the rich Emperor and Loloma gold reefs in Fiji. Most of these adventures in the balmy tropics were highly profitable, but investors waited in vain for the discovery of rich surface outcrops on Australian soil.

It seemed that the sharp-eyed prospectors of the previous century had found almost every lode that rose above the ground and some that didn't. Only a few new deposits were developed in the years 1910 to 1930. A few gold reefs were opened in the West, and some of them led to the foolish Bullfinch-Chaffinch boom. On the flank of Ben Lomond in north-eastern Tasmania the Aberfoyle deposit, rich in tin and wolfram, was opened in 1926, only two miles from a mine that had been worked for years. Tasmania also had its brief osmiridium rushes in the 1920s. On the sandy beach at King Island, between Victoria and Tasmania, that same Tom Farrell who had found the rich Farrell silver-lead field found in 1904 the mineral scheelite, and when the demand for tungsten rose in World War I a small Melbourne company, King Island Scheelite, opened glory holes in the scheelite by the sea and after paying a few dividends fell asleep until World War II. At Mount Isa in north-west Queensland silver and lead were found in 1923 but that field, like the new Tasmanian discoveries, waited long before it was profitable to investors. So the new fields gave no incentive to mining investors who had been suckled for the last eighty years by a run of rich discoveries around the continent.

Mining needed a new type of investor and a new type of prospector. As Australia had found most of the outcrops of common metals it had to search for those deposits that were buried beneath a thin layer of soil or clay, or were deceptively poor at the surface and rich below, or slender on the surface and massive at depth. Many such deposits had already been found. The blasting of the roadmakers had found the rich North Lyell copper mine and the burrowing of the wombats had found Wallaroo, and prospectors blindly driving tunnels into barren-looking hillsides had chanced across rich reefs. But most of these chance discoveries had occurred in mining booms, when a colossal sum of money and an army of men were engaged in the search. Money was even more essential if similar hidden deposits were to be found in the 1920s.

A new way of prospecting was also essential; the search had to be scientific. It seems strange that an industry that had gained so much from applying science to its problems should so long refuse to apply science to its most crucial problem, the finding of new lodes. There was a crust of prejudice against geologists. It could be traced to the famous prophecies of Professor McCoy of Melbourne and Sir Roderick Murchison of London who predicted in the 1850s that gold reefs would cease to pay a few hundred feet below the surface. Their mistake was derided throughout Australia for two generations. But it was forgotten that other geologists such as Selwyn of Victoria had not endorsed Murchison's theory. It was conveniently forgotten that mine managers and metallurgists had made immeasurably more mistakes, and more costly mistakes, than the geologists. The prejudice against the geologists therefore went deeper. As nearly all geologists were employed by governments, they were suspected by many mine owners. They were condemned as impractical and in part they were, for they were absorbed in fundamental problems as well as practical problems; they were trying to understand how orebodies were created. The few geologists employed by mining companies in the nineteenth century spent most of their time reporting on leases that were about to be floated into companies; in making their reports they were invariably on new fields where few workings had uncovered essential evidence below ground level, and thus they made many mistakes. Only the optimistic geologist, likely to give a favourable report, was engaged by the mining promoters, and if his report was lukewarm it was not published. Not surprisingly, investors came to scorn geologists.

The geologist did not deserve all the blame. Predicting whether ore would be found in a certain place he had an unenviable task, because much of the evidence he needed was buried. His profession was necessarily inexact and imaginative, and so was often scorned by mining engineers and metallurgists and assayers. Nevertheless, the fact was inescapable that in the business of prediction the geologist was usually the most skilful man. But no Australian mining company employed a geologist permanently until the eve of the depression.

Mining had revived at the turn of the century from a complex array of interacting causes, and likewise decline seemed to gather its own momentum. Just as technical brilliance and optimism was a highlight and a vital cause of the revival, so technical sluggishness and pessimism fed the decline.

Kalgoorlie, once renowned for ingenuity, reflected this decline. Immersed in a litter of broken machinery and scrap iron, the old mills lost money and gold and self-respect. Most Kalgoorlie metallurgists did not think the flotation process was even worth testing in a laboratory, and not until 1930 did the Lake View flotation plant open a new era in metallurgy on the field. Kalgoorlie was a monastery, walled from the world of new ideas. An expert from South Africa, Kingsley Thomas, toured the mines as a royal commissioner in 1925 and deplored their inefficient steam engines and the hand trucks that carried the ore underground and the drays that carted it from shaft to mill. He saw seven companies each living in isolated cells within the golden walls: 'Seven managers, seven mine offices, seven London boards of directors, seven, in fact, of everything that goes to the working of a mine, except modern appliances and labour-saving machinery, and an appreciation of up-to-date mining practice.'

Dearer wages and higher costs of mining had cut profits; that was the first step in decline. It could have been met by increasing output and introducing the latest machinery, but once Kalgoorlie failed to meet this challenge the vicious circle of decline spun faster. As profits fell, the London companies refused to provide more capital. Without capital the mines could not try new methods and machines. Without new methods the profits continued to slide. Gold mining became an old man's industry and therefore conservative. The old men who had risen with the industry to its pinnacle had to fend off defeat in old age. What happened in Kalgoorlie happened on most goldfields in the 1920s.

25

END OF REDSHIRT CAPITALISM

ON SUNNY DAYS in 1953 Jimmy Elliott sat on the bench in a tiny park at Mount Lyell, stooped in his tweed overcoat, a wiry man of eighty-six with sunken eyes and long fingers. His stepfather had been a goldminer at Barry's Reef in Victoria and had coughed black spit and wasted of miner's disease in the 1870s. He said his stepbrother had gone to the Klondike rush and, climbing from a frozen shaft, had thawed legs and feet by the fire; gangrene set in and a leg was amputated. Jimmy himself had worked or sought mines from his twelfth year. In his middle age he was trapped underground for five days in the smoke-sealed North Lyell mine with a host of dead men. All his life he had a child-like faith in his luck but at the end he had few possessions. He had married the sister of Ironbark Jimmy, the pugilist, and had two daughters, but even they vanished from his world and he did not know whether they were alive. But like thousands of Australian workingmen of his age he remained infatuated with mining, for all its hazards and disappointments.

Thousands of workingmen had lived through the same changes and were not pleased. They started their working life with the freedom of the independent gold digger and closed it working for wages; they began mining with the hope that they would become rich and instead mined wealth for others to spend; they first mined with the open sky overhead and later mined in the darkness of a deep mine. The period of the late 1860s was probably the watershed between the era of the independent digger and the era of the wages miner, and from that time a majority of Australian metal miners worked for companies and wages.

The generation that changed from the romantic life of the digger to the discipline of a company mine did not overnight become militant. Most men at first may have preferred the new order, for undoubtedly they earned more money than they ever earned by digging. Married men usually preferred the regular weekly wage. Men who had paid for the privilege of sinking a barren shaft in a co-operative venture came to like a regular wage. Men could earn wages in a company mine and so finance their own mine or visit the latest rush. Many men alternated the seasons between mining and sheep-shearing and harvesting, and found a miner's wage the highest. Many owned an acre or two and cows and poultry and gladly worked intermittently in the mines. Many had been wages miners in Cornwall or Northumberland and knew no other life. Lads from Australian cities who wandered from home and worked three or four years on mining fields, and migrants fresh from Birmingham or Venice, could not share the lament about the new order of company mining for they did not know the old order. Above all, thousands of miners blended the benefits of the old and new orders; they retained the optimism of the old order by speculating in mining shares and they gained the regular wage of the new order.

It is true that many foremen in company mines were taskmasters. Richard Thomas, one of innumerable Cornish miners with Welsh names, was underground manager of Block 10 at Broken Hill and noticed some miners were loafing. 'Are you new men?' he asked. 'I think the sun has been shining on your heads all your life'—meaning that they were new to underground mines. He warned them to work. Some weeks later he approached their distant working place, heard whistling and singing, and sacked them. 'No man,' he told a Royal Commission at Broken Hill in 1897, 'can strike a drill into hard rock, and whistle and sing at the same time. It takes a man all his wind to fulfil his duty.' However, the nature of work in large mines was such that miners were comparatively free from close supervision. Working in isolated parties in scattered places, they were rarely visited by a foreman or shift boss more than once a day. Moreover, they worked no harder in a company mine than they would have worked in their own shallow hole or mine. Nevertheless, unless they were tribute-miners, they were on longer their own masters. In many minds that rankled.

In the diggings a Limerick man often worked with a Limerick man; Californians worked together, Durham men together. But in deep mines the manager chose the miner's workmates, and that made for discontent. Some managers avoided that problem by employing only Catholics or Primitive Methodists; many employed only men from certain towns in Cornwall and many refused to employ any Cornishmen. These prejudices and antipathies stand out in the answers that an old Broken Hill Proprietary foreman, William Harry, gave on oath to a Royal Commission in 1897.

In the open cut that he supervised, hand trucks were pushed along rails into a small tunnel where an overhead shoot and trap door released the ore mined above. The men thought such a shoot was dangerous, and so William Harry was questioned:

Q. Is it called all sorts of names?

A. Yes; 'Cousin Jack', 'Irishman', and such like, just according to what the navvy takes it into his head to call it. He generally names it after the man on whom he has the greatest set; and I suppose that is the reason why it has been named here the 'chinaman'.

Q. You want competent men, and you do not care what nationality or association they belong to?

A. No; I would not care if they were Chinamen; but I bar 'Cousin Jacks' sometimes.

Q. Very good; but would you ask them if they were 'Cousin Jacks'?

A. No; I know them by sight.

Mustering all his prejudices to resist the cross-examination he announced that the miners forty years ago did not need to be watched or supervised 'and there never was any accidents to speak of'. The man working for an early company in fact had too many accidents. He certainly was more prone to accident than the gold digger. Working near deeper shafts he could fall further; usually working in quartz mines he could be hit harder by falling rock than falling earth. He used explosives and worked in poor light and often breathed foul air. Going down shafts that had no safety cage he climbed vertical ladders for three hundred feet or more. The ladders were often fastened so close to the side of the shaft that the wooden rungs had no space for a firm foothold and a man held his life in the grip of his hands. 'It seems like the ticking of the everlasting clock, that noise your feet make, as you go down that horribly straight ladder, pasted up against a wall 300 feet high,' said a Victorian politician in 1873. In Victoria in each year of the 1860s about twenty men fell down shafts and were killed. Even miners who descended in swift cages were sometimes in danger. Engine drivers drowsed after twelve hours by the controls, or the engine clutch slipped, and the cage crashed to the bottom or soared above the mouth of the shaft, smashing the trapped men.

Most fatal accidents were underground. Dynamite or powder misfired, blowing men against the wall. A rock face that seemed safe suddenly crumbled. Heavy logs that propped the roof fell without warning. Men suffocated in the foul air in deep alluvial mines. Negligence or ignorance caused most deaths, reported a Victorian Royal Commission in 1863.

Few miners wore protective clothes, though some Cornishmen wore a hard resinous felt hat and iron-capped boots. Old miners became too accustomed to danger and were careless. 'You have almost to knock them down with a stick to keep them from the [powder] magazine with a pipe stuck in their jaw,' said one Broken Hill manager, and his comment caused scant surprise. Managers themselves were often careless, sometimes criminally irresponsible in their neglect of safe methods. The states for long did not bother to enforce precautions in mines or to inspect machinery. The one exception was the parliament of New South Wales which passed a lax law to regulate its few coal mines in 1862.

Angus Mackay of Bendigo was Minister of Mines in Victoria in 1873 and he piloted through parliament the Regulation of Mines Statute, one of the most advanced social laws in the English-speaking world. It enforced an eight-hour day for miners and engine-drivers. It prevented women and thirteen-year-old boys from working in mines. It regulated ladderways and appointed inspectors. And it copied the recent British law that made mine owners prove that accidents were not due to their neglect.

The Chief Inspector of Mines reported after one year that his criticism of dangerous mining practices was accepted by managers, owners, and men with a fairness 'that might have caused surprise if the general character of the mining population was not already known'. Even the Chinese alluvial miners who burrowed so hazardously read a Chinese translation of the Act on the doors of their joss houses. But ninety Victorian miners were killed in that first year of inspection. The quartz miners might have been more prudent if they had known on New Year's Day that one in every three hundred would be killed underground that year. Rashness was so widespread that the comment of the chief inspector was understandable: 'Strange to say there have been no deaths during the year from the bursting of boilers.' Although rigid inspection cut the accident rate, other colonies were slow to imitate Victoria's regulation of metal mining. And, as mining education was weak, no law could have prevented some disasters in deep gold mines.

The New Australasian Company at North Creswick worked a buried river of gold. As the custom was, it sank a vertical shaft through the basalt rock and the buried river and down a short way into the Silurian bedrock. From the bottom of the shaft men drove horizontal tunnels known as reef drives through the solid bedrock, and these tunnels drained the water from the overlying alluvium and gave the miners access. The company eventually exhausted all the golden gravels that could economically be reached from that central shaft. In 1880 it allowed the old workings to fill with water and sank a new shaft two thousand feet down the course of the buried river. The manager who directed the work was William Nicholas; his nephews who lived at the gold town of Majorca nearby were to create the international firm of Aspro but Nicholas himself, like so many good mine managers, was not good at sums. He made an error of fifty-five feet in measuring the surface distance between old shaft and new, and so he did not realize that the old flooded workings were close to the new drives his miners were extending. Nevertheless, he was observant and experienced, and all the signs decreed that his new workings were draining the water that seeped from the old workings. As the pumps in his shaft were powerful he was not alarmed.

At sunrise on 12 December 1882 two miners were extending a drive in the half light of candles. Suddenly they heard a roar of water and saw slurry pour from the roof of their narrow tunnel. They ran five hundred feet back to the shaft and warned Michael Carmody, who ran one-third of a mile along a tunnel in the opposite direction to warn other miners that a wall of water was rushing their way. He passed on the warning and fought his way back through the rising water towards the shaft that was the mine's sole exit. There were twenty-nine men at the remote end of the mine and only two succeeded in wading through the water to the shaft. The other men were trapped in a dead end of the mine. With the water rising above their waists in the tunnel they climbed the ladders that led to the main workings overhead, and hoped that the water would not reach them. The water followed them up and continued to rise until it reached their necks. They stood in water and darkness, lighting no candle for fear of polluting the small layer of air between their heads and the rock ceiling. Some took courage from the soft singing of hymns. Some hoped that the pumps and baling tanks, working with all speed in the distant shaft, would lower the water. The water ceased to rise the final foot to the overhanging roof, but the breathing of so many men in such a narrow space vitiated the air. Men lost awareness and slumped into the dark water. Their mates did not have the strength to drag them from the water and support them. Twenty-two men were drowned. Five survivors waited two days and two nights for the rescuers to reach them. This worst accident in the history of Australian gold mining impaled the imagination of the nation, and the hymns which the dying miners sang were sung again with vicarious emotion in gospel halls and sewing bees for months afterwards.

A miner's funeral tramping past shuttered shops or the idle miner with bandaged limb was too often seen on the Victorian goldfields. The families of these men got poor compensation. Some companies gave them a cottage or a purse of sovereigns or a job for the eldest son. Some widows got gratuities from the friendly societies that were strong in the gold towns, the Rechabite Ark of Safety Tent at Scotchman's Lead or the Manchester Unity Gold Miners' Pride at Bendigo. Some were helped by the Miners' Accident Societies and some were helped by a hat passed round at the shafthead of the mine. Miners had more social insurance than any other group of workingmen, and needed more. The state, of course, paid no compensation nor did it compel companies to compensate men who were injured through the neglect of managers or mine owners.

The gold mania of 1871 was the immediate spur to the creation of miners' unions. Many mining towns were prosperous and company dividends were high, so miners agitated for the right to work eight hours daily instead of the ten-hour shifts that were common in deep mines. Robert Clark, Bendigo quartz miner, unlettered, alert and sympathetic, formed the Bendigo Miners' Association in the summer of 1871-2. Similar associations were formed at Stawell and Clunes and the deeper fields. In 1872 the wages men of Bendigo and Stawell won the eight-hour day. At Clunes in 1874 they defeated the attempt of the Lothair company to employ Chinese miners underground. The unions influenced the Regulation of Mines Act that became law on the first day of 1874. And the climax was a meeting at Bendigo in June 1874 of leaders of twelve miners' unions representing 1,800 men and the creation of the Amalgamated Miners' Association of Victoria. Then came anticlimax, for the new A.M.A. weakened as goldmining became less profitable. Then came depression in 1878 and gold mines revived; the unions had more bargaining power and a deep alluvial miner from Creswick, William G. Spence, revived the A.M.A. In 1883 it had over seven thousand members and was Australia's largest trade union, and was moving up the continent. Its generous accident and funeral and sickness benefits attracted miners, and in its first eighteen years it paid members and their families £100,000 in accident and funeral fees and £6,600 in strike relief. But for all the success of the Amalgamated Miners' Association most metal miners belonged to no union as late as 1900.

The miners' union was watchdog rather than bulldog. John Sampson, a working miner of high skill and character, was first president of the A.M.A. branch at Creswick (the town known as the holy ground of Australian unionism) and his political principles were not unlike those of his grandson, Sir Robert Menzies, Australia's prime minister. The A.M.A. was rarely militant. It knew that men who risked their savings in mining shares deserved a high reward, and begrudged no company its high dividends. It did not demand excessive wages, knowing that they might close useful mines or reduce the dividend in successful mines. There was no sharp rift between shareholders and wages miners. Thousands of miners themselves were shareholders, for goldmining was the one industry in which the humble employee was in the best position to speculate. If a new reef was found, underground miners often sent secret messages to a stock-broker in the town and bought shares before they rose dramatically. As the price of gold was fixed, the working miner could not be out-witted by city speculators who understood prices. Thus share registers of Victorian companies were sprinkled with the names of miners. The classes of occupation that provided most insolvencies in Victoria in some years were publicans and miners. Some of the insolvent hotel keepers no doubt had been miners who made enough from shares to buy the pub.

Goldminers did not use their trade union as a battering ram against company mining, because they still hoped to make a fortune. Most underground men did not think they were doomed to be wage-earners for the term of their working life. Share speculating was not their only avenue of hope. On Ballarat and many declining fields hundreds worked as tributers in the Cornish manner, and rented gold mines from moribund companies and paid them as tribute a share of the gold they mined. Some made fortunes from rich patches of gold, and bought white villas in Bendigo or chocolate-soil farms near Ballarat. Others retired with gold they had stolen. The free gold in Victorian reefs and the loose gold in the deep alluvial mines was easy to gather, secreted in boots or hat or mouth, and taken home with faint chance of detection. A Cornish tributer was once asked by a Bendigo visitor to define a 'tribute' and he replied that he kept the gold he could carry and shared the rest with the mine owners. A mass of evidence points to the frequency of gold-stealing in company mines and critics even blamed the gold-stealers for the collapse of Bendigo; but this was one of the amenities that made goldminers more than wage-earners. The mildness of unionism reflected an industry where the humblest man was not yet divorced from excitement, hope, and profit.

While most goldminers still hoped to retire on a fortune, many lost hope of surviving until the age of retirement. The dust drifting from the rock drills underground carried a new hazard, more danger-ous than physical accidents because it was invisible. The machine miners' main job was to drill holes into the rock and fill the holes with explosives, and the drilling and breaking of the rockface and the shovelling of the broken ore created millions of particles of fine dust. Dust in many working places was so thick that a lighted candle was dimmed as if by fog.

The slow advent of the mechanical rock drill from the 1870s had intensified the fine dust in mines. Its drilling speed created far more dust than the old hand drills that were hammered into the rock with the slow blows of a miner's hammer. One American rock drill was aptly named 'the widow maker'. The increasing depth of mines made ventilation difficult, and often no current of air blew away the dust from dead ends where men worked. The practice of employing three shifts of miners, a practice not so common in the smaller mines of earlier years, gave the dust only Sunday in which to rest and settle.

In quartz mines the fine particles of silica slowly damaged men's lungs and bronchial tubes. Such men continued to work vigorously for years, then got a recurring cough and mild bronchitis. They had to spit, a complaint popularly called the black spit. In time they were short of breath and tight in the chest and reluctantly retired from the only living they knew; and doctors could not walk along the streets of Bendigo or East Ballarat without seeing men of fifty with stooped shoulders and walking stick. Their weak lungs were now vulnerable to tuberculosis. The tubercle germ spread rapidly in the quartz mines, for many men in the first stages of the miner's disease passed on the germs by spitting or by shouting close to their mate's face in order to be heard above the roar of the rock drill.

The tragedy of the disease was obscured for nearly a generation. Even after enquiries in Cornwall and the Rand isolated the causes and stressed the mortality of the disease, Australia was slow to fight it. Then in 1907 Dr Walter Summons discovered that ten Bendigo miners were dying from lung diseases for every miner killed in an accident. On the Western Australian goldfields the rising toll of tuberculosis and silicosis was blamed on miners who had caught the diseases in Victorian goldfields, but in 1910 Dr J. H. L. Cumpston estimated that, amongst machine miners who had worked only Western Australian gold mines, one in four had signs of silicosis of the lungs. How many lives had been cut short by the fine dust of Australian mines cannot be counted, but the number must have exceeded ten thousand. Ironically once the menace was seen it could easily be curbed by playing water on the dust and by ventilating the deeper workings. However, many miners ignored their water jet and many companies owning deep shafts on narrow leases recoiled at the expense of providing ventilation.

The black spit was most common in the gold towns and seemed likely to affect the amicable relations between mine owners and men. Moreover, as gold mines became poorer the opportunities for gold pilfering decreased; strong police precautions partially restricted them at the greatest gold city, Kalgoorlie, for the telluride gold ores of that field were distinctive and miners who illegally acquired specimens could not plead the old alibi that they had found the gold in the bush. The goldminers too had less opportunity for speculating in the large low-grade mines that became the norm in Western Australia, for directors tried to steady the output and therefore the share prices; most share fluctuations were now dictated by London. The tribute system too was less widely practised. Nevertheless, the waning of flannel-shirt capitalism was never complete and so did not seriously impair the relations in the gold mines. Unions were stronger but rarely militant; the men they elected to parliament were rarely radical; the working man in the gold towns was—and still is—more inclined to be independent and individualist.

Broken Hill was the first Australian metal field to experience that sharp rift between companies and miners that had long distinguished the coalfields. Significantly it was probably the first metal field in which miners lost their hope of rising within the economic system. The largest group of men at Broken Hill had migrated from the sick copperfield of Moonta and Wallaroo. Whereas those South Australian fields had racial and religious unity, the paternalism of Cornish managers, and the gamble of the tribute system, Broken Hill lacked those solaces. Many Broken Hill managers were Americans who knew no Cornish ways and enforced more discipline than the Australian managers. The Moonta tribute system was not introduced to Broken Hill, and so the working miners were not their own masters and had no chance of making a fortune if their section of the mine proved rich. As the main Broken Hill mines were directed from Melbourne, the miners did not get those welfare amenities which were once common in South Australian mines but unknown in Victoria.

The one hope of the ambitious miner at Broken Hill was to gamble in shares or peg one of the hundreds of silver outcrops in the district. Hundreds of miners held shares in local mines that were refloated in southern cities at handsome profits during the silver fever of the late 1880s. Crowds of dusty miners gathered at the open-air exchange in Argent Street, jumped on to stationary carts and waggons to see over people who obscured share quotations on large blackboards and faces of haggling sharebrokers. A surviving photograph entitled 'Share hawking in Argent Street, 1888' shows such a large crowd that the popular author, Ion Idriess, seemingly thinking his readers would not believe so many workingmen could be engrossed in shares, published the photo in his recent book *The Silver City* with new caption 'A fight on the street corner'.

After the silver boom subsided, miners lost even this avenue of hope. Shares in outside syndicates became unsaleable, local sharebrokers vanished, and many miners lost heavily. Above all, the Broken Hill lode was so massive and assured that silver shares were relatively stable compared to shares in the erratic Victorian gold reefs; the underground miner thus had few chances to exploit inside information on the prospects of the mine. Moreover, the fluctuations in Broken Hill shares were determined more by investors' predictions of the likely price of silver and lead on the world market, and the Melbourne or London investor was more capable of predicting than the Broken Hill miner. Silver miners lost their privileged seats as speculators. As an ounce of silver was worth less than one-twentieth of an ounce of gold they could not even profit much by stealing rich specimens from the mine.

Miners who had relied on their own initiative and judgment to improve themselves became more sympathetic to collective steps. The Amalgamated Miners' Association enrolled more members. As most miners were now gathered into the few large surviving mines they could more readily be marshalled into the union. A seven-day strike in November 1889 persuaded the mining companies to employ only members of the union. After a strike in the following September the men's working week of forty-eight hours was cut by two hours, and they became the most privileged workers in Australia. The union at Broken Hill grew stronger. It gained militancy from the national unrest in shearing sheds and on the waterfront in 1890 and 1891 and from the pessimism of its own members. The sharp fall in the price of silver in 1892 forced mining companies to economize, and it was clear to unionists that they might be forced to work harder for lower wages. At Broken Hill the forces of friction were far stronger than on any other Australian field.

In July 1892 there was bitter dispute over contract mining. The companies insisted that all ore should be mined on the contract or piecework system, arguing that miners would then work hard. The unions opposed the system, claiming that it encouraged dangerous mining practices, that it penalized older miners and might lead to the dismissal of slow workers. Neither side budged. The unions withdrew all men from the mines and picketed the leases. Richard Sleath, the union secretary, who had been a company promoter and provisional director of the stillborn Broken Hill Smelting & Refining Company as recently as January 1890, was now bitter in denouncing the greed of capitalists. He bought a share in B.H.P. and thundered at the shareholders' half-yearly meeting in Melbourne. The large force of police mustered at Broken Hill could barely prevent violence between unionists, outside miners, and company officers. When a trainload of strike-breakers arrived from Victoria the city almost rioted. After eighteen weeks the strike was defeated. Many unionists were not employed again and seven strike leaders were taken to the distant pastoral town of Deniliquin, tried for conspiracy, and imprisoned for terms ranging from three months to two years.

The contract system nearly doubled output in the mines. Victory emboldened mining companies, as their profits fell lower in 1893, to reduce wages and increase the working week to forty-eight hours. Broken Hill men were still amongst the nation's highest paid workmen but their union became weak and their memory of 1892 became sour. Above all, they had lost the miners' traditional hope of rising from the ruck. The mines did nothing to restore amity, the unions did nothing. The discord festered, but so long as the national economy was sick and unemployed men valued any job the unions on Broken Hill and every Australian metal field had small bargaining power.

Broken Hill presaged troubles that would worry most base-metal fields. So long as a copper or lead field was new or the price of its metals high, there was small danger of a sharp rift between owners and men. But once workers' capitalism declined and metal prices fell, tensions quickly marred personal relations. As Broken Hill revealed, the economics of base-metal fields were such that workers' capitalism tended to decline more quickly than on a goldfield. Moreover, unlike gold, the price of which was fixed, base-metals on the world market behaved erratically, and thus a sudden fall in price imposed sudden tensions on the mining fields. When companies suffered a drastic fall in profits they tried to cut wages or exact more work from men. The men resisted the attack.

The nation recovered a taste of prosperity with the new century and men no longer flocked from miserable cities to booming mining fields. As the queue of work-seekers shortened outside the mine offices the unions regained their power to organize and bargain. The periodical slump in metal prices stung them into aggression. In Tasmania the Hercules miners struck for nearly two years on windswept Mount Read. The Magnet silver miners struck for twenty-eight weeks. The copper miners at Mount Lyell were out for eight weeks. On the mainland Broken Hill became the stormland. In 1909 some of the mines were so hit by the falling lead market that they tried to cut wages by one-eighth; a lockout ran for twenty weeks and all the bitterness of 1892 floated again. Men struck for eight weeks in 1916 and the bitterness grew until it exploded in 1919 in a strike that lasted over six hundred days—remembered acidly as 'the spud and onion days'. The men eventually returned to the mines with many gains—a 35-hour week underground, a high minimum wage, and compensation for all suffering from dust on the lungs. Other copper and lead fields from the Gulf of Carpentaria to the Southern Ocean shuddered during slumps in metal prices. The decade 1911 to 1920 was turbulent for the base-metal industry and the flag of the International Workers of the World replaced the old dividend flag on the poppet head of many mines.

Many mine managers and directors became aggressive in resisting union demands that endangered their mines. Many unionists were firebrands in oratory and action. Anger fanned mysterious fires. In the huge North Lyell mine a pump house caught fire in October 1912 and the fumes killed forty-two men and trapped a large party of miners underground for five days. The cause of this worst disaster in an Australian metal mine was obscure. 'Forty-two men,' reported a Royal Commission, 'are said to have lost their lives in various parts of the mine; and, with so many voices lost to us in the silence of death, the evidence is necessarily incomplete, and we can only deplore the fate of those whose testimony concerning the happenings in the mine on the fatal 12th of October will never be given before an earthly tribunal.' Even so, the strongest evidence suggested that the fire was deliberately lit to embarrass the company, not to kill men, in a year of unrest. At Broken Hill on 30 July 1919 in the large South mill 'fires were maliciously started by some person or persons unknown' and the mill burned to the ground with the onlookers refusing to fight the flames. In the next summer the oregon timber in the C.S.A. mine at Cobar caught alight inexplicably some hours after all miners were believed to have left the mine, and the fire was so fierce that the mine was sealed and closed. That fire possibly sparked spontaneously in combustible ore, but the serious underground fire that closed Mount Morgan's mine in 1925 appears to have been one more skirmish between men and bosses.

In the long decline of mining, companies had tried to survive by demanding sacrifices of the men. These demands, and the monthly prospect that so many mining towns might close, unsettled men who faced the loss of not only their jobs but their homes. As mining towns vanished the local union leaders who had lost their jobs seemed often to go to those surviving mining fields where unionism was strongest, and they descended mines by day and ascended soapboxes by night. Some of the biggest base-metal fields reflected the new waves of opinion that were breaking on the trade union movement in many Australian workshops and workplaces. The I.W.W. was rising, the Australian Communist Party was born, and the Labor Party adopted its policy of socialism.

The gold towns remained placid in their slower but more serious eclipse, and a few base-metal fields avoided the strained era, while some fashioned strain into triumph. At Mount Lyell the bitter strike of 1911 was followed by the underground disaster and the mine had to be flooded to quench the fire and for months the bodies of men could not be recovered. Disaster scarred the mind of the community; the landscape had long been bleak and scarred. Misted mountains guarded eroded hills and the black-tarred roofs of houses in the valleys. Sulphur from the copper smelters shrivelled gardens and heavy rain scoured even the soil from long-bare gardens. The climate and terrain were forbidding; isolation made for dear living. In this crisis the mine manager, R. M. Murray, persuaded the company to improve the town's life and living conditions. By 1920 the company was providing cheap houses for many employees, turning old hotels into good boarding houses, opening clubs and libraries for employees, giving men holiday passes on its two railways to the coast, building holiday cottages at the beach, offering employees cheap firewood and probably the cheapest electricity in Australia, subsidizing local bands and amenities, and opening shops to lower the cost of living. This plan closed the rift between managers and men; it stabilized the cost of living when it was rising elsewhere and so saved the company from those legitimate demands for higher wages which closed so many fields in the recurring slumps in the price of copper. The bold scheme was not simply benevolence in the mind of its creator; it was self-preservation.

Wallaroo and Moonta had been paternal companies in the Cornish tradition as far back as the 1860s, and when they united they slowly enlarged the benefits they gave to employees. The Wallaroo and Moonta Mining & Smelting Company was not rich, but the manner in which it treated its men sixty years ago would be accounted visionary even today. 'Probably nowhere in Australia is the welfare of the employees, both at work and after working hours, provided for in such a generous manner,' wrote an industrial officer of the Port Kembla refinery after inspecting Wallaroo in 1918. Australia's three most important metallurgical plants, Port Pirie, Risdon, and Port Kembla, emulated Wallaroo. By the late 1920s every big base-metal field or plant had welfare schemes, with the exception of Broken Hill, and there W. S. Robinson of the Zinc Corporation introduced the new concept in the 1930s. Thus the mines triumphed over tensions, pioneering a brand of industrial relations that was so advanced that the new social benefits conferred by parliaments often seemed mean or irrelevant.

Paternalism became so common in the mining industry that it multiplied the expense, and therefore the risk, of opening new mines.

Whereas in 1880 workmen flocked to a new mining field and built their own houses and financed their own streets, and promoters came and built railways or gasworks or iceworks or theatres, now new mining companies had to entice men to the outback with the promise of cheaply-rented houses and cheap lodgings for bachelors, had to supply electricity and water and ice and often sewerage to the whole town, had to foster social clubs and bands, provide pension and accident benefits, and make or subsidize hospitals, streets and fire stations. That expense of planting a community was one of the burdens of opening new fields such as Mount Isa and Mary Kathleen, or of reopening old fields. Governments or municipalities or the people themselves provided these amenities in the cities but not in the outback. Investors who opened a new mine therefore took not only the risk inseparable from mining but had to create a community and all its amenities—an additional expense with which virtually no other industry was faced.

The famous lead or copper bonus was another answer to the tensions on base-metal fields early in this century. Like the welfare scheme it broadly derived from Cornwall and was hammered into shape at Wallaroo and Moonta; it was also like the welfare scheme in that it rebounded sometimes on its creators. The Wallaroo and Moonta company had devised the copper bonus for employees in 1905, promising to pay them a sliding bonus whenever the price of copper was high. Their copper bonus resembled the tribute system that had almost vanished from the field. The tribute miners had backed their judgment that their small part of the company's mine would be rich and they gambled too that the price of copper would rise, for they shared with the company, in varying proportions, the actual revenue that came from their strip of ground. Now, under the copper bonus, they shared part of the additional profit the company earned whenever the price of copper increased. In 1906 the price of copper soared unexpectedly, and in September all employees got a bonus equal to 20 per cent of their wages and in November the bonus reached 25 per cent, falling away again as copper fell.

In prosperity Wallaroo treated its men generously and in bad years the men were generous in accepting lower wages to enable the company to survive. At Cobar a less generous copper bonus was awarded to miners by the Arbitration Court of New South Wales in the boom of 1906-7. The copper bonus ceased in Australia when all the fields that paid it died soon after World War I.

The bonus emerged again as a lead bonus at Broken Hill during the lead boom in 1925. Popularly believed to be a solace for the hazard of lead poisoning, it was in fact a small venture in profit sharing.

After World War II the price of lead soared with the general inflation and the bonus was for a time double the Australian basic wage. The Broken Hill miner and millman became an aristocrat.

The Queensland Industrial Court had awarded Mount Isa miners a lead bonus of sixpence a shift in 1937, and after the war the bonus jumped with the price of lead. Shareholders whose willingness to invest had alone made Mount Isa profitable waited twenty-five years for the first dividend, and then in the following decade they got less in dividends than the employees got in lead bonuses. The weakness of the lead bonus was that it had no automatic relation to a company's profits or production; and as most union leaders had ceased to understand the peculiar economics of mining when they ceased to be share speculators, they believed that, if Broken Hill and Mount Isa had a bonus, their own field should have a similar bonus. The fact that most mines would have to close if they had to pay a high bonus was not readily or willingly grasped, and so the bonus bred content on a few fields and discontent on those fields that lacked a bonus. The bonus also disturbed mining investors. They already had to take a high risk in investing in a new mining field; they had to take an additional risk by planting a community that would yield them no adequate income and could easily be a hopeless loss if the mine failed prematurely; and after taking these risks the profits from their mine, if any, might go in large part as bonus to employees and as tax to governments.

It is easy to concentrate on all the gains the working miners have made in the last half-century, the increased safety underground, the comparative freedom from occupational diseases, the higher standard of living, more leisure and infinitely more social security. But old Jimmy Elliott, sitting as an old man on the park bench at Mount Lyell, knew that all these gains did not atone for the loss of the individual miner's hope and optimism that he would strike his own fortune by his own skill and labour. He once said, if he had life afresh, he would be a stockbroker instead of a miner.

26

GOLDEN EAGLES

On 16 January 1931 a car drew into the kerb by the stately Kalgoorlie post office. Inside the car a Coolgardie police constable and two Larcombes, father and son, guarded the largest nugget of gold ever found in Western Australia. A crowd gathered round the car to see this lump shaped like a golden eagle, and when at last the Larcombes carried their prize into a bank it would not fit on the gold scales. People eager to see climbed on to counters, packed the banking chamber. Finally the gold was carried to another bank and lowered on to the scales, and after much manipulating of weight-pieces it balanced the scales at 1,136 ounces, about half the weight of Australia's largest nugget. For weeks men streamed south of Coolgardie in motor cars, bicycles and horse jinkers to root the red earth where the Golden Eagle had been found. That gold was like the nearby Londonderry of 1894, an isolated patch of treasure, but the knowledge that heavy nuggets could still be found was a tonic to prospecting.

As if prospectors needed a tonic in 1931. The world depression had already reached Australia, and thousands of men looked for jobs and thousands looked for gold. The price of gold, static for centuries, leaped in 1931 with the depreciation of Australia's pound and then Britain's pound. Within three years the price of gold more than doubled to £8. 10s. an ounce and all the glamour of the yellow metal returned.

Western Australia's old goldfields throbbed once more. Storekeepers came to occupy shops that had long been windowless. Coolgardie, owning the saddest set of ruined buildings in the land, sold paint and nails and new corrugated iron again. Ora Banda and Norseman had Saturday-night dances again in renovated halls. There were marches on Anzac Day in towns that had lost men at Gallipoli but were deserted in 1915. Bakers and butchers created busy delivery rounds in places long abandoned. Around Coolgardie and Broad Arrow and Menzies, countless teams of men reopened little mines and built themselves humpies of salvaged iron or camped under old water tanks. At night the smell of frying bacon or hot stews came from fires blazing at a thousand points of darkness, and men walked to neighbours' camps to argue who caused the depression or why there was no gold in their small shaft.

Men sank shafts until the water filled them, and they had no money for pumps. Men worked months at hammer and tap, extending a drive in an old mine believing that gold lay inches out of their reach. Some found patches of stone so rich that they returned to the cities or went on sprees; the Denver City Hotel at Coolgardie had its best year in 1933, the worst year of depression in Australia. There were men who averaged £8 or £10 a week throughout the depression from their little mines when £3 a week was to be treasured, and there were men who found no gold after spending all their savings and months of sweat. On the mining charts small mines were everywhere: Tom Thumb and Stray Shot at Marble Bar; Swan Bitter and Fanny Bay at the Black Range; Mug's Luck and Garibaldi at Cue; Kelly's Eye and Donegal Sligo at Menzies; Minnie Palmer and Grace Darling and several thousand more. Motor truck and railway made provisions cheaper than in the 1890s, the state batteries treated small mines' ore, and the small teams that worked for themselves were probably as successful in making a steady living as they had been in the 1890s.

Most of the reefs and lodes that jutted above the surface had been found and partially explored in the 1890s. Experienced prospectors now sought lodes that did not protrude above the soil. In much of Western Australia auriferous rocks were hidden by a blanket of red loam or soil, but careful sifting and washing of the loam often revealed strong specks of gold. With water more plentiful on the goldfields, prospectors developed a skill in 'loaming' that few men of the 1890s could equal. Finding traces of gold in the soil they systematically sampled a wide area, panned the samples for gold, and carefully marked the amount of the gold and the spot where they found it. Near the hidden lode that shed the gold, the surface soil often contained more gold. There the prospector dug a trench or shallow hole through the soil to bedrock and often he found a lode. This method required plenty of water, and the goldfields pipeline and numerous surface dams now supplied it. Most new lodes found in Western Australia in the 1930s and after were discovered through 'loaming'. Some lodes were large like the Comet at Marble Bar, and many were small but handed £5,000 or £20,000 clear profit to the discoverers.

Most men had no money to go prospecting, and worked instead on the goldfields for wages. When goods trains reached Wiluna, Norseman, Kalgoorlie, and Leonora, a few men would emerge from beneath tarpaulins and throw down swag and billy and disappear. They found

20 A gold dredge near Araluen, N.S.W., about 1900
(N.S.W. Mines Department)

21 Hammer and tap miners at Wyalong, N.S.W., in the late 1890s
(N.S.W. Mines Department)

22 Zeehan, a Tasmanian silver town, was a symbol of the decline of Australian mining in the 1920s

23 Jimmy Elliott (1867-1956), famous Tasmanian prospector
(*Mercury*, Hobart)

the gold towns the most prosperous in Australia. More men arrived than the expanding mines could employ, and many had to live on the generosity of the towns, getting a flap of mutton or soup bones from a butcher and loaves from a baker.

Unemployed men resented the hundreds of Italians who had long lived at Kalgoorlie and who got high wages in the mines. On 20 January 1934 in the 'Home from Home Hotel' in Kalgoorlie a naturalized Italian barman and a customer fought over a debt. The customer was knocked down and killed. On the following night a mob wrecked the hotel with axes and iron bars and set it afire, burned a wine saloon and the All Nations boarding house, and looted foreigners' shops. Mobs went on the electric tram to Boulder City and burned and wrecked more buildings on two successive nights. Two men were killed and sixty buildings were destroyed in two days of rioting, the first serious disturbance on an Australian goldfield since the deep-alluvial troubles at Kalgoorlie in 1899.

The return to the goldfields was not confined to Western Australia. Even in Victoria, where goldmining had almost ceased, the government gave 16,000 unemployed men a tent, mining tools, prospector's guidebook, and free railway pass to the goldfield of their choice. Thousands more found their own way to the old goldfields and fossicked for alluvial, opened small mines, or worked in new mines for wages. Nevertheless in all Australia the prospectors found only two valuable goldfields in the 1930s: Tennant Creek in the Northern Territory and Cracow in Central Queensland. The courageous prospector was trying to find lodes without the aid of capital in an age when capital was essential for prospecting.

The rising price of gold attracted capital. Even if the price had been static money would have flowed into goldmining for want of alternative investments in the depressed nation, but now it poured. In Victoria only twenty-eight new mining companies were registered in 1930, but there were 160 in 1934. So many mining companies were listed on Australian exchanges that the mere reciting of their vital statistics filled nearly five hundred pages in mining handbooks.

Game investors took the prizes in the gold revival. In May 1932 a few Melbourne sharebrokers, amongst them Wallace H. Smith and L. G. May, formed a tiny syndicate known as Australian Gold Development No Liability to seek gold mines. They allotted only a hundred shares and raised only £5,000 for their initial search. Almost as soon as the syndicate was formed, their consulting engineer Tommy Victor went to the Dawson River in central Queensland where several prospectors had developed a gold reef amongst the spindly gum trees on the slopes of Cracow. His Melbourne syndicate bought the reef and floated the Golden Plateau mine and it paid £800,000 in dividends by 1939. Victor's alertness got his syndicate groundfloor shares in Cocks Pioneer and Cocks Eldorado, valuable deposits in north-eastern Victoria. In 1934 he went to Fiji and reported favourably on the Emperor and Loloma gold deposits, and when these wonderful producers were floated into companies his Melbourne syndicate had priority to buy shares. Later his syndicate got a cheap bundle of shares in the magnificent gold mine at Noble's Nob near Tennant Creek and the rich King Island Scheelite. Victor's ability to select good mines was almost uncanny. Only Brookman's Adelaide syndicate of 1893, which pegged the south end of Kalgoorlie, was more successful in this form of promotion, and their success came from a rich bracket of mines that were huddled together, not thousands of miles apart.

Nearly every other company and exploration group floated in the 1930s explored old fields. That was not always a hardship. The long string of mines that had closed after World War I now were promising with the higher price of gold. Fields that had never succeeded gave hope to companies that used the flotation process and diesel power stations and electric pumps and large-scale mining methods and the advances of the last few decades.

In Queensland the Mount Coolon field south-west of Bowen was worked vigorously, and old Mount Morgan was reopened after a long struggle. In New South Wales old Cobar, dead as a copperfield, was resurrected for gold. In the Victorian mountains deeper reefs were found in the old Morning Star at Woods Point and the A1 at Gaffney's Creek and lorries used old pack-horse trails. Electric dredges chewed fortunes from the old river flats of central Victoria. Money was won and lost in the second attempt to revive Bendigo, and Western Australia continued to dominate goldmining, partly by opening large deposits at Wiluna, Norseman, Kalgoorlie and the old towns.

The reworking of old mines called for heavy outlays of capital to build new power houses, treatment plants and often townships, to sink new shafts and provide modern machinery. Perhaps London would supply the capital, just as it had supplied it in the 1890s. Claude Albo de Bernales went to London to see.

De Bernales was prince of gold promoters in the thirties. He had magnificent physique, long black hair and dark Spanish complexion, dressed like an ambassador and spoke like a statesman. On documents he simply signed his name de Bernales in the style of nobility. In a hotel in Kalgoorlie or a boardroom in London his physique and presence commanded notice. The man was flamboyant and so were his ideas. He lived in a Spanish palace in the Perth suburb of Cottesloe and built in the heart of Perth his famous London Court, a sixteenth-century street lined with shops. His name was reverenced in Western Australia where his gold mines sustained thousands of families in the depression, and scorned in England where the Stock Exchange of London delisted his companies and Scotland Yard studied his activities.

De Bernales had reached Coolgardie in 1897, a young Londoner who loved the ambitious mining city then at its pinnacle. He sold mining machinery, cycling from mine to mine along smooth camel tracks. They said he carried a clean collar and shirt in his kit and changed his dusty shirt before calling on each manager. He was personable and smooth, always a salesman. He married in 1903 a Kalgoorlie widow who owned a local foundry, and the marriage improved his prospects. He bought and sold boilers and pumps and cyanide vats and speculated in mining shares and mining leases, a rising young man in a failing industry. With a mining engineer named Henry Urquhart he acquired leases on hot plains at Wiluna, a goldfield far from the railway in the Murchison. Patient experiment was essential at Wiluna to extract the gold from an ore that contained much pyrite and arsenopyrite, and experiments called for capital. As the lodes were massive and poor they required mining and treating on the big scale and that too called for capital. De Bernales failed to raise the capital in France on the eve of World War I, but in the 1920s his company finally found that a complicated process entailing grinding, floating, roasting, cyaniding, and smelting would recover most of the gold.

When the mining captains in the West were timid de Bernales was brave and optimistic. In 1926 he went to London and floated the Wiluna Gold Corporation at much profit to Australia and ultimately to himself. The state government extended the railway from Meekatharra to Wiluna, and the company had its own shining oil tank on the waterfront at Geraldton and twelve railway tankers to haul oil 445 miles to its vast powerhouse at the mine. State and Commonwealth knew the importance of this new mine to a sick industry and guaranteed a loan of £300,000 that the company needed for new plant in 1930. In April 1931, as the depression settled on the land, the Wiluna mill began to win gold.

Thousands of unemployed men and defeated farmers paid their fare or jumped the rattler to reach Wiluna, and men and women who lived there in the 1930s seemed to believe there never was and never would be a town so vigorous and warm-hearted. Investors were as enthusiastic about the mine. On the stock exchange it was valued at more than £6 million before a dividend was near. And in time it earned £12 million revenue from ore that averaged only 4½ pennyweights of gold to each short ton.

De Bernales had the foresight to see that gold mining would boom again. In addition to Wiluna he had acquired the leases of scores of mines that were once rich or promising but were closed by rising costs or metallurgical woes. His family company, known as Australian Machinery & Investment Company, held these leases, and in June 1932 he formed twelve new holding companies and sold leases to each company, in return for shares. These twelve apostles issued to his family and friends shares with a face value of £1,361,450. This transaction did not involve the public. De Bernales was merely shuffling figures, creating the impression that his companies with their large paid-up capital had spent heavily in acquiring or developing their mines. He hoped to persuade British investors that his mines merited a high price.

De Bernales was not only a shrewd accountant, he was also a shrewd historian. He examined old records of his abandoned mines to find which had the most prospect of success under changed economic conditions, and he employed their history to raise capital for them. Between 1933 and 1935 he floated eight big companies in London on pleasing terms.

Southern Cross Gold Development Ltd was typical of his London companies. De Bernales sold it five abandoned gold mines at the old field of Southern Cross including Fraser's mines that long ago had won 193,000 ounces of gold. His own geologist reported flatteringly that the mines had a rich past and rich future. His decision to issue 5s. shares rather than £1 shares attracted the small investor, and they eagerly paid £150,000 for shares in the new company. De Bernales' family trust collected £24,000 in cash, £75,000 in fully paid shares in the new company and an option to buy more shares cheaply should the mines prove rich.

Mines that have once been rich in gold are not necessarily still rich; they may be worthless because all accessible gold has gone. De Bernales had such an array of once-famous mines that he could allot them by the half dozen to each London company. Thus the Commonwealth Mining and Finance Limited took the old Sand Queen Gladsome mine at Comet Vale, the old Bayley's Reward at Coolgardie, the Eureka gold mine at Pine Creek in the Northern Territory, and a few obscure mines for good measure. All the mines he owned had produced over £10 million of gold, but what was left in them was for the new London shareholders to discover. They eagerly accepted the risk.

De Bernales found it easy to float companies in London by employing the service of Great Boulder Proprietary, the richest company in Kalgoorlie's history. Great Boulder's old chairman, John Waddington, sat with de Bernales on the board of Wiluna and became a close friend. He slid neatly into the de Bernales web. The promoter nominally sold some of his abandoned mines to Great Boulder which in turn sold them to new London companies, thus lending its magic name and an implied guarantee that the mines were valuable. In fact Waddington did not bother to ask the Great Boulder staff to inspect the mines on which the company was staking its name. In return for these favours the shareholders in Great Boulder were granted the 'privilege' to buy new shares in de Bernales London companies. Some found it was not a privilege.

De Bernales became managing director of Great Boulder in 1936. A bower bird of the mining world, he applied at Great Boulder the practices so effective in his other companies. One family company, Mercer's Trust, took over the share register for a fee. Another provided office and secretary for a fee. Another family firm acted as technical managers for Great Boulder at £2,400 a year. And when Great Boulder wanted new machinery, de Bernales' Kalgoorlie foundry supplied the very items for £121,000 and saved the bother of calling tenders.

De Bernales' reward for floating his London companies was £1,012,000, of which one-fourth was cash and the remainder fully-paid shares. The reward for managing the mines and selling them machinery was also generous, one of his colleagues noting unkindly that 'he charged to the hilt except for the string that held the machinery together'. In his early sixties the tall charming adventurer had mines from Marble Bar to Boulder City, a mansion and art treasures in England and a long Perth arcade modelled on the Elizabethan style that he admired in architecture and business. His assets probably were near a million. The assets of the mining companies he had floated were rather less, for though he had rich orebodies at Yellowdine and Marble Bar his companies had paid no dividend.

The marble palaces began to crumble in 1939. Some Great Boulder shareholders led by Wilfred Barton, K.C., London barrister and son of Australia's first prime minister, accused de Bernales of corruption. The Beaverbrook press echoed the charges. De Bernales replied with long newspaper advertisements under the implied imprint of the Adelaide stock exchange until Adelaide cabled its protest, 'We consider it most misleading and objectionable'. Joe Thorn, general manager of the leading Kalgoorlie mine Lake View and Star, publicly praised the way Great Boulder was managed, and evidence suggests it was well managed. In June 1939 Great Boulder held its annual election of directors and de Bernales' team retained office.

The campaign turned to the new de Bernales' companies. They had paid no dividend and each five shillings share was now valued at an average of only eightpence. The Board of Trade investigated these companies and the London stock exchange suspended trading in their shares in July 1939. De Bernales resigned from the chair at Great Boulder and slowly his new companies collapsed. After the war the British Board of Trade sent detectives to Australia, but they did not prosecute the man who had attracted nearly £2 million of English money to Western Australian goldfields.

The floating of several thousand independent mining companies in London was the cardinal feature of the 1890s gold boom but, except for de Bernales' companies, that was rare in the 1930s. The flow of foreign capital to Australian mines resembled a quiet creek rather than rapids. Some South African gold companies dabbled in Victorian goldfields. A company that had owned the celebrated Camp Bird mine in Colorado invested in Victorian deep-lead mines and Wiluna. American Smelting and Refining Company opened Big Bell mine in the Murchison district of Western Australia. W. S. Robinson, one of the few Collins House leaders who still loved the risky task of finding new mines, united London, Melbourne, and Johannesburg interests in a powerful group that sought mines; and Western Mining Corporation and Gold Mines of Kalgoorlie were two fruits of this alliance. More important than the gold these companies won was their emphasis on scientific exploration and the employing of geologists. They employed more geologists than all other Australian mining companies put together, and at the old goldfield of Norseman their geologists found, by deduction, one of Australia's richest gold deposits.

Other ideas raised the output of gold. One idea was the big mine. It seemed that each generation had to learn for itself the economies and security that came from large-scale mining and milling. A big mine had to spend money generously on developing its orebody well ahead of that year's call for ore; it had to invest heavily on new plant to maintain the economies of large-scale working. Australia's gold leaders had learned that idea at the turn of the century, only to forget it when it became difficult to raise capital for new plant. In the depression the idea was revived, and the big gold mines began to dwarf the largest gold mines of a generation previously. Lake View and Star mined 620,000 long tons for the year 1937-8 and Wiluna more than half a million, and such companies could make profit on poor ore that would have once been rejected. The idea of the big mine enabled Kalgoorlie to survive.

The depression, the higher price of gold, and the technical awakening raised Australia's gold output from 427,000 fine ounces in 1929 to 1,646,000 in 1939. In one decade the annual value of the gold output increased from less than £2 million to £16 million. Gold had become again an industry of national importance, and the number of men that gained jobs in gold mines was probably comparable with the number that gained relief from public works. The spur to gold mining even touched base-metal mining. Tennant Creek was opened and Mount Morgan was re-opened, and both fields became large producers of copper as well as gold. The revival of mining investment and the rising metal prices stimulated silver-lead-zinc mines, and in the second half of the decade New Broken Hill Consolidated and Lake George in New South Wales and Rosebery in Tasmania began to mine on the large scale.

It was customary to blame World War I for the decline of gold mining, but World War II was more damaging. Rapid expansion of gold output was turned into rapid decline, and across the continent resurrected gold towns died again.

27

LASSETER'S LAND

LEWIS HAROLD BELL LASSETER was like de Bernales, a legend of the thirties. He was one of many Australians who believed that rich gold reefs were hidden in the heart of the continent. He had even seen one with his strange staring eyes. He recalled that as a young man he had gone to the Arltunga goldfield near Alice Springs in Central Australia early in this century. Wandering west with horses he found an amazing gold reef amongst sandhills and bare cliffs near the unmarked borders of the Northern Territory, Western Australia, and South Australia. He would have perished, he said, but for a passing Afghan who guided him towards civilization.

Again he returned to his reef with a surveyor named Harding who carefully defined the position of the reef with instruments. Unfortunately their watches were seventy-five minutes slow, so in calculating the position of the reef they made an error of a hundred miles. Lasseter explained that in desert without landmarks the reef could only be rediscovered with charts and compass.

Lasseter told his tale in Sydney in 1930, when the world depression was deepening and the fever for gold was reviving. He spoke intensely and plausibly and said he could rediscover his reef if he had a well-equipped party. In Sydney the Central Australian Exploration Company financed a party, and from the small town of Alice Springs Lasseter guided the expedition west. He pegged a lease on the Blacks' Ceremonial Ground near the Petermann Ranges and another near Mount Lasseter, but he had not found the reef when he suddenly left his expedition and disappeared with camels. Then his camels strayed and Lasseter was alone.

Early in May 1931 a bushman named Bob Buck reached Alice Springs and informed the authorities that natives had led him to a new shallow grave in which he found the body of a man. The diaries and charts and false teeth which he carried to Alice Springs proved that the body was Lasseter's.

The story that Lasseter had again found his reef while wandering alone spurred more expeditions to enter the search. Ion Idriess, experienced prospector, wrote in 1931 a popular book, *Lasseter's Last*

Ride: An Epic of Central Australian Gold Discovery, that made Lasseter's reef a legend. Investors bought shares in companies that sought Lasseter's reef or claimed to have found it. A long line of parties went west to Lasseter's country with camels or motor trucks equipped with coir mats to lay on deep sand drifts. Adventurers told city syndicates that they too had seen Lasseter's reef, but when they approached the vicinity of the reef they quarrelled or refused to go on or quietly disappeared. Some mysteriously produced specimens of gold which they vowed were chipped from the reef but none could produce the reef itself.

It was widely believed that the Petermann Range was mysterious, forbidding, rarely visited by prospectors and therefore likely to conceal the Coolgardie and Kalgoorlie of the twentieth century. In fact, many skilled explorers and geologists and prospectors had gone through or near the ranges. Ernest Giles had passed by in 1874 and those gallant Western Australian prospectors Frost and Tregurtha had passed by before they vanished to the Klondike, and Tregurtha had carved his name and the date 1896 in the quartzite at the Piltardie waterhole as evidence for the later parties that they were not the first. Expeditions led by W. H. Tietkins, F. Hann, Dr Henry Basedow, Michael Terry, and many other skilled men, had penetrated the ranges near the borders of the three territories and found only meagre signs of minerals.

Even while Lasseter was alive his story puzzled many men. Dr L. K. Ward, South Australia's geologist, talked with Lasseter and could not glean where he had gone on the trip that led to the finding of the gold-studded reef. When Lasseter reached Alice Springs to begin his last journey he could not answer the questions asked by suspicious old bushmen; he did not seem to know the rockholes and mountain ranges in the area he said he had visited. Western Australia's government geologists, Talbot and Blatchford and Ellis, went with expeditions in the 1930s and could not understand how a prospector could fail to rediscover a rich reef in country that was studded with distinctive scalloped ranges visible from afar. They could not fathom why a prospector who survived in such dry country with horses should prove such a poor bushman on his last expedition. As for Lasseter's tale that the slowness of his surveyor's watch had mislaid the site of the reef by a hundred miles—the time on the watch in fact would have created an error of 1,100 miles in longitude, and surely not even a bush surveyor would have cheerfully made that error. 'It is wrong,' wrote geologist H. A. Ellis, 'to think that because a large portion of "Lasseter's Country" is desert nothing is known of it. The natives have been for centuries and still are, hunting over it, and over thousands of square miles of it the natives will tell you there is not even a pebble of white quartz.'

It is almost certain that Lasseter found no reef and it is possible that he was mad. But there was method in the madness of the hundreds of thousands who believed in his reef. The generation who remembered gold saving the nation in the 1890s hoped that gold again might save it. The price of gold began to rise in the month that Lasseter was lost in the desert, and his death in a glamorous cause gave him the halo of the mythical hero. He was the Patrick Hannan of the 1930s depression, and Central Australia was to be the mecca of the poor just as Western Australia had once been. 'The stage is set for a big discovery,' wrote Ernestine Hill in 1937 in *The Great Australian Loneliness*. 'Where will it be?'

About halfway between Alice Springs and the Kimberley gold diggings of 1886 were sand and spinifex and low scrub and low hills and a few patches of gold. Allan Davidson, a bold prospector working for an English syndicate, had found gold in 1900 near tors of granite 360 miles north-west of Alice Springs, and though the gold was poor and the water foul a few prospectors came down from the top corner of Western Australia and from the cursed Tanami goldfield to win alluvial gold with their dry-blowers. Stewart, a prospector, was murdered by a black in 1911, and soon the Granites were abandoned. Schultz, a prospector, found more gold and died in 1925, reputedly from foul water. The discovery of more alluvial gold at the height of the depression in 1932 drew hundreds of unemployed city men and bushmen to Alice Springs in the fortnightly train from Adelaide, and promoters came to collect leases and journalists to collect news. A dozen companies were floated in Adelaide and Melbourne. At the end of 368 miles of wheeltracks stretching north-west of Alice Springs, over countless sandy creek crossings, was a cluster of tents and humpies spread along five miles of outcrop. Charles Chapman, a well-sinker near Roma, came west with a Queensland party of twenty men and paid £5,000 for the Burdekin Duck mine and won 130 ounces of gold bullion from 70 tons of ore in 1932, taking the gold to town in golden syrup tins. His wireless provided contact with cities and doctors, the holes he bored provided water for the thirsty and his gold was food to the hungry. With small quartz reefs around the Burdekin Duck at one end of the field and large ironstone lodes at the other end, near the Golden Shoe and Bullagitchie leases, some said the field resembled Kalgoorlie; but the geologists and engineers said it didn't. The gold was poor and patchy and the companies were not rich enough to drill the big lodes, and soon the procession of motor trucks with boiling radiators went back the track to Alice Springs, leaving Chapman to guard the Granites.

Some who left the Granites went east to Tennant Creek. For sixty years it had been a lonely repeating station on the overland telegraph line from Adelaide to Darwin, and the little community at the telegraph and the few travellers had occasionally observed signs of minerals on the hot plains and flat-topped hills. H. Y. L. Brown, South Australian government geologist, spent eight days in the district in 1895 and washed one colour of gold in a creek. Allan Davidson, finder of gold in 1900 at the Granites and Tanami, rested his horses while he sampled the many quartz reefs. A linesman from the telegraph station found alluvial gold in 1926 and a few men sank shallow shafts. They hoped for gold in the quartz rock and the red soil, but in both places the gold was very weak. The field was to prove rich in gold and copper, but the gold lay in ironstone that most prospectors shunned and the copper was invisible near the surface. Water was scarce, the sharp-pointed spinifex on the plains vexed the horses, the cost of stores carried a thousand miles from Port Augusta was exorbitant, so few prospectors stayed long at Tennant Creek.

Early in 1932 a blackboy tending the telegraph station's cattle found gold about seven miles south of the station, near the present site of Tennant Creek town. His master pegged the Peter Pan lease. Malachi Noble, Ralph Hadlock and Bill Garnett were camped in the vicinity and pegged Wheal Doria lease. Noble loamed the surface soil and, finding gold, sank holes until he found the reef. Hadlock drove his waggonette three hundred miles south to Alice Springs and returned after six weeks on the track with £70 of stores. The depression and the gold boom had drawn hundreds of men to Central Australia, and some hurried to Tennant Creek. The depression made that field.

The richest stone had to be packed in bags and sent by truck and train to the government battery at Peterborough in South Australia for crushing. The Wheal Doria got forty-eight ounces of gold from one load of twenty-eight hundredweight, and other loads were nearly as rich; they had to be, to pay the freight. An Afghan came with van and trailer and built a little battery with two stampers, and crushed ore carted from surrounding mines for £2 a ton. The mine owners stood near the noisy stampers, some carrying guns, to make sure they got their gold from the crushing, and the Afghan would sell them silk hat or shirt to add to his profits. The town of Tennant Creek developed itself by the dirt track from Adelaide to Darwin, a tough town with brawls and gunplay and a publican who sold bottled beer at a bar partitioned from the customers by wiremesh. More than a hundred separate mines operated before World War II, and the gouger with his share in a small mine and rattletrap truck was king. The common complaint of the Australian investor in the 1930s was that there were few new goldfields to invest in, but Tennant Creek justly complained that investors were timid.

Tennant Creek's peak yield of gold in the first decade of active mining was a mere 16,000 ounces in 1942. All the shafts were shallow. Not one large shoot of ore had been found. Without an intensive search the field could not live beyond 1950.

Malachi 'Jack' Noble, tall wiry prospector of the Wheal Doria, was so taken with the goldfield and its patchy richness and size that he sent a message to old friends in the Kimberley district of Western Australia. Bill Weaber, an elderly cattleman, came and formed a partnership with Malachi Noble to search for mines. It was a wonderful alliance. Weaber was quite blind and Noble had one eye, so Weaber gave money and Noble his eye in the search. Noble went east to the flat-backed hills and found four deposits, Rising Sun, Weaber's Find, Kimberley Kids, and the jutting bluff named Noble's Nob, and in a small stamp mill the syndicate crushed the rich ore gouged from the Rising Sun. Weaber went to live in a large iron bungalow at the mine. He felt the golden ore with the sensitive touch of the blind, questioned the miners on what they saw, liked the rushing sound of the stampers and the voices of the men, and so created his own fantasy of the rich mines he controlled.

Jack Noble was more restless and, selling his shares for about £9,000, bought a hotel in the town, returned to his native Queensland for a spree, and is said to have once hired a taxi from Townsville to Tennant Creek. He was soon penniless.

In Adelaide in 1947 drowsed a small company that had spent a quarter of a century in vain search for mines in the western half of Australia. Owning the grand title of Australian Development, but no mine, its main asset was a few Commonwealth bonds. In 1947 it received favourable reports from Charles Blackett, an experienced gold metallurgist, on blind Weaber's small mine at Noble's Nob. Although the mine had yielded only 2,500 ounces of gold Blackett suggested that Australian Development should get an option to buy the mine. A secondhand mill began to produce gold at the end of 1948, and the company struggled to gather money to exercise the option. In July 1949, with time running away, they sank a winze or shaft at the 135 foot level. The shaft ran into the richest ore yet seen on the field. For the first forty feet of sinking the ore averaged the astonishing assay of fifty to sixty ounces to the ton. The company bought the mine and began high dividends, sparing money to build a new plant and powerhouse and small town of dazzling corrugated iron, painted white and silver, one street riding barebacked on the hill. The ore was so rich that the company mined only 14,000 tons in the year 1950-1 but paid £315,000 in dividends. Fourteen years after taking the option it had sold £8·1 million of gold and paid nearly £4·5 million in dividends—at least forty times the actual money its shareholders had invested in the mine. Two men with one eye had opened this mine that yielded more gold than any single mine of Ballarat or Bendigo.

At night Noble's Nob looks out on a splash of light on the plain. The mine on the plain is Peko, being so named by a barrel-chested Russian who was one of a small team that gashed into the hematite outcrop for gold. After the war a small Sydney company explored the deposit a little deeper, and at the 210 foot level found oxidized copper. On the roof and walls of the underground drives the beautiful mineral salts, coloured rich pink and blue and yellow and green, were quick to form once the ore was exposed to the air. Assays showed payable copper. The Bureau of Mineral Resources drilled a few holes and found the copper extending deeper. Like Mount Lyell and Mount Morgan and many Australian deposits, Peko was strong in gold in the oxidized zone and strong in copper in the deeper sulphide zone. As the oxidized zone goes deep on the Tennant Creek field, and as few companies had money to probe deeply, the copper was not appreciated until the field was nearly twenty years old.

Eighty years after its first mining boom the Northern Territory at last had a rich mining field, but Tennant Creek was a mere pinpoint on the plains of the tropics. Four hundred miles to the east an immeasurably richer field was rising in red hills that had long taunted prospectors and gougers and the Afghan camelmen that had carted their treasure.

28

A MINE IN AGONY

JOHN CAMPBELL MILES was one of those jacks of all trades who wandered the outback. In 1923, the most noted year of his life, he turned forty. He had spent most of his previous years working along the eastern half of Australia, ploughing wheatlands in the Wimmera, driving a haycart near Melbourne, carting sugarcane and repairing fences and windmills and shooting wild pigs in Queensland. He was wiry, quiet, and temperate, loved horses and quiet campfires and did not stay long in the noisy towns. He had once worked underground at Broken Hill at hammer and tap but stayed less than a year. Word came in 1908 of a gold rush at the Oaks in north Queensland and he bought a bike and rode with a mate 1,500 miles across the continent, his swag strapped to the frame and the punctured tyres stuffed with grass. His father had been a gold digger who had arrived too late at most rushes in the 1850s, and Miles arrived too late at his first gold rush. He was always patient and leisurely, liked to watch the billy boil or the sun rise.

'In the year Sister Olive won the Cup' he was working on a sheep station in Queensland and decided to go to the Northern Territory and maybe prospect for gold. Next year, 1922, he started with his stallion Hard Times and two mares and three packhorses, and travelled slowly through Hughenden and Mount Elliott. The plains were cracked with drought and he moved cautiously, camping long at each waterhole, waiting for his mares to foal or his packhorses to rest. He had been almost a year on the track when he reached the dead copper town of Mount Elliott and saw a billiard table standing in the sun where once a hotel had stood, and streets of deserted houses in the mining region he was about to revive.

On a hot morning in February 1923 he led his horses along the dry valley of the Leichhardt River, the powdered dust rising in red clouds from their tracks. In the dry riverbed the smooth trunks of the gum trees glistened ivory white and on both sides of the narrow plain the brown contours of low ranges pranced in the heat. Miles looked for a camping place and suddenly his packhorse smelled water and ran down to the watercourse and wallowed in the wet sand by a small waterhole. Miles followed her and made camp, fastening bells on the horses and filling his waterbags. He liked to prospect when he was in mineralized country, so about midday he took his farrier's hammer and rode to the low ranges half a mile from the track. Tapping his hammer on the yellow-brown boulders on the hill he chipped a lump of rock that was black and honeycombed and unusually heavy. He thought it must be a mineral but he did not realize it was carbonate of lead. Wandering over the hills that were prickly with spinifex, he saw in several places parallel bands of the same ore crossing the rock like a zebra skin. He recognized one specimen he chipped as galena, the sulphide of lead he had once mined at Broken Hill. That night in his camp on the dark plain he wondered if he had found another Broken Hill.

The Camooweal mailman passed along the road in his lorry and took ten specimens of the ore to Jim Tregenza, the government assayer in Cloncurry, who noted by return post that the samples contained from 49 to 79 per cent lead and much silver as well. Tregenza advised Miles to sell a few tons of ore to the metal buyers to see if it paid in bulk. Four cattlemen who were working the Native Bee copper mine a few miles to the south helped Miles to bag sixteen tons of ore. Two carriers generously lent waggons to cart the bags to the railway station at Duchess, and Miles stood with his little box camera and snapped the long string of horses.

News that he had got a good cheque from the sale of his ore attracted men from copper mines and cattle stations to peg out leases at 10s. an acre and gouge the richest lead from countless narrow veins. In shelters of boughs or hessian they shovelled the ore into bags, loaded the bags on to camels and horse waggons and lorries, and eased their thirst in the cruel summer with water bought for sixpence a gallon. They christened their leases with sweat and imagination, Ace of Diamonds, and Joker, Durban Angel and Queen of Sheba, Silvery Moon and Silver Dollar, Black Star and Bright Star and Star of Bethlehem and seventeen other stars. It required the same vigorous imagination to see the lead in many of the leases, and the earnings of the entire field in the first year were £700 from the sale of ore.

Australian mining was so pessimistic that few promoters bothered to visit the new Mount Isa field, and those who bothered came like snails on the long railway trip through Brisbane and Townsville and Cloncurry. William H. Corbould, aged fifty-seven, white helmet shading his crusty face and wiry body, came at the cabled summons of a Cloncurry publican to see the field. He had seen many new fields—Broken Hill where he was an assayer and trustee of the first iron church (which he never entered), Kalgoorlie where he managed

Hannan's Reward, and the copperfields of Burraga and Mount Elliott—and he remained an optimist and even sentimental about mines, and he had the vision to see that Mount Isa might prove rich. He got option to buy four hundred acres from the prospectors, and floated Mount Isa Mines Limited in Sydney in 1924.

Randolph Bedford, a giant of a man with red face, long boomerang moustache and broad brimmed hat, who delighted in red hills and hot sun and the colour of rare minerals and the taste of rare whisky, was in Brisbane when he heard of the new discovery. Individualist, journalist, novelist, and politician, he had been on Broken Hill in the boom of 1888, and had swelled the noise at every mining boom thereafter. He hurried to Mount Isa and signed with sweating hand options to buy lead leases from Arthur Campbell the cattleman. Old William Orr, who had floated more good copper companies than any other Australian, bought leases and floated his company too. Like Corbould and Bedford, he knew Broken Hill in its roaring infancy and hoped that Mount Isa would be as rich. These three promoters were optimists in an era when mining had lost its optimism, and they floated three companies, but raised only £85,000. Soon two of the companies sold out, and Corbould's company controlled the field. No other virgin Australian mining field had ever before been owned by the one company.

The prospectors got shares or cash for selling their leases to the companies, and few thought the shares—or the money—were worth keeping. Bill Simpson, a ginger-bearded convivial bushman who had become partner with Miles, owned shares worth nearly £30,000. He hired a taxi to carry him a thousand miles to Brisbane, celebrated his luck for a few days in the bars of the city, and was killed by a car while crossing a street. Three men who had sold the Rio Grande for £5,000 boarded the train for Townsville with an ice chest crammed with beer and spirits and were the only passengers who cheered when floods delayed the train four days at Richmond. Their celebrations at Townsville were remembered for years. Their money miraculously disappeared, except that of Condamine Davidson who left enough to buy a shining seven-seater Studebaker but soon he was driving a lorry, for wages. And Campbell Miles, the discoverer, frugal and abstemious, vanished into obscurity, returning a generation later to find the field he had founded and named one of the world's greatest.

A line of windlasses capped the lodes at Mount Isa in 1924 as miners explored the ground. Those who descended the holes in a bucket praised the size and richness of the lodes. Sitting on packing

24 Part of the Golden Mile at Kalgoorlie, today
(W. A. Chamber of Mines)

25 Headframe of New Broken Hill mine, seen across parklands that have replaced sand drifts.
(Conzinc Riotinto)

26 Australia's tallest chimney towering above Mount Isa
(Mount Isa Mines)

cases in fresh white tents they wrote reports likening it to Broken Hill. The price of lead was pushed high by the rising automobile industry, and that spurred the optimism. The company's first manager, E. C. Saint-Smith, was a good geologist and predicted that the lodes would be massive and payable, and easy to treat in a simple gravity mill. His shallow shafts found that the lodes were not as massive as he had predicted, and not as rich. And when his hands were festered with Barcoo rot and he resigned, he admitted that the field was still 'in the lap of the gods'.

Corbould refused to admit it. He spent his company's capital in exploring the lodes and when the exchequer was low he personally invested £5,000. He rode his horse around the leases, dismissing any man who paused to roll a smoke and preaching frugality. He pleaded for the Queensland government to extend its railway fifty-two miles from Duchess to the mine, and promised that his company would pay up to £130,000 as compensation for losses in running the railway. His company had barely gathered half that sum from investors during its life; critics said it might never possess £130,000. Even so the government decided to build the railway, linking the mine with Townsville 603 miles away.

A few engineers still compared the field to Broken Hill. Both had silver-lead, and there the likeness ceased. Broken Hill was only half the distance from a deepwater port and its companies did not have to underwrite a railway. Their surface ore was so rich that it financed the mines, while Mount Isa had no rich ore. Their deeper sulphide ore was half as rich again as Mount Isa's average ore. They had built no township to house the miners whereas Mount Isa would have to create town and amenities and dams to attract the men. Their ore was easier to treat than Mount Isa ore and released more of the imprisoned metals. If all Mount Isa's disadvantages had been appreciated in 1926 the field would almost certainly have been abandoned.

Corbould wanted half a million to make his mine produce lead and silver. Most engineers who toured the mine said the investment was too risky. The Australian stock exchanges agreed. W. S. Robinson, a leader of the Broken Hill group, had an option to lend £500,000 and did not persevere. Corbould hawked his mine in London in 1926 and finally interested the Anglo-American Corporation of South Africa, rich in gold and diamonds. That company would provide £600,000 and take a controlling interest in the mine, if its experts ruled that the mines were valuable. The experts came, and again a cloud of red dust settled on the despondent town.

Another financier spoke with Corbould. He was John Leslie

Urquhart, aged fifty-three, forthright and exuberant, an international mining financier who spoke seven languages. Born near the biblical city of Ephesus, trained in engineering and chemistry in Scotland, he bought liquorice root in Turkey for manufacture into chewing tobacco and then turned from that waning industry to the rising oil industry. At the age of thirty he managed four British oil companies at Baku on the Caspian Sea, but, narrowly escaping assassins in 1906, he moved to metal mining in Russia and Siberia. He floated in London the Anglo-Siberian, Kyshtim, Tanalyk, and Irtysh corporations to develop large mines and consulted leading American mining engineers, amongst them Herbert Hoover and D. P. Mitchell who had first managed mines in the Western Australian gold boom. In 1917, Urquhart's companies owned twelve metal mines, coal mines, 250 miles of railway, 32 riverboats, smelters, steelworks, sawmills and dynamite plants and all the amenities and adjuncts of an industrial kingdom. Then came the Russian revolution of 1917, and his work was swallowed. He lodged a claim for £56 million compensation and united his companies into the Russo-Asiatic Consolidated to press his claims and search for other mines outside Russia. The Soviet refused to compensate him for the loss of his mines, and the only new ones he opened were poor zinc mines in France. 'It is now,' he told his followers in 1926,

> almost twenty-one years ago since I obtained an option over the Kyshtim property, which laid the foundation with British capital of the greatest mining and metallurgical enterprise ever created in Russia . . . you will therefore share my bitter disappointment that, despite all our endeavours, our coming of age should see us with our properties confiscated, the work of a lifetime lost, and with no apparent desire on the part of the Soviet Government to take the only honest course.

In the following year, Urquhart flung his company's dwindling assets into Mount Isa.

More than four years after its discovery Mount Isa at last had promise of the essential capital. The town grew. Churches and ice-works and hotel and courthouse and houses were carted from the deserted copper towns to the east, the shops with earthen floors were rebuilt, and in the houses ice chests replaced Coolgardie safes, and for entertainment the silent movies were flashed on to screens beneath the stars, with violin music to herald the climax. With money from London the company created a town for its employees on the other side of the low range, the first attempt to plant a model town in tropical Australia.

The steel skeletons of powerhouse and mill and smelters and mine headframe stood hot in the sun. The first train arrived through a suburb of tenthouses. Water flowed along a pipeline from the company's reservoir. Urquhart visited the mine and vowed that it was larger and richer than any of his Russian mines and predicted a profit of £570,000 a year. Like so many mining financiers, his mind was a pendulum, swinging between extremes of pessimism and optimism. His moods perhaps had more reason to swing. After losing his Russian mines and spending a decade seeking new mines he had got in quick succession Mount Isa and rich gold mines in New Guinea, and in his optimism he began to commit his company beyond its resources.

Mount Isa was gluttonous for money. When Corbould was hawking the mine he sought no more than £600,000 to prepare the mine for the production of bullion. When Urquhart became chairman he planned more ambitiously. He wisely spent heavily on drilling the lode to see how many million tons of ore he could reasonably expect. He built the town and hospital and amenities which Corbould in his frugality did not envisage. And he built far larger mills and smelters to gain the economies of large output. He spent the original budget three times over, and in 1930 the plant was still not complete and neither Urquhart nor his burdened company had more money to spend. The world depression was beginning, and some critics thought Mount Isa might have to be abandoned before it had sold a ton of silver-lead bullion.

In the European summer of 1930 Leslie Urquhart invoked New York's aid. He called on Senator Simon Guggenheim, one of seven sons of a Jewish migrant who had risen from peddling lace and trinkets on doorsteps to the control of the world's largest smelter and refiner of metals, the American Smelting & Refining Company. The senator was president of the company in 1930, controlled mines and furnaces from Newfoundland to Chile and Peru, and was persuaded to provide half a million to complete Urquhart's project at Mount Isa.

The mine soon swallowed much of the money. The engineers and geologists did not know that a few hundred feet below the parched surface were porous masses of rocks holding hidden reservoirs of water. In 1930 new underground workings pricked the reservoirs, and the mine did not have the pumps or power to master the water and one of the main shafts was flooded. After the shaft was pumped dry another inrush of water drove out the miners. Stronger engines and pumps were installed, but the engines failed just when one of the main underground haulage ways entered another zone of saturated rock.

When the new American general manager, Julius Kruttschnitt, arrived in Christmas 1930, he found mine and treatment plant and power house still being equipped. In the accountant's office he saw the sheaf of unpaid bills and was told that there was no money to pay them. He cabled the directors for more money, but Urquhart's company had none to spare. Again the American Smelting & Refining Company had to lend nearly £600,000 to save the money it had invested only six months previously. In all £3,500,000 had been invested when the roar of the treatment plant muffled the town in the blue winter days of 1931.

Urquhart waited in London for the exultant cables reassuring him that the plant ran smoothly. At Mount Isa the blast furnaces were temperamental, the sintering plant and mill were inefficient. Urgent cables summoned experts from America, who ordered new machinery for almost every step in the treatment process. The grains of silver and lead in the hard carbonate ore were fine and intergrown, and too many of them defied the bubbles of the flotation process. The ore from the mine was poorer than the careful diamond drilling and sampling had originally suggested—this was a surprisingly common hazard for companies that worked massive low-grade mines.

The company in its early years extracted from each ton of ore far less metal than the careful tests in laboratories and pilot plants had originally predicted—that was also a common hazard for companies that worked low-grade ores. The price of lead had halved since the company planned its venture. Month after month the working losses grew. In two years another £900,000 was borrowed to keep the mine alive and 1,200 men in their jobs. Urquhart had lost his Russian empire and now his Australian mine and its shackles of debt were slowly passing to the Americans. The brave optimist caught pneumonia and died on 13 March 1933.

Seven months later two miners were suspended by the company for blasting rock in a dangerous manner. The episode was magnified and the miners went on strike. The company closed the mine. Two thousand people left the town, the goods trains crowded with men singing to the music of violin and saxophone in the close tropical nights. In New York, H. A. Guess confessed that the decision to finance Mount Isa was 'the one serious blunder I have made in my long career of over thirty years with American Smelting & Refining Company'. He advised his company to sell the mine—if it could find a buyer. Instead the company borrowed £500,000 in London to prepare mine and mill for the working of the sulphide ores that were strong in zinc. After three months the pennant of smoke streamed from the chimneys, but the sulphide ore proved one of the most intractable lead-zinc ores in the world. Moreover, much of the zinc output was dumped at Mount Isa because it could not afford the journey by railway and steamer to overseas smelters. Not until the sixth year of production did the company make a profit, and even then the shareholders in New York and Paris and London and Sydney got nothing. The company had debts of about £3,500,000, and those debts and the annual interest had to be repaid first. It was safe to predict that the mine would be exhausted and the works dismantled before the shareholders got a penny.

A few men who clambered over the hills of Mount Isa saw green stains in cavities which the wind had hollowed. The splash of green signified copper but on the rocky outcrops they found no copper lode. In 1930 a deep hole drilled on the hanging wall of the Black Star lode pierced poor copper for several hundred feet and the engineers were surprised and delighted. They drilled a few more holes but the copper was deep and very erratic. As it did not come within eight hundred feet of the surface of the rocky valley its size could not be accurately or cheaply defined. The few holes they drilled were like arrows from a long bow; most missed the distant target.

The miners found more copper just when Japan entered World War II and made Australia aware of the decline in its copper industry. Copper was urgently wanted for war, and by 1942 Mount Isa had found so much copper that it was persuaded to switch from lead to copper mining. Luckily the copper lode lay so close to the Black Star lead lode that it could be worked from the existing shafts. The existing mill and flotation plant could treat copper ore. A new smelting plant alone was necessary and scavengers went to the ruined smelters of the abandoned Mount Elliott and Mount Cuthbert and Hampden mines and found electric crane and converter vessels and thousands of odds and ends lying in the sun or leaning buildings. The ghost copper smelters were resurrected at Mount Isa and worked as efficiently as the streamlined lead smelter. For three years Mount Isa became the nation's largest copper mine, but won no more profit from copper than it had won from lead. At the end of the war the company reverted to lead and zinc and the debts stood like a gigantic boulder above the main shaft.

So far Mount Isa's achievement was merely to survive. It had been for a decade and a half one of Australia's largest underground mines, producing about half a million tons of ore a year, but in that time its debts had increased rather than fallen. And all the time the company seemed to be approaching the day when all the ore would be exhausted. Roland Blanchard, the geologist, wrote a confidential and careful report in April 1943 and suggested that Mount Isa could be a ghost town within another decade. The logic of his report was cold and irrefutable.

The ambitious scheme to make Mount Isa a rich mine had been thwarted by the unexpected, the strange behaviour of ores, the erratic price of metals, the unexpected poverty of the lodes. Unpredictable events could still occur. After the war, in two years, the price of lead leaped from £25 to £91 a ton. The company paid its first dividend in 1947 when William Corbould, who had fathered the mine nearly a quarter of a century previously, was an octogenarian in Monaco.

The old miners and millmen who had clung to Mount Isa in its lean years got prosperity. The company paid a bonus that varied with the price of lead, and during the Korean war the average employee earned three times as much as wage-earners in Brisbane and Melbourne. In ten years the men got more in bonuses than the shareholders got in dividends. Hotels and betting shops were jammed, trains carried in the latest cars and refrigerators. So many men wanted the most prized jobs in the land that the company had to announce in southern newspapers that no jobs remained. Still they came with battered suitcases and toolboxes in the trains from Brisbane and Sydney and slept in the open air and queued each day at the company's office for the magical jobs that no longer existed.

Annual profit passed the million and all debts and debentures and loans were redeemed. With more money for exploring, the company found new treasures of ore. It had that rare alliance, a large lead-zinc lode and a big copper lode side by side, only feet apart in places, and side by side from 1953 were separate treatment plants for lead and copper, an insurance against fickle metal prices.

The American Smelting & Refining Company now owned half the shares and earned half the dividends from the mine it had nursed. But the Spanish mansions on the mine-leases passed from American to Australian engineers, and George Fisher came from Zinc Corporation at Broken Hill to guide the expansion. He became chairman just before the Australian uranium boom of 1953-4 when Mount Isa clicked with geiger counters, and landrovers radiated out the dusty roads with geologists, and all the hills were believed to be luminous at night with uranium, and even a Mount Isa pig was said to be radio-active. In July 1954 a syndicate of eight men from Mount Isa found uranium in a grassy valley near the ruined Rosebud smelters, and their Mary Kathleen became the richest uranium mine in Australia, and in the excitement in geologists' camps and city stock exchanges the big mine at Mount Isa was almost forgotten. But while new mineral lodes were being sought far from bitumen and milkbars, richer lodes were being discovered far beneath the schoolground and streets and oleander trees of Mount Isa itself.

In the ten years after the dividends began, Mount Isa Mines earned net profit of £21 million and £100 million of revenue. High profit financed the search which the company had long been too poor to make. The diamond drills soon pointed to huge reserves of lead. More and more copper was found in the Black Star and in areas where no copper had been seen. Near the small Rio Grande mine geologists found a little copper in the dolomite rock. As dolomite was the parent rock of the copper in the Black Star the company drilled deeper in the hope of finding more dolomite and more copper. Just as the Black Star copper lode was buried far below the surface, so copper at the south end of the field might also be buried deep. The first drill holes found copper, poor but plentiful. New holes were drilled deeper. From the depth of half a mile the drillers hauled up a core of rock rich in copper. Another deep hole penetrated even richer copper for a distance of five hundred feet, one of the most sensational holes drilled for metal in any country. Hidden from the sight of man was the most important Australian discovery since Kalgoorlie and Iron Knob were opened in the 1890s. Years would pass before the immensity of the new copper lode far beneath the schoolground was accurately plumbed but even in May 1956, Fisher wrote to New York: 'I have no hesitation in saying that the recent developments at Mount Isa have been so extraordinarily good that this field could be stated to combine the lead existing at Broken Hill and also the potentialities of Rhodesian copper.'

Mount Isa began to plan to double and treble output. If the mine had been near a city the challenge would have been formidable, but being in parched land, six hundred miles to the nearest port and linked only by a weak railway, the challenge was almost forbidding. A stronger railway was essential and would cost £30 million, and it took the government nearly four years to promise the money. Power and water were wanted. The company plugged a rocky gorge to impound a sweeping sheet of water, and built Australia's first outdoor power station with turbines and boilers hauled by locomotives and lorries through central Australia. In effect it had to create a new industry from the underground crushers and ore-winders to the surface silos and mill and smelters and offices and laboratory at Mount Isa and the copper refinery and fabricating plant at Townsville. The company spent about £33 million on this expansion in six years alone.

Out amongst the hard red ranges the horsemen mustering cattle could see an orange and white beacon rising far away like a barber's pole, the tallest chimney in Australia standing near hills where black men were slaughtered and whites were murdered in their sleep in years which the oldest men of Cloncurry and Mount Isa still recalled. Travellers flying over the black empty earth saw at night the luminous glow of Mount Isa from forty miles away. On the thin band of bitumen running into the Northern Territory motorists saw the long trail of white smelter smoke. The mines that had so long been in agony and the town that seemed certain to die produces today possibly more export income than any city in Australia.

In the final chapter all sums of money are given in dollars; two Australian dollars equalled one Australian pound when decimal currency was introduced in 1966. The book would have appeared more consistent if the £ had been converted automatically to the $ in all the chapters, but such a conversion could be misleading. In 1900, for example, £1 was closer to $10 than $2 in purchasing power.

29

NEW AGE

THE MINING FRONTIER no longer seemed boundless at the end of World War II. It seemed that the prospector with his pick and tin dish and waterbag and dollypot had found nearly all the valuable outcrops in the land. Most surviving prospectors of the last great era of discovery were old, confined on hot days to their camp of bags and boughs on the fringes of ghost mining towns in the inland, or confined by rheumatism to their hut in the forests of Gippsland or Tasmania. They were always ready to name a hill or plain where they believed they could find a rich reef or lode if only they had the strength of their youth. They did not concede, as so many men did, that the richest surface lodes had all been found. Their optimism seemed eccentric and was not.

One path of expansion lay with the 'new metals'. Atomic powe needed uranium, the high speed cutting tools needed tungsten, air craft were made with titanium that was strong and light in weight, the uses of aluminium were numberless. Australia was thought to lack Africa's or North America's diversity of payable minerals, but it had ores of all these metals.

Australia had long mined the tungsten ores, wolfram and scheelite, and about 1905 Queensland was mining about half the world's supply of tungsten. But the mines languished in the long decline. When the Japanese overran the world's great tungsten region in south-east Asia in 1942 Australia was urged to provide tungsten for electric lamps and radio valves and certain steels. Six hundred Chinese labourers evacuated from the phosphate mines in Nauru were sent to mine wolfram in Central Australia, and they failed. In the Tasmanian mountains Aberfoyle and Storey's Creek produced wolfram, but not enough. Two Commonwealth officials, M. A. Mawby and P. B. Nye, went to the windswept beach on King Island to see how much scheelite could be wrung from the small struggling mine found thirty-eight years before, and sensed quickly that it was one of the world's most promising scheelite lodes. The diamond drill proved their prophecy, and when tungsten became even scarcer after the war King Island Scheelite became one of Australia's most profitable companies, and the Australian mines produced annually about 4 per cent of the world's tungsten.

From the golden beaches between Sydney and Brisbane came another metal. The surf of a million stormy nights had dumped heavy particles of mineral on the changing beachline, and those sands of rutile and ilmenite were the source of the glamorous lightweight metal, titanium, and of titanium paints and pigments. The rising demand for titanium was met by a few Australian companies before the war and many after the war. In one year, 1957, output of these beach sands was worth nearly £10 million. More important for some, the heart of the sandmines was on Queensland's playground coast, and old engineers who began their mining career in Western Australian desert could end it amidst the dancebands and neon lights and seawinds of Surfer's Paradise.

The mines of uranium, unlike those of the other new metals, were in the hot, remote regions to which Australian mining engineers had long been exiled. The oldest uranium mine was at Radium Hill, South Australia, about eighty miles over the dry plains from Broken Hill. Abandoned by a Sydney company in the first war it was worked by the South Australian government after World War II in the hope that nuclear power would revolutionize a state that was weak in thermal fuels. Radium Hill's second life ceased in 1961.

The forgotten Northern Territory provided richer uranium only fifty miles from the port of Darwin. Gold seekers hurrying to Pine Creek in the 1870s had passed Rum Jungle's uranium. A decade later Chinese coolies had built a railway past the uranium deposit and, even later, miners had worked nearby for poor copper ores. The Commonwealth tried to settle farmers at Rum Jungle before the first war and it built a bitumen airstrip and army camps there during World War II, but no enterprise seemed to flourish at Rum Jungle. And when the atom bomb had conquered Japan and the troops had dispersed and the grass crept close to the bitumen airstrips, Rum Jungle had promise of another undisturbed sleep.

A forty-two-year-old prospector named Jack White lived in a bush humpy near Rum Jungle in 1949 and read a pamphlet issued by the Bureau of Mineral Resources to guide prospectors in identifying uranium ores. He soon had occasion to read it again. Walking one day with his gun past the old Rum Jungle copper workings he halted for a few minutes in the shade of a tree, and as he crouched in the shade he noticed a green mineral a few yards away. Thinking it was unlike the green copper carbonate that littered the area he took it home and read his pamphlet and decided that it must be the radio-active ore, torbernite. Next morning he dug a trench and found green ore and some yellow ore that he thought was the radio-active ore, carnotite. He took samples to Darwin but the mine officials, having no geiger counter to test the ore, advised him to send them to Alice Springs. He packed them in a Sal-Vital tin and posted them away. Soon his peace was disturbed by officials who came to inspect his find. As the Commonwealth had in effect a monopoly of all uranium, and as the deposit was on freehold land, he got no mining lease. His ultimate reward was £25,000, which he distributed freely far and wide.

The government drilled the lode and found it large and rich. Consolidated Zinc developed the mine for the Commonwealth, and in the hush of secrecy created a deep open cut, power house and treatment plant and workshops, and a beautiful garden township. From Darwin they shipped a yellow powder, the uranium concentrate, and copper as well, and though no balance sheet has ever been released, Rum Jungle appears to have been the most productive mine the Northern Territory has known.

Rum Jungle inspired no boom and no swift search for new uranium deposits. But when the Commonwealth relaxed its monopoly of uranium, private companies joined in the search, and when oil flowed from a wildcat well on the Western Australian coast in November 1953 a speculative mania infected investors, and the infection spread from oil to uranium shares. In the winter of 1954 the quest for uranium grew intense and the northern centres of Darwin and Mount Isa and Cloncurry had most of the country's geiger counters and geological teams, and their hotel bars were high-powered transmitters for rumours of new uranium finds and fabulous prices paid for old. In July 1954, at the peak of the excitement, a syndicate of eight searchers found genuine cause for excitement. The unofficial leader of the eight, Clem Walton, was a taxi driver in Mount Isa, and an alert prospector in his spare time, and specimens of the uranium ore called pitchblende that he was shown reminded him strongly of ore which an old copper gouger named Jack Fountain had mined in the red ranges towards the Rosebud copper smelters where he had lived as a lad. With members of his team he prospected in the hills where he believed the old copper gouger had long ago found the unusual ore, and on a steep hill overlooking a wide valley they found grey ore that made their geiger counter race.

The beauty of a large uranium deposit is that its potential can often be measured by instruments before even a trench has been blasted on the outcrop. The richness of this virgin Mary Kathleen mine was so apparent that companies vied with one another to buy it. Barely a month after the discovery Walton's syndicate accepted £250,000 in cash and a high royalty on all ore mined; no Australian prospectors had received a comparable price for an untested mine since the famous Londonderry Golden Hole near Coolgardie was sold exactly sixty years previously. But Mary Kathleen was more than a golden hole, and when the neat township had been built in the valley and the open cut gouged into the hill and the treatment plant built, it began to earn huge profits for the discoverers, the Australian shareholders, and the Rio Tinto company of England.

The new uranium fields had the glamour of a royal birth, but by the early 1960s they were dying. The stockpile of uranium in Britain and the United States was adequate. Until nuclear power was used more for generating electricity, the market for freshly-mined uranium ores would be small. The closing in 1961 of Radium Hill, in which little ore seemed to remain, was followed two years later by the closing of Mary Kathleen, which still retained more ore than it had yielded in its seven fat years. After the uranium mines near the gorges of the South Alligator River sent the last drums of uranium oxide through Darwin to Britain, only the treatment plant at Rum Jungle continued to work. Elsewhere caretakers of new towns waited for, or diamond-drillers worked for, the revival of uranium prices expected in the 1970s.

The burst of uranium discovery between 1949 and 1954 had helped to swing exploration to tropical Australia, for the three new fields lay near the northern coast. The deep copper found at Mount Isa in 1955 quickened the swing to the north. And in the same year the discovery of bauxite deposits, capable of nourishing a great aluminium industry, suggested that tropical Australia might soon rival southern Australia in mineral wealth.

When the first aluminium was smelted at Bell Bay in northern Tasmania in 1955, the raw material came from Malaya. In retrospect the carrying of bauxite to Australia seems almost as odd as the idea of carrying gold to Kalgoorlie, for Australia now holds about one-third of the world's known reserves. Knowledge of the existence of payable bauxite in Australia, however, came slowly. The richest deposits lay on the secluded tropical coasts, in the path of those monsoons whose moisture had helped to concentrate the alumina in the surface rocks. The only European settlers in the vicinity were missionaries. The nearest mining fields were far away. Even if the bauxite had been seen in the 1890s, when the surface of Australia was scoured by thousands of sharp eyes, it would have aroused no interest. In 1899 the world consumed no more aluminium than Bell Bay now produces in a month. Even when the demand for the ligh metal soared in World War I, the main smelters of aluminium—th United States, France, and Switzerland—used alumina from nearb deposits. The bauxite on the Gulf of Carpentaria was too remote f export to the northern hemisphere and even too remote from T mania, the source of the cheap power so essential for the electrolytic method of producing aluminium. There was faint incentive to search for bauxite until World War II when the scarcity of aluminium in Australia began the slow steps which ultimately set up the government industry at Bell Bay.

The shoreline of the Gulf of Carpentaria resembles a mis-shaped horseshoe, each end of which holds bauxite. On the western end, Captain Fred Wells and Fred Waulkes were members of the crew of a yacht sent in 1949 to investigate a find of bauxite in Arnhem Land. The find was unpayable, but they continued the search; next year they were successful on the Wessel Islands. The news attracted the Aluminium Production Commission, the agency of the Commonwealth and Tasmanian governments which was setting up the plant in Bell Bay. Drilling on the tropical islands revealed about 10 million tons of bauxite. None has yet been mined, but the discovery led to another find at Gove, on the mainland almost opposite. Curiously, red pebbles of bauxite at Gove had been seen but not recognized during World War II. At a time when Australia was anxious to find bauxite, a construction gang had carved an airstrip in a promising deposit; the runway at Gove was red with it. In the 1950s a procession of companies inspected or explored Gove. After long delays, a syndicate known as Nabalco agreed in 1965 to spend about $100 million on an open cut, an alumina refinery, a township and port. Swiss Aluminium, the biggest fabricator of aluminium in Europe, was supplying half the capital, and from 1971 the red pebbles from Gove, it seemed, would be shipped as alumina for final treatment in European smelters.

Across the warm water of the Gulf, three hundred miles to the east, the bauxite was richer and more massive. In places the red seacliffs were so conspicuous that they had been observed by Captain Matthew Flinders when he was making the first circumnavigation of Australia in 1802; a study of his charts reveals that all the red cliffs which he mapped were bauxite. Towards the end of the nineteenth century some of the cramped ships which steamed to Normanton, the small port for Cloncurry's copper and Croydon's gold, may also have passed within sight of the red cliffs, though usually they kept clear of the desolate coastline. Few explorers and prospectors landed on the shores of mangroves and mudflats and tidal creeks. Presbyterians founded the Weipa aboriginal mission within a pebble's throw of a bauxite deposit, but their mission was to gather converts, not minerals. In 1902 a government geologist collected a sample of low-grade bauxite, but understandably he was not excited. In 1947 a university geologist collected samples of high-grade bauxite, just when the economic conditions were ripe for discovery. An official request from the Mines Department in Brisbane to three remote mission stations in the area to collect samples of bauxite was faithfully answered, but in Brisbane the samples proved to be low grade.

Of the Broken Hill companies, the one most interested in exploration was Consolidated Zinc, now known as C.R.A. As early as 1953, Maurice Mawby suggested that his field-geologists should be alert for phosphate and bauxite in northern Australia; he thought bauxite could possibly be found in Cape York Peninsula. Two years later one of his geologists, Harry Evans, was leading a small party in search of oil-bearing structures in the Peninsula when he noticed boulders and blocks of bauxite between a remote cattle station and the Weipa mission. He found that the deposits were extensive, and the samples which he carried south were promising in quality. It was July 1955, the very month in which the Bell Bay plant began to produce aluminium ingots from Malayan bauxite.

Evans returned to the Cape York Peninsula in October, landed in a small aircraft at the ghost gold town of Coen, and hired a Landrover and a fibreglass dinghy at a cattle station. He drove over the rough track to the Weipa mission and hired the help of an old aboriginal named Matthew. They hoped to trace the estuaries and coastline in the dinghy, landing wherever bauxite was visible or feasible, but on the first day the plan was thwarted. The outboard motor on the nine-foot dinghy was swamped by choppy waves. Then, in cleaning the motor, they accidentally let the throttle slip overboard and failed to find it in the thick mud around the roots of the mangroves. After rowing the dinghy back to the Weipa mission, they could find no gadget or spare part that would act as a throttle. 'The trip down the coast seemed now to be out of the question', wrote Evans. Eventually the handle of a native spear, carefully carved to the dimensions of the carburettor, served as a throttle, though it had the effect of literally throttling the engine at the wrong moments. Using most of the space in the dinghy for carrying fuel and only a small space for bedrolls and food—they carried only tea, dried milk and sugar—they set out again.

'As the journey down the coast revealed miles of bauxite cliffs', Evans confessed, 'I kept thinking that, if all this is bauxite, then there must be something the matter with it; otherwise it would have been discovered and appreciated long ago'. There was nothing the matter with it. The payable bauxite in fact covered several hundred square miles, the largest area of payable mineralization ever found in Australia. By 1964 systematic drilling suggested that Weipa held about one-quarter of the 'known potential resources of bauxite in the world'.

Comalco—a partnership of Conzinc Riotinto of Australia and the Kaiser Aluminium & Chemical Corporation of U.S.A.—developed Weipa. No mining project in Australia had been based on such a variety of tests. Ten thousand samples of bauxite were assayed before the mine was planned. A water-filled model of Albatross Bay was made at the University of Delft in Holland to guide the dredging of a channel for the big bauxite ships which arrived from 1963. A meteorological station was set up to provide accurate knowledge of the heat and wind and rain before a township was designed. Colonies of Weipa white ants were deported to Canberra where their appetite for building materials could be measured and controlled.

The main destination for Weipa's bauxite was the huge cooperative refinery opened in 1967 at the central Queensland port of Gladstone. From Gladstone the alumina (aluminium oxide) went to aluminium smelters in the United States and to the smelter at Bell Bay which Comalco now controlled. The costliest venture in the history of Australian mining, the aluminium empire based on Weipa had absorbed a capital of nearly $300 million by 1968 and is still expanding.

The bauxite on both sides of the Gulf of Carpentaria was far from plants where aluminium was smelted or saucepans shaped. Only thirty miles from Perth, other patches of bauxite had long been known. After World War II they were examined as the possible base of an aluminium industry, and rejected. While some patches carried 47 per cent alumina, they were not as rich as the 49 per cent at Gove and the 58 per cent at Weipa. Moreover they contained four or five times as much quartz or silica as the Weipa deposits. The silica, it was believed, would clog the extraction process. In the late 1950s a syndicate led by Western Mining Corporation, the largest gold miner in Australia, examined the shunned bauxite in the Darling Ranges. They were not so sure that silica was a metallurgical poison and they made careful tests. The result was surprising; this silica aided rather than impeded the process of extracting the alumina. The problem of finding adequate ore was solved by June 1960 when 37 million tons had been proved. The problem of raising a huge sum for the building of an alumina refinery near Perth and an aluminium smelter near Geelong was met by forming a partnership with the Aluminium Company of America under the name of Alcoa of Australia.

This rising demand for relatively new metals—aluminium and

tungsten, uranium and titanium—had stimulated much of the successful prospecting in Australia in the 1940s and 1950s. The following decade was marked by the discovery of minerals for which the world's appetite, even half a century ago, was high. The belated finds of oil, iron ore, nickel and phosphate rock in the 1960s seemed, at first sight, out of sequence. Australian prospectors had sought nickel before they were even sure how to spell 'aluminium', and companies had sought oil before they had even heard of uranium.

Eighty years separated the sinking of the first oil bore and the finding of a payable oilfield in Australia. The search for oil began late in 1881 on the sandy Coorong, not far from the South Australian port of Kingston, after a substance on the surface was mistaken for evidence of oil seepage. The Salt Creek Petroleum Company raised the necessary capital during Adelaide's share boom of 1881—that same boom which financed the early Silverton mines and the prospecting which led to Broken Hill. As it happened, Adelaide investors would have found a shorter cut to oil if they had bought new shares in 1885 in the Broken Hill Proprietary and then waited for eighty years. For the bore on the Coorong went down more than 600 feet, and produced mere salt water.

In the following twenty years the probe for artesian water in dry pastoral regions offered some hope that oil might be accidentally found. Artesian bores were sunk in sandstone, limestone or mudstone, and those sedimentary rocks were also likely places for oil. Indeed, an artesian bore at Roma in Queensland discovered natural gas in 1900 at a depth of 3,700 feet and the gas was piped in 1906 to the main street to illuminate the town. For ten nights the lamps burned, and then went out; water had probably flooded the gas sands in the borehole. A wide search for oil and natural gas in the district yielded a few traces of oil, fitful flows of gas, and a long line of red figures in the ledgers of the oil-drilling companies. Success did not come until the 1960s when a series of gas wells ultimately justified the laying of 276 miles of pipeline from Roma to Brisbane.

The search for oil in the 1920s extended from the Kimberleys and Papua to the coast of Victoria. There, the small fishing port of Lakes Entrance, a base forty years later for the discovery of a major oilfield beneath the ocean, had a tantalizing glimpse of oil. William Baragwanath, the most experienced gold geologist in Victoria, had argued by analogy that east Gippsland could contain oil; on his initiative the Victorian government in 1922 began to sink a line of bores near the western end of the Gippsland Lakes. At the same time G. W. Shirrefs moved from his small silver-lead deposits in the ranges near Buchan to look for oil by the coastal lakes, and on

25 July 1924 his bore was down just over one thousand feet when it found the first accumulation of oil in the continent. The rush of peggers and promoters to Lakes Entrance was reversed when his second and third bores found more artesian water than oil. Their flow of oil was easily measured; it averaged a steady pint a day. The monthly flow from those two wells was just sufficient to drive one T-model Ford from Lakes Entrance to Melbourne.

All kinds of arguments were conjured to explain why Australia had to import all the fuel for T-model Fords, motor vessels and diesel powerhouses. The search for oil in Australia had failed, it was said, because foreign companies sabotaged the search; they had found oil and then sealed it from sight. It was argued that oil would not be found south of the equator. It was argued that Australia's rocks were too old to contain oil. These myths were pricked by Dr W. G. Woolnough, geological adviser to the Commonwealth, when he reported on his tour of oilfields in the Americas in 1931. He was one of a few geologists and mining men who believed that oil would be found—if only more brains and money were employed in seeking it.

The fitful search for oil was in one sense a reflection of the attitude of the entire mining industry towards exploration in the years 1900-45. As few new mineral discoveries were made, the search for most minerals was lethargic. As the search was lethargic, the chance of a discovery was small. The vicious circle of doom was not easily snapped. The rich mining companies in Australia, between the world wars, were not interested in the risky search for oil in the ground, though some Broken Hill companies did experiment with the idea of winning it from coal or shale. The quest for an oil well therefore rested on little wildcat companies floated on Australian stock exchanges and, occasionally, an overseas oil company. Private spending on the search in Australia in all the years to 1945 had not even reached $3 million. The scarcity of capital for drilling was matched by the scarcity of geologists who knew where to drill; some oil companies consulted rain-makers, water-diviners and blindfold seers. The reluctance to enlist every geological clue when planning the site for a drill-hole was a reflection of a mining industry which was slow to enlist the aid of geology in seeking metals. But in the search for oil, the neglect of geology was more costly. For if mineral detection is likened to criminal detection, then those seeking oil rather than metals usually lack even the evidence of fingerprints to guide them.

Oil was scarce in Australia during World War II and petrol was rationed until 1949. The national incentives to find it were strong.

The Commonwealth's Bureau of Mineral Resources, created in 1946, facilitated a more scientific search by mapping sedimentary basins and making geophysical surveys in favoured areas. One of those areas, Rough Range in Western Australia, yielded an exciting flow at the end of 1953. But the pool of oil was too small, as the following eighteen holes slowly and painfully revealed. The search in the 1950s was many times more strenuous than in all the previous decades added together, but at the end of the decade Australia and Antarctica still remained the only continents without an oilfield.

A commercial oilfield was at last found in December 1961. A Californian-Australian partnership struck oil at Moonie, 192 miles west of Brisbane. They found a second small field sixty miles further west, at Alton, in 1964. In the same year, on the other side of the continent, another Californian-Australian partnership, which had suffered defeat on Rough Range, went 150 miles north to Barrow Island and found payable oil. In remote parts of the continent discoveries of natural gas coincided with the discovery of the small oilfields, and were perhaps more significant.

Meanwhile the oldest mining company in Australia began to seek oil less than two hundred miles from its own head office in Melbourne. It was a strange ricochet of events by which Broken Hill Proprietary had moved from the mining of silver in the dry inland to the search for oil beneath the sea. Its silver smelters at Port Pirie had needed ironstone, so the company acquired the nearest iron ore deposit—Iron Knob in the Middleback Ranges. Possession of the iron ore in turn induced B.H.P. to make steel at Newcastle and that forced the company to mine its own coal. As the coal lay in sedimentary basins there was a chance of finding natural gas or oil in the vicinity, so B.H.P. began surveys near its Port Kembla steelworks and coal mines in 1954. After one well had been drilled the company consulted Lewis G. Weeks, a petroleum geologist from the United States, and in 1960 he reported pessimistically. He did add, as a bonus, that if B.H.P. were interested in finding oil elsewhere he could recommend a promising area. Recent advances in offshore drilling convinced Lewis Weeks that for the first time a search in the sedimentary basins near Bass Strait was feasible. B.H.P. accepted his advice and began the long and expensive survey of the geology of the sedimentary rocks beneath the sea. By 1964 the geological pattern pieced together—by aircraft equipped with magnetometers, and ships which measured underwater explosions—encouraged B.H.P. to enlist as drilling partner Esso Exploration, a subsidiary of Standard Oil of New Jersey. A drilling-vessel sailed from Texas to the eastern end of Bass Strait, and two days after Christmas 1964 spudded in Australia's first offshore well, nearly forty miles from the long beach where oil had been unsuccessfully sought during the Lakes Entrance boom.

The drill churning below the seabed found a rich natural-gas field named Barracouta. The next wells were dry, but the fourth found the Marlin oil and gas field early in 1966. The oilfields of Kingfish and Halibut were found in 1967. Broken Hill Proprietary, already the nation's largest miner of coal, builder of ships and maker of iron and steel, now had half of the shares in underwater reservoirs which held many times the oil and gas previously found in Australia. And Victoria had its first valuable mineral discovery—excepting the brown coal developed after World War I—since the gold era of the nineteenth century.

On the glassy seas off Barrow Island the oil-tankers sometimes pass larger ships carrying iron ore from new ports in the same region. More than two thousand miles away, on the windswept waters of Bass Strait, men on the oil-rigs towards the northern shore sometimes see a ship on her way to load iron ore from a new port on the southern shore. These adjacent bonanzas of the 1960s—oil and iron ore—had been found in the face of very different obstacles. While governments consciously encouraged the search for oil, they unconsciously discouraged the search for new iron deposits. They eased the difficult task of finding oil and made difficult the easier task of finding iron.

On the eve of World War II the main iron ore region was the Middleback Ranges. Most of its ore was shipped from Whyalla to the Broken Hill Proprietary Company's steelworks on the Pacific coast, but occasional shipments went to steelworks in Japan, the United States and Belgium. Australia was only a minor exporter of iron ore, but in 1938 the traffic seemed likely to expand. The Nippon Mining Company, eager to draw Australian ore to Japan, was opening a deposit of iron ore on Koolan Island in Yampi Sound. Of the seventy men who worked on the rocky, sweltering island in the remote north-west of Australia, only a few were Japanese technicians; but their presence on an unguarded coastline apparently worried the federal government. Japan was already at war with China and her territorial ambitions were widely feared. To Joseph Lyons, the Prime Minister of Australia, the problem was how to remove the Japanese from Yampi Sound without creating a diplomatic incident. The simplest answer was to ban the export of iron ore not only from Yampi Sound but also from every Australian port. He placed the embargo in July 1938, and no iron ore left Australia for foreign ports for a quarter of a century.

Lyons publicly justified the embargo by emphasizing that the Australian continent was meanly endowed with payable iron ore. Dr W. G. Woolnough had elaborated that fear in a pessimistic report to the Commonwealth in May 1938: 'it is certain that if the known supplies of high grade ore are not conserved Australia will in little more than a generation become an importer rather than a producer of iron ore.' He argued that if Australia did not conserve her iron deposits for her own needs, the steelworks at Port Kembla and Newcastle in the 1970s or 1980s would have to import iron ore from overseas.

Today the reserves of iron ore are at least fifty times greater than they were in Woolnough's mind; and therefore it is deceptively easy to dismiss his prediction simply because it proved to be wrong. But his report was dubious even in the light of the facts on his desk in 1938. It is fair to suggest that his report carried a superstition about the causes of mineral discovery which was commonly held during that generation when few new mineral fields were found in Australia. Woolnough believed that payable oil had not been found in Australia because it lay far below the surface and therefore could not be seen; only expensive drilling could find oil, but not enough holes had yet been drilled. In contrast, a mineral such as iron ore outcropped on the ground and therefore could be easily seen by prospectors, drovers, timber-cutters, rabbiters or anyone who passed by. He believed that every iron outcrop near the coast or railway lines—in short every outcrop which was favourably placed—had already been seen. His belief was reinforced by the experience of most mineral searchers who went outback in the 1930s. Almost wherever they wandered they found tins, tentpoles, discarded shovels and tin dishes, to prove that prospectors had been in the area ten or twenty or fifty years earlier. They therefore assumed that the chance of easy discoveries was now slight.

The flaw in this assumption was that early prospectors had been interested only in gold, tin, silver-lead and copper—the only minerals which seemed likely to reward them in return for a small outlay of capital. Iron ore, however, rarely interested them. They could see no way of making money from it. Even in 1938 their chance of selling iron ore to the sole Australian steelmaker was faint; B.H.P.s own mines in the Middleback Ranges could supply all that its blast furnaces required. There remained the prospect of selling iron ore to foreign steelworks. The deposit in Yampi Sound was the first to be developed solely for export, but until it proved profitable a prospector had no incentive to search for iron. Moreover, even if he happened to find a rich deposit, he had small incentive to pay for a mining lease and so make his discovery public knowledge.

One consistent lesson in the history of Australian mining is the correlation between intelligent searching and discovery.* An awareness of that correlation was absent from Dr Woolnough's report. Indeed Australia's known reserves of iron ore in 1938 were surprisingly high—if measured against the weak incentives for seeking it. If Dr Woolnough was entitled, on the available evidence, to be optimistic that oilfields would be found in Australia, he was no less entitled to be optimistic that more reserves of iron ore would also be found in Australia.

Another lesson in the history of Australian mining is the frequency with which unpayable deposits have been made payable by new technology. The history of Mount Lyell, Bendigo, Kalgoorlie and nearly every large field illustrates that truth. 'Very large tonnages of iron ore', Woolnough admitted, 'are known to exist in the interior of Western Australia'; but he explained that they were too far from the coast to yield a profit from existing markets and existing methods of mining and transport. This was true in 1938, but he assumed that it would also be true in 1978. Even if Dr Woolnough had seen the massive iron deposits at Mount Tom Price—deposits which today are very profitable because of advances in mining, metallurgy and transport—he would have given them no place in his predictions.

One critic of his reasoning was Western Australia's Minister of Mines, A. H. Panton, who reminded parliament on 6 September 1938 that in the next half century improved means of transport 'might make very inaccessible ore bodies very accessible'. He reminisced how he himself had come west as a gold seeker in 1900 and how, walking the dusty gold trail from Cue to Peak Hill, he had passed a signpost pointing to the small gold town of Wiluna, 185 miles away. On the signpost, beside the '185 miles', someone who had walked the whole way had scratched the comment, 'Every Inch of It'. The Minister did not need to remind his listeners in 1938 that advances in transport and metallurgy had recently transformed sleepy Wiluna into one of the most prosperous gold towns in Australia. In contrast the signposts that Woolnough saw on the tracks to his ironfields were pessimistic and permanent. That he was not

*The correlation can be stated in several ways. It can be stated in the negative: if a mineral has not been vigorously sought, it is not logical to argue that the mineral is rare in a country. It can be stated also in the positive: the more mental and physical effort spent in searching, the greater the chance of discovery. The halo of romance, of luck and unpredictability, which traditionally surrounds the discovery of a mining field obscures this correlation.

alone in his pessimism is shown in the support which his opinions received for more than twenty years.

In isolated places around the Australian coast or far inland, valuable iron deposits lay untouched throughout the 1940s and 1950s. Visible on the ground, they were invisible in Canberra. In years when many Australian economists and politicians were justifiably perturbed that the nation's exports were increasing too slowly, iron ore offered a valuable treasury of export income. The treasury, however, was locked.

Successive governments reaffirmed in the 1940s and 1950s that iron ore was relatively scarce in Australia. By retaining the embargo on exports, however, they closed the only market for new-found deposits. They thereby erected a fence against searchers; they promoted the very scarcity which they feared. Canberra's policy was complemented by Perth's. The government of Western Australia, which in 1938 had criticized the embargo on the export of iron ore from Yampi Sound, virtually erected their own fence against searchers. Until 1961 they refused to grant prospectors or companies a title to explore for iron ore. They also gave no assurance to a discoverer that he could keep what he found. Deposits were to be awarded, not to the finder, but to the company which promised to bring manufacturing industries to Western Australia. So Koolan Island—scene of the Nippon Mining Company's frustrated plan—was awarded to B.H.P. in return for a promise to erect a steel-rolling mill near Perth. Just as the search for gold in eastern Australia in the 1840s had been retarded by the English common law that all gold belonged to the Crown, so the search for iron ore in Western Australia more than a century later was retarded by the new law that all iron ore belonged to the Crown. Canberra and Perth thus ignored another lesson which the history of Australian mining seemed to furnish—the obvious one that laws discouraging the search for minerals also discouraged discovery.

Meanwhile the glow of Japanese steelworks was visible in the 1950s from many ports along the Indian and Pacific oceans. As a steel-maker Japan passed Great Britain and eventually won third place behind the United States and the Soviet Union. Japan increasingly imported iron ore from India, Malaysia, Canada, the United States, Chile and Peru, but not from Australia. Ministers in Canberra from time to time reaffirmed their pessimism that Australia was deficient in iron ore. The official count of the reserves in Australia had increased from about 260 million tons in 1940 to about 370 million tons in 1959, but then the Australian steelworks' appetite was growing at a faster rate than the ore reserves. Although the embargo on exports of iron ore was increasingly criticized, it would probably have survived longer but for the recession known as 'the Credit Squeeze'. Late in 1960 Australia's overseas funds were low. One way of increasing them was to sell iron ore to Japan. In December 1960 the federal government cautiously offered companies the chance to export from either second-ranking deposits or such deposits as exploration might reveal. Such a rash of discoveries followed the lowering of the barrier that it was lowered further in 1963.

The sudden boom in iron-ore exports in the 1960s embraced a string of harbours spaced along one-quarter of the coast of Australia. The iron ports ran from Geraldton on the Indian Ocean to Darwin on the Timor Sea; they extended from a port so far south that it shipped wheat, to a port so far north that it shipped crocodile skins and pearl shell. Iron ore was shipped 4,000 miles to Japan from ports which had been the springboards for the Indian Ocean Gold Trail of the 1880s and from ports which had never harboured a ship until the 1960s. The Western Australian town of Dampier is not yet printed on most maps of the continent, and yet in October 1967 it sent to sea the largest cargo ever to leave Australia—$1 million of iron ore in the 104,500-ton ship *Sigsilver*. She sailed for Rotterdam, the only European port which could accommodate her. Although the dredged port of Dampier was only opened in the previous year, it will soon handle more cargo than either Sydney or Melbourne, and virtually its only cargo will be billions of lumps or pellets of dark iron ore. Dampier is one of the ports which is changing the seven hundred miles between North West Cape and Yampi Sound into the Iron Coast.

Most of the new iron deposits appeared on no mineral map in 1960. In Geraldton's hinterland, Tallering Peak was believed to be the best deposit of iron. Accordingly the government of Western Australia called tenders from companies willing to mine the ore for export and accepted a tender from Western Mining Corporation. But even before Western Mining explored the deposit, its geologists found a more profitable deposit only twelve miles from a railway line which carried the wheat harvest to Geraldton. The new deposit in the Koolanooka Hills, the basis of the first of many contracts signed by Japanese steel mills and Australian iron mines, showed how perfunctory had been the earlier search for iron ore in Western Australia. If an iron deposit in the settled corner of the State could remain undiscovered until 1961, the chance was high that other deposits would be found in the remote corners.

In the Pilbara, an old gold region north of the Tropic, the Western

Australian government invited tenders in 1961 from companies willing to mine at Mount Goldsworthy, which official records said was the most promising iron deposit. It certainly was promising, and after the new owners had built seventy miles of railway to the old gold port of Port Hedland in 1966, they began the regular shipments which would win an estimated $144 million from Japanese steel mills in seven years. But before one mile of railway had been laid, Mount Goldsworthy was dwarfed by discoveries in the same region. Ironically some of the discoveries had been made in the 1950s and kept secret because there was no reward for discovery.

Lang Hancock, a Pilbara pastoralist, was flying his small aircraft low over the tumble of red-brown ranges in 1952 when he saw the black outcrop of mineral below. Next year he investigated the spot, gathered samples, and found that the mineral was iron. He said nothing and did nothing, until the repeal of the restrictive laws gave him an incentive to act. In his own aircraft he flew mining men to the secret site in 1961. The ore proved to be limonite—too poor to pay. Asked if he knew where the richer hematite could be found, he led the visitors to a hill not far from his own Hamersley homestead. Neither outcrop has yet been mined, but Hancock was clearly the discoverer of a new iron-ore field. One of many Australian pastoralists who found minerals on their own grazing run, he was to outstrip them all in the money he made from mining.

Conzinc Riotinto of Australia and the Californian firm of Kaiser, the partnership which was already developing bauxite at Weipa, began a systematic search of the hot ranges which Hancock had shown them. Bands of iron ore were found to run across country for several hundred miles like the stripes of a zebra. The stripes were poor, averaging about 30 per cent iron in an area where twice as much was essential to pay the costs of mining and transport. It was feasible, however, that somewhere along those bands an enriched area could be found. One man who held that theory was Haddon King; a North American geologist who had come to Western Australia during the gold boom of the 1930s, he now directed exploration for Conzinc Riotinto. To test the theory, field-parties systematically walked across the banded iron formation as it crossed the waves of rocky hills. To speed the work a helicopter carried the searchers to the top of the ranges. They then walked downhill, testing the bands of iron as they went, and from the foot of the hills the helicopter carried them to the starting point of the next traverse. The area marked for intensive searching was almost covered when the enriched area—now known as Mount Tom Price—was found in September 1962. It was four miles long and in places four-fifths of a mile wide. Systematic drilling by 1963 revealed at least 500 million tons of rich ore and hundreds of millions of tons of poorer ore. Mount Tom Price held more ore than was known to exist in Australia only three years previously. That was only one of many massive deposits found in the Pilbara where, by 1968, seven mining syndicates held big deposits of iron.

The Pilbara iron mines were hungrier for capital than any other mines opened in Australia. The isolation of the ports, the double isolation of the mines, the massive scale of operations and the dearth of government aid all increased the cost. Between the new port of Dampier and Mount Tom Price, 182 miles of standard-gauge railway were laid in exactly a year. A longer railway of 265 miles is about to link Port Hedland and Mount Newman; hitherto the longest railway built by a mining company in Australia was 103 miles of narrow track linking Mareeba in Queensland and the Chillagoe mines and smelters in 1901. In the open cut at Mount Tom Price the iron ore was carried in 100-ton loads by lorries, then hauled in mile-long trains to Dampier, where conveyor belts loaded some of the biggest ore-carriers afloat in less than twenty-four hours. It is likely that in its first ten years Mount Tom Price alone will yield as many tons of ore as all the mines of Broken Hill have ever yielded—and they had eighty years' start. The ranges had become a huge offshore quarry for Japan, a loading station on a conveyor belt moving towards the Rising Sun.

The south-east of Australia had its own iron-ore fleets. The Middleback Ranges, which in the half century from World War I to the mid-1960s was the main and often only source of iron ore, continued to feed Australia's blast furnaces with the aid of Yampi Sound. And from the Savage River in the rain forests of western Tasmania—an iron deposit that was valueless in Dr Woolnough's day—the first crushed iron ore was pumped in 1968 along fifty-three miles of pipeline to the new Port Latta where it was concentrated into pellets and shipped to steelworks in Japan. Although the south-east of Australia was producing more ore than ever before, it was surpassed by the new Iron Coast.

Western Australia was recovering from a slow start as a mining land. Slow to mine iron ore, it now produced more than the rest of Australia. The last Australian State to find a goldfield, it has mined, over the years, more gold than any other except Victoria. Traditionally the most backward in base metals, it revealed in the late 1960s a nickel belt whose wealth seemed likely to surpass that won from the nearby Golden Mile.

Near the shores of a salt lake, thirty miles south of Kalgoorlie, once stood the gold camp of Red Hill and the surveyed townsite of Kambalda. In 1898 it had a puff of fame when a magnificent shoot of gold was found in the shallow mine of the Red Hill W.A. Syndicate; 550 ounces of gold were extracted from about 3 hundredweight of stone, and a specimen known as the Golden Butterfly was one of the most photographed gold specimens found in the West. The mine, however, proved patchy and did not last long. Its life would have been different if the company had continued one of its workings for perhaps one hundred feet, for evidence now suggests that one of the nickel lodes passes close by. Only obstinacy, however, would have made the company explore further. It was interested only in gold, and the indications of gold were no longer promising.

In 1947, long after the town was deserted, two prospectors spent several months seeking gold in the area. A long thin outcrop of dark gossan took their eye. It had no gold but it still intrigued them. When the uranium boom came in 1954 they remembered the rock and wondered if it might be radio-active. They took samples to the laboratory of the School of Mines in Kalgoorlie and heard, to their disappointment, that the samples held only nickel. 'In view of the meagre Australian resources of nickel and the relatively high prices', added the assayer's report, 'it might be worthwhile to explore the extent of the occurrence and to submit a sample for nickel assay.' Apparently the prospectors, John Morgan and George Cowcill, did not think it worth while until ten years later, when they showed four specimens to Roy Woodall, a talented young geologist with Western Mining Corporation in Kalgoorlie. The new samples were assayed and revealed an average of 0.7 per cent nickel and 0.5 per cent copper. Even if one billion identical specimens could be obtained, the prospect was still unpayable. Nevertheless Western Mining investigated the area near the abandoned Red Hill mine where some of the nickel specimens had been collected. In the following summer they employed two university students on vacation to map the narrow gossan outcrops in the area. The gossan was proved to be extensive and nickel-bearing, though not payable. Tests by geophysical and geochemical methods—the testing of hundreds of samples of soil for nickel—in the winter and spring of 1965 pinpointed areas which justified a search for buried lodes of nickel. The first hole was drilled in midsummer near the old Red Hill mine on the shores of the shimmering white lake. On 28 January 1966 the drill, down nearly 500 feet, penetrated nickel ore for eight feet—only a small intersection, but unexpectedly rich. Further drilling found more ore. The drills went on finding ore, proving 10 million tons in the first two years. On the other side of

Kalgoorlie two of the old gold companies—Great Boulder and North Kalgurli—pierced the blanket of leached rock and found the hidden Scotia nickel lode in 1968. It was now not unreasonable to predict that Western Australia in the 1970s might become third to Canada and New Caledonia as a source of nickel for stainless steel, nickel plating, and the numerous nickel alloys.

The town of Kambalda was reborn. The deserted townsite, with its streets long obliterated by brushwood, was redesigned after the solving of the legal tangle created by the dispersal long ago of those who owned street frontages. A bitumen road, a power line from Kalgoorlie and a water pipeline from the old Coolgardie scheme quickly arrived, giving the new town those amenities which the old had never possessed. The town of Kalgoorlie, its population slipping below 20,000 as the output of gold waned slowly, gained even more. Unsaleable land and cottages were sold, the sale of paint soared. Hotels and boarding-houses which had scrounged a living found their bars and bedrooms full. The colonnade of the stately public offices was plastered with the marriage banns of mining—the formal applications for new mining leases over ground which had not been pegged for decades. Kalgoorlie deserved its reprieve. No mining town in Australia had retained such faith in the importance of searching, or such a regiment of weekend or part-time prospectors, or that symbol of hope—a resident stockbroker.

When, at last, Kalgoorlie seemed encircled by nickel lodes and rumours of lodes, many people began to wonder why the nickel had not been found in the 1890s, when the ground was scoured with tens of thousands of eyes and probed with thousands of shafts and potholes? Even then Kambalda was payable; moreover many Australians knew the value of nickel, for they had managed or invested in nickel mines in New Caledonia or had treated their nickel ores in smelters at Newcastle and Dapto. The first obstacle to finding payable nickel near Kalgoorlie was simply the fact that visible evidence was weak. Long ago the outcrop at Kambalda must have been rich in nickel, but ages of weathering by sun, rain, wind and frost had dissolved most of it. In lodes within several hundred feet of the present sandy surface most of the nickel had vanished. In gold lodes in the region, however, the same process of oxidation had had the opposite effect; the proportion of gold in lodes near the surface had been increased by the dissolving of sulphur and other constituents of the lodes. Thus surface prospecting found payable gold but not payable nickel. As the gold lodes were distinct from the nickel lodes, the warren of underground workings which penetrated the gold-bearing rocks did not find nickel. A rare exception was the discovery in 1910 of 'bunches of nickel minerals' at the 250-foot level of the Star of Fremantle gold mine at Kunanalling, a small town not far from the Scotia nickel discovery of 1968. These bunches, however, were seen as curiosities, not as clues.

As the available evidence in 1900 or 1950 convinced miners that this was simply a vast goldfield, they sought only gold and followed only the clues of gold. This was understandably the pattern in nearly all mining fields. Thus on the Cloncurry copper belt the outcrop of silver-lead at Mount Isa was discovered late because all eyes and minds were focused on copper. Then later, at Mount Isa, strong evidence of copper at depth was regarded sceptically because silver-lead mining had come to dominate those ranges. In the history of mineral discovery the eyes in the mind are as important as the eyes in the eye-sockets. The finding of the clues near the old Red Hill goldmine had to be accompanied by an alertness to the meaning of these clues, and it is doubtful if that alertness existed amongst any mining engineers and geologists until the 1950s. Geologists needed a hypothesis to fit the clues, the hypothesis that the vast goldfield might also be a base-metal field. A few geologists were playing with that hypothesis in the 1950s, for they realized that the pre-Cambrian shield of Canada could throw light on rocks of similar age on the West Australian goldfields. In Quebec Province base metals had belatedly been found in a region worked only for gold; perhaps base metals could also be found below the blankets of surface sand and oxidized terrain near Kalgoorlie. Western Mining Corporation had already investigated base-metal clues in the West and had a deeper knowledge of the geology of the area than any other company when the two prospectors produced the meagre samples of nickel in 1964. It was mentally prepared for the clues, and therefore willing to risk money in evaluating them. Appropriately this company which under W. S. Robinson in the early 1930s had pioneered the scientific search for minerals in Australia, and under Lindesay Clark had continued to search, made three of the most significant finds of the decade 1956-66: the bauxite in the Darling Ranges, iron ore in the Koolanooka Hills and nickel at Kambalda.

In the year in which the first nickel lode was found, the first payable deposit of phosphate was found in parallel circumstances, 1,200 miles across the desert lands. The phosphate rock, like the nickel, lay in a region which had long been traversed by prospectors; it was only twenty miles from the ghost copper town of Duchess, a siding on the Mount Isa-Townsville railway. Moreover it was a mineral which had long been wanted in Australia. The wheatlands were fertilized by superphosphate, and it was made from phosphate rock shipped from Nauru in the Pacific and from Christmas Island near Java, islands cut off by Japan in World War II.

The main impediment to finding phosphate was the absence of eye-catching clues. The specimens found in 1966 near Duchess had no distinctive feel, texture or colour; only a chemical test could show whether they did contain phosphate. Bores put down in search of oil in sedimentary rocks in north-west Queensland actually passed through phosphate rock, but the rock was not recognized. The companies were searching for oil, not phosphate, and they failed to test the rock. Similarly, geologists officially mapping the surface of the ground must have walked across outcrops of phosphate and missed their significance.

The first step towards 'discovering' what had been seen but ignored was the systematic search for phosphate in the 1960s by the Commonwealth's Bureau of Mineral Resources and the old mining company, Broken Hill South. They decided, by analogy with other lands, which areas in Australia seemed most likely to contain phosphate. As the evidence pointed to north-west Queensland they tested cuttings collected and preserved from unsuccessful oil bores. On 27 February 1966 they found that one borehole, unknown to the drilling company, had passed through 190 feet of low-grade phosphate. Broken Hill South sent a party to the abandoned area near Duchess and their field-work easily located beds of phosphate rock on the surface. More discoveries followed over a wide radius. Here was a major field—if cheap transport could be provided.

The finding of Australia's largest deposit of manganese, on Groote Eylandt, in the Gulf of Carpentaria, suggested that in the search for minerals it was easy to ignore strong clues as well as weak ones. 'Ores of manganese are very common and, being jet-black, are conspicuous. These very facts perhaps explain', wrote Sir Harold Raggatt in his valuable book, *Mountains of Ore*, 'why, until 1960, nobody thought to test the small outcrops of manganese ores scattered over the coastal plain of Groote Eylandt.' One may suggest that the delay in appreciating the manganese was also prolonged by the island's inaccessibility, its cloistered role as a reserve for aboriginals, and above all its position near to a corner of the continent which had seemed deficient in minerals. Two Commonwealth geologists, P. R. Dunn and P. W. Crohn, were the first to sense the promise of the manganese outcrops lying less than a mile from an Anglican mission station on the island. If the deposit proved to be large, the same manganese oxide with which islanders coloured their bark-paintings could be used in the making of special steels. In 1962 Broken Hill Proprietary conducted a search, sinking nearly a thousand pits and drill holes to explore the massive bed of ore which lay just beneath the soil and tree-roots. Four years later ships were regularly carrying the manganese ore to Japan and to B.H.P.'s own ferro-manganese plant in Tasmania.

While new discoveries had the glamour, the traditional metals remained important. With copper from Mount Isa, Mount Morgan, Mount Lyell, Tennant Creek and the reborn Cobar field, copper output was multiplied by six between 1950 and 1966. In the same period Australia more than doubled its output of tin and increased that of silver and zinc by about 80 per cent and lead by 60 per cent. Even gold production was slightly higher, though some famous mines—Sons of Gwalia, Wiluna, Big Bell (Western Australia) and Morning Star (Victoria)—dismissed their last men and auctioned their machinery away. Gold, however, was clearly losing its gilt.

The long boom has made the mining industry again one of the dynamos of the economy. Minerals seem likely to surpass wool as the most valuable export in the early 1970s; and it is iron ore which seems likely then to earn more than any other mineral.* Australia's long dependence on many imported minerals, ranging from petroleum to aluminium, has either diminished or ended. Not since the first years of the twentieth century have Australian mines been so important. Never before have they been so productive.

Tropical Australia is at last becoming a great mining region. Though it occupies four-tenths of the continent, its mines had not been the main Australian source of even one important mineral in any one year until the 1950s, when they began to yield more than half of the nation's annual production of copper. By 1968 the tropics led in six of the fourteen main mineral groups. The tropics supplied most of Australia's copper, iron ore, bauxite, manganese, oil and uranium; another mineral, phosphate rock, will probably be added in the 1970s, but before then the tropical zone will have lost its short-lived supremacy in oil.

The new mining fields were ravenous for capital. Between 1955 and 1968 the aluminium industry—from mines to smelters—had used about $600 million of capital. In four years Hamersley Iron invested more than $200 million on equipment and amenities in the Pilbara, ranging from a long railway line to streets of brick houses which were air-conditioned and supplied with water purified

* Iron ore actually reigned until 1975-6, in which year the main mineral exports were: black coal $974m., iron ore $770m., aluminium $480m., pig iron and steel $200m., copper $140m., zinc $135m., lead $130m., mineral sands $126m. In the year 1975-6, three times as much export income came from mineral products as from wool. Likewise black coal was a more valuable export than wheat, and iron ore exceeded meat.

from the sea. The investment in seeking and equipping oilfields in the period 1946-1966 exceeded $400 million. To a mining promoter of the nineteenth century such figures would have savoured of astronomy, not mining. The two richest Australian companies of that century, B.H.P. and Mount Morgan, held such rich ore beneath the grassroots that they financed their own expansion after tiny injections of outside capital; Mount Morgan needed only about $4,000 and B.H.P. $36,000. In contrast, many of the discoveries of the mid-twentieth century were so low-grade that they could not earn one cent of profit until an enormous sum had been spent in preparing them for large-scale production. Most of that money came from overseas. Whereas the mining industry was largely owned by Australians before 1890, and largely owned by British investors on the eve of World War I, it was now a medley of Australian, United States and British interests with small pockets of Canadian, Swiss, French, German and Japanese capital.

The importance of a new mining field could no longer be measured by the population which lived near by. Of the mining fields created or extended since 1945, only Mount Isa so far supports a small city; it is populous because it has a higher proportion of married men than most fields, because it concentrates and smelts its ores on the field and because it is the largest underground mine in Australian history. Elsewhere there was a marked swing from underground to open-cut mining. The narrow vein-type deposit dominated nineteenth-century mining and had to be followed underground; but most of the new deposits were massive and shallow and lent themselves to mechanical quarrying by a small team of operators. The new miners sat in cabins and drove machines in quarries which would have been swarming with men and horses in the days of pick and shovel.

The most puzzling facet of the post-war boom, to most Australians, is the series of important discoveries. They wonder why so much has been found in a continent whose great mining days were believed to have passed. They wonder why so much should have been discovered recently in regions which had once been thick with prospectors.

Some of the seeds of the long mining boom are easily labelled. Australian mining would have expanded less swiftly but for the relatively new metals—mineral sands, uranium, aluminium. It would have expanded less swiftly but for mechanical advances. The development of offshore drilling techniques made possible the discovery of oil in Bass Strait. The use of heavy machinery in open cuts cut the cost of labour, which was higher in Australia than in most mining lands. The building of ocean bulk-carriers gave Australia for the first time cheap access to distant markets.

The powerful revival of Australian mining clearly reflected international events. If an Australian mining leader had been told in 1945 that Japan, in the next twenty years, would become a major industrial power he would have felt uneasy. But Japan's demand for huge tonnages of iron ore, alumina, coal and other minerals, and its proximity to Australia's tropical ports, was a sharp spur to Australian mines; Japan eventually replaced Britain as their main market. Political events in the post-colonial era—the retreat of European powers and the rise of independent governments in Africa and south-east Asia—also spurred Australian mining. European mining investors who, before World War II, had preferred to invest in mines in the colonies and dependencies because they had political stability, cheap and docile labour and low taxes, began to shun those lands; their advantages were whittled away. A lot of mining capital was diverted from tropical lands to Australia in the 1950s and 1960s. Without it many of the recent mineral developments in Australia would have been stillborn; indeed they probably would have occurred in other lands. The oil rigs would probably have been moored not in Bass Strait but in the Macassar Strait; the 'Savage River' iron mine might have been in Manchuria and the 'Gove' bauxite project might have been in central America.

Between the world wars Australia had been shunned by most overseas mining investors. Taxes were denounced as too high; trade unions were denounced as the most aggressive in the world. Many American and British mining companies preferred exploring for minerals in lands where, if they succeeded, they could fully enjoy their success. After the war, however, Australia slowly became less forbidding to mineral explorers. Taxes were modified, mining leases became larger, and industrial unrest seemed to diminish at the very time when Indonesia, Bolivia and many of the former havens of international mining capital became stormy.

Another barrier was overcome in the post-war years. The large mining companies began to search continuously for new deposits. Before 1939 most of the capital for exploration had come from the floating of new companies on the stock exchanges, and that capital was usually raised in the periodic mining booms. Exploration therefore was spasmodic. Two years of vigorous exploration might be followed by eight years of lethargy. In contrast, the continuous investment of money and talent by the established mining companies in seeking new mines gave them possession of most of the new post-war fields. Whereas in 1939 the giants of Australian mining confined themselves to one mining field, the giants of 1969 are nation-wide.

The highest barrier to expansion of metal mines in Australia was simply pessimism that Australia's rich era of mineral discovery had vanished for ever. That pessimism was dissolved in the 1950s by the chain of discoveries around the rim of the continent from the tropical tip of Cape York Peninsula to Perth. The old gold trail that stretched late in the last century from the Pacific to the Indian Ocean was followed again, promising copper and bauxite and nickel and uranium and iron. The rush to Ophir had often faltered but never ended.

POSTSCRIPT: THE 1970s

Near the desolate gold town of Laverton, on the outer fringes of the West Australian goldfields, a percussion drill began to drill a hole in search of nickel on 23 September 1969. The hole was begun at midday, and by nightfall it was 140 feet deep. Most geologists thought the prospect of finding nickel in this particular area was unfavourable: hence this ground was held by a small exploration company.

Next morning the drill noisily resumed work. A geologist was present, and he inspected the wet sludge which came up from the deepening hole. After a few minutes—when the hole had reached 145 feet—he noticed a change in the colour of the sludge. The drill, he realized, was now entering a sulphide mineral deposit. Examining the sludge he saw copper and—he suspected—nickel in promising quantities. For several hours the same rich sludge continued to emerge as the drill pounded into the rock. By 5.15 in the afternoon it had passed through at least forty feet of mineral deposit. Samples from the 'discovery hole' were then sent to a laboratory in Kalgoorlie where a few days later the new-found deposit was confidentially said to be rich.

The shares in the small company had been bought and sold for about $1 in the week before the discovery. Each day the shares jigged up and down by five or ten cents, depending on rumours and dealings. On 1 October 1969 the Adelaide Stock Exchange was officially informed that the small Adelaide mining company had discovered a massive lode of nickel. Until that day the average share speculator barely knew of the existence of that little company formerly registered as 'Poseidon, No Liability'. Within weeks, Poseidon was known throughout Australia. The 'no liability' became known later.

The price of Poseidon shares jumped. They quickly rose from $1 to $10 and then to $25, and many shareholders began to wonder whether they were now too high. The shares continued to be driven up by maniacal bids. They passed $50, moving upwards in unpredictable spurts. The daily movements of Poseidon shares displaced the weather, sport and politics as the universal topic of small talk. The promise of Poseidon elevated scores of other mining shares: not only the shares of nickel mines but of producing mines which had no nickel and were a thousand miles away from the Poseidon leases. On the Sydney Stock Exchange in the last three months of the year the price of the average mining share rose by about 50 per cent, and much of that increase was spurred by Poseidon.

The annual meeting of the small Poseidon company was to be held in Adelaide six days before Christmas. The same meeting a year earlier was considered, even by Poseidon's own shareholders, to be so insignificant that a quorum could not be raised; but now in 1969 the meeting created such intense interest in London, New York and every city in Australia that shareholders hurried to Adelaide to hear the latest news from the chairman's own lips. The large hall was packed to the doors, and those who could not squeeze inside heard the speeches through a loudspeaker. In London an afternoon newspaper, reporting the exciting meeting, announced on its front page that Poseidon was 'the most valuable hole in the world'. The chairman, closing the meeting, seemed to confirm the highest hopes: 'I see no reason why any shareholder in Poseidon should not have a wow of a Christmas.'

The first slice of Christmas cake came almost at once to the shareholders, for that day the price of Poseidon rose from $100 to $130. By the end of the year they had reached $200. The share market seemed to lose touch with reality in the summer of 1969-70. Perhaps this was the last fantasy of a market which had long been floating in the clouds. Between 1967 and 1970 the price index of the mining shares had multiplied by five on the Melbourne Stock Exchange, and most of the increase came from discoveries that took place in the minds rather than the mines of shareholders.

By the first week of February 1970 the profits made by the shrewd speculators in Poseidon—those who 'sold too early'—must have been enormous. A clerk or widow who had chanced to invest, say, the sum of $400 in Poseidon five months previously now owned shares worth more than $100,000, and that parcel of shares was still rising in value by thousands of dollars a day. On 10 February the price peaked. On that day the mine—a line of trenches and a row of drill holes and a cluster of sheds and camps—would hardly have been worth more than a glance from the pilot of a small aircraft flying overhead, but it was valued on the stock exchanges at more than $700 million. Those who said the mine was over-valued had no idea that the overvaluation was on such a colossal scale.

Exactly a year after the discovery of nickel at Poseidon, uranium was found at Nabarlek, about 170 miles east of Darwin. Said to be the highest-grade deposit of uranium in the world, its importance was prematurely inflated by its owners and supporters. Shares in the main company, Queensland Mines, leaped from $7.50 at the end of July 1970 to $37 at the end of November 1970. Even more important were the nearby Ranger deposits, which were first detected in an aerial survey in June 1970. On either side of the Ranger deposits, were soon found the uranium of Jabiluka and Koongarra, making Arnhem Land perhaps the most important uranium region so far found in the world.

The boom in shares and the boom in exploration were intimately linked. The inflow of foreign capital accelerated exploration and discovery; the discoveries lifted the price of shares; and the rising price of shares enabled more companies to be floated and the search for minerals to be intensified. In this merry-go-round of excitement part of the motive power was rational but part was irrational. New discoveries were hailed and puffed up long before their importance could be known, long before there was any chance of their minerals reaching a market, and long before transport and metallurgical or political obstacles could be sensibly assessed. Thus, from the most dramatic discoveries of the exciting years 1969, 1970 and 1971, no dividends have yet been paid. At the end of 1977 Poseidon was in the hands of the receiver; the rich uranium region between the East and South Alligator Rivers had produced only token shipments of uranium ore; and the continent's largest field of natural gas—first located in 1971 in the North-West Shelf, and explored in the following three years at astronomical cost—was not yet exploited.

The share boom ended in 1971. The big Sydney company, Mineral Securities, which had traded largely in insecurities, crashed early in 1971. The price of shares fell, and fell again. The share news began to desert the front page, people no longer knew the latest price of Poseidon, that 'wow of a Christmas' seemed long ago, and the regiments of clerks working for the stockbrokers, and stock exchanges were slowly demobilized.

The boom and slump had happened many times previously: it had been experienced often in the 1860s, 1870s, 1880s, 1890s, and at least once and sometimes twice in nearly every decade of this century. The tipsy share boom of the years 1967-70, however, had been followed closely by a Senate Select Committee on Securities. Its enquiries uncovered fraud, malpractice, and many varieties of trickery in the conduct of company directors, managers, geologists, stockbrokers and share speculators. In a mining boom the opportunities to defraud can be large even if the regulation of the share market is strong: in Australia, the Senate Committee reported, the regulation of the share market was feeble or ineffective. It may well be true that, in the last decade, the armed bank-robberies in Australia have taken a total sum no larger than that which 'changed hands' through unarmed robberies in some new mining companies during the share boom of the years 1967 to 1970. In 1977 the share markets in Australia were still so loosely regulated and supervised that many of the scandals of the last boom will—unless drastic reforms are made—be repeated in the next boom.

The mining industry had been increasingly criticized because of the chaotic way in which its camp-followers raised capital in the cities. It was criticized even more in the 1970s because of the chaos it was said—wrongly and rightly—to create in the environment. The conservation movement had become a crusade, and mining was one of its largest targets.

The opening of many mining fields now was postponed by the question: will the mining operations harm the environment? The same question had sometimes been asked a century ago but the answer was usually ignored. Pollution was then widespread on many mining fields. The fumes from copper smelters destroyed patches of vegetation. The wood-cutters chopped nearly every large tree within a radius of several miles of many mining towns. When alluvial goldfields were busy, the sludge discharged from the puddling machines flowed into the rivers and creeks, discolouring them far downstream. In the 1850s the Barwon River at Geelong changed colour after the gold rushes set in to Ballarat, more than fifty miles to the north-west, and at Bendigo the slimy sludge from the gold-treatment plants buried roads, fences, farms, and bridges. When quartz-mining became common, the hard gold-bearing rock was crushed by iron stampers; and the noise of the stamp mills was likened by the poet Edward Dyson to 'an earthquake in a boiler-metal town'. The English novelist Anthony Trollope, visiting the Gippsland gold town of Walhalla in the 1870s, stayed in a hotel near the great Long Tunnel stamp mill: 'Sleep', he wrote, 'was out of the question.' Most Australian pioneers, when faced with the alternative of whether to preserve their own jobs or the environment, understandably sacrificed the environment. They had crossed the world to improve their lot, not to protect strange landscapes and creatures.

Conservation criteria were to affect new mining ventures occasionally in the 1960s and powerfully and frequently in the 1970s. In the mineral fields of tropical Australia, more and more reports called for the safeguarding of the Aboriginals' sacred grounds, cave paintings, archaeological relics and tribal lands. The right of Aboriginals to veto new mining projects or to share in their profits was increasingly affirmed. The idea that the natural environment was a national asset, to be prized as highly as a rich mine, was increasingly supported.

In the 1970s the drilling for petroleum near the Great Barrier Reef was halted for fear that, if payable oil were found, accidental leakages might endanger the sea life and the coral reefs themselves. The dredging of rutile and zircon on Fraser Island—the largest sand island in the world—was banned by the Commonwealth in order to protect a wilderness which many naturalists regarded as one of the quiet wonders of the Pacific. The Fox inquiry into the merits and dangers of mining the Ranger ore bodies in Arnhem Land was strongly influenced by the argument that the plants, insects, birds, swamps and sandstone escarpments of a biological wonderland should be protected from the hazards of the mining and treatment processes. The possibility of opening new lead mines was restrained by fear that one of the main markets for lead—as an additive in petrol—would be eliminated. The future of uranium mining in Australia seemed to be pinned to the prospect that if uranium was deemed by public opinion to be dangerous to the peace or the health of the world, it would remain unmined.

The conservation issues had become one of the tightest brakes on the expansion of mining. Whereas the prospect of opening a new mine had been almost unanimously applauded in the 1950s, the news that a major mine might open was now viewed by many with alarm. For the first time in Australia's history a small, ardent minority was virtually arguing that the long rush at last should end.

And yet Australia was, by the mid 1970s, more important in world mining than in, probably, any previous decade. In bauxite Australia was the world's leading producer, and in iron ore it was exceeded only by the Soviet Union. Australia ranked third as a producer of lead, fourth in zinc and nickel, and probably ranked fifth or sixth amongst the world's miners of coal and silver. The discoveries of the 1950s and 1960s were now in full fruition, and few facets of national life were not affected by the controversies provoked by mines or the wealth produced by mines, or the sense of national independence subtly fostered by the abundance of minerals.

GLOSSARY OF MINING TERMS

Alluvial gold or tin: ranging in size from a pinpoint to a small boulder, this metal lies amongst sand or clay or gravel. Originally the metal was in the solid rock, but over the ages the rock eroded. Nature in effect crushed the rock and the released mineral gravitated into watercourses. Thus it was easy for a poor man to find and mine.

Basalt: a tough dark rock, formed by hot liquid from a volcano. It covered many rich deposits of alluvial gold in Victoria and New South Wales, and some deposits of alluvial tin in New South Wales and Tasmania.

Battery: a machine for crushing metal-bearing rock and extracting the metal.

Claim: a square of Crown land which a gold digger held and worked by virtue of his licence or miner's right. A claim was smaller, cheaper, and less secure than a mining lease.

Cradle: a wooden box which, rocked manually to and fro, separated the heavy alluvial gold from the lighter sand and gravel.

Deep lead (pronounced 'leed'): A stony course of a river which once drained the country but has since been covered and concealed by layers of clay or basalt rock. These deep leads often contained gold or tin.

Drive or *Crosscut:* an underground passage, usually narrow and always horizontal, that connects the shaft with the place where the metal is being mined.

Dry-blowing: a way of separating grains of alluvial gold from sand on arid Australian goldfields. The dirt was dropped from a height, and the wind blew away the lighter dust but not the heavier gold.

Fool's gold: iron sulphide, valueless but rather like gold in appearance.

Lode: a deposit of mineralized rock lying in a crack or opening in the rock.

Metallurgist: a specialist in extracting pure metal from mined rock.

Mill: a house where ore is crushed or milled as preparation for extracting the mineral.

Mullock: rock which is mined but contains no payable metal.

Nugget: a large piece of alluvial gold, as distinct from a lump of gold embedded in solid rock.

Outcrop: the part of the mineral deposit which is visible on the surface of the ground.

Orebody: a payable deposit of mineral in hard rock.

Oxidized zone: the upper part of a lode that has been altered by oxygen and rain. For example, the oxidized zone of a deposit of zinc-lead-silver has no zinc and much silver.

Peg: to take an area of Crown land for mining purposes by driving marker pegs into the ground, usually on the boundaries.

Prospector: a man who went into new terrain to search for new lodes or mining fields.

Quartz: a mineral of crystalline silica, usually white and shining, that contained much of the gold mined in Australia.

Reef: a lode or vein of quartz, parts of which contain gold.

Shaft: a vertical, or nearly vertical, passage going from the surface into the ground. It serves the same function as the lift-well of a skyscraper, a vertical road to the deeper workings of the mine.

Sluicing: a method of extracting alluvial metal from sand and gravel by shovelling the paydirt into a wooden flume of running water.

Smelting: using fire and a furnace to extract metal from an ore; the process was much more important in treating base-metals than gold.

Stamp mill: a machine consisting of heavy cast-iron rods or stampers that pulverizes ores; also called a battery.

Stope: an underground rock-walled chamber or cavity in which men mine the ore. It is usually reached by descending the shaft, walking along the drive, and then climbing up or down a ladder.

Sulphide zone: the unweathered, usually larger, part of a mineral deposit. It is deeper, and its ores are harder to treat than those of the oxidized zone. Sulphur, in the form of pyrite, is one of the main constituents of sulphide ore.

Tunnel: an underground, horizontal passage. As the word is well known, I have used it more freely than an engineer would use it. For example, I often use it instead of such words as drive, adit, level, or crosscut.

Tribute: a contract under which a party of working miners give the mine owner a fixed proportion of all the metal they mine.

Windlass: a wooden roller with a handle which a miner turns to pull up his rope and bucket of ore from a shallow shaft.

Winze: a shaft which connects underground workings but which does not reach the surface.

ACKNOWLEDGMENTS

On and off for about ten years I gathered material for this book. Then through the generosity of sponsors I was able to travel extensively to most Australian mining fields and work full time on the research. I am grateful to Mr George Fisher, chairman of Mt Isa Mines, who began to arrange support from mining companies while he was president of the Australasian Institute of Mining and Metallurgy and who supported the venture in every way. I am indebted to Sir Ian McLennan and the Broken Hill Proprietary, to Sir Maurice Mawby and the Broken Hill Mining Managers' Association, to Mr G. Lindesay Clark and the Chamber of Mines of Western Australia, for their support and for checking errors in the manuscript. I am also indebted to Professor W. Woodruff and the University of Melbourne for a research grant and for much help, and to Miss Beryl Jacka of the Institute of Mining and Metallurgy. I should add that this is not an official history of Australian mining, that many of the sponsors no doubt disagree with my judgments and opinions on the past and present, and that no other Australian industry has been so liberal and co-operative in sponsoring a history.

Many people kindly gave me information or guided me to it. I mention some in the text and some in the bibliography and the following list neglects many others whom I would like to have thanked by name.

In the Northern Territory I was grateful to John Tonkin and the staff at Rum Jungle, H. E. Edwards and Australian Development, and R. E. White and staff at Peko.

In South Australia—W. A. P. Phillips, Malcolm Hill, Gerald Fischer, F. A. Green and staff at Broken Hill Associated Smelters, and A. L. Keats.

In Tasmania—O. H. Woodward and John Reynolds.

In Queensland—J. Malcolm Newman, F. L. Hennessy, B. Lennon, and Mount Morgan Limited, and many of the staff of Mount Isa Mines.

In Sydney—the Mitchell Library and Dr John McCarty.

In Broken Hill—R. Pitman Hooper and Zinc Corporation, M. Howell and Broken Hill South, the late A. R. West and North Broken Hill, A. R. 'Floss' Campbell, and A. Coulls.

In Western Australia—L. C. Brodie Hall and Western Mining, Ray Simpson and Gold Mines of Kalgoorlie, L. Edgar Elvey and Great Boulder, F. A. Davis and North Kalgurli, R. C. Buckett and Lake View and Star, Mr Jennings and the Chamber of Mines, B.H.P. (Yampi), Central Norseman, Sons of Gwalia, Kevin Finucane, G. Spencer Compton, J. W. Cranston, George Beresford, Keith Quartermaine, H. H. Carroll, and Miss M. Lukis of the State Archives of Western Australia.

In Victoria—Haddon King, G. B. O.'Malley, C. E. Blackett, M. R. McKeown, E. A. Beever, Professor H. H. Dunkin, Dr John McAndrew, Nils Nilsen, the late Austin Edwards, Harry Stacpoole, E. F. Semmens, H. W. Nunn and the State Archives, Dr Geoffrey Serle, the authority on the 1850s, and Mrs Moira Willes who typed the manuscript.

BIBLIOGRAPHY

Most books and journals and papers used in writing this history are listed below on a state or regional basis. Those sources of information which concern mines and metallurgy in more than one state are listed under the heading of 'National'. The bibliography is divided into nine parts:

1 National
2 Victoria
3 New South Wales
4 Broken Hill
5 Queensland
6 Northern Territory
7 Western Australia
8 South Australia
9 Tasmania

1. NATIONAL

Amongst the autobiographies of mining men four books are fascinating, though all are prone to exaggeration: Randolph Bedford's *Naught to Thirty Three* (Sydney, 1944), G. D. Meudell's *The Pleasant Career of a Spendthrift* (London, 1929), Frank Penn-Smith's *The Unexpected* (London, 1933) and Herbert Clark Hoover's *The Memoirs of Herbert Hoover: Years of Adventure 1874-1920* (New York, 1952). My article discussing Hoover's memoirs and his Australian mining career is in *Business Archives and History*, Sydney, February 1963.

Many other travellers wrote books which described Australian mining towns: J. A. Froude in his *Oceana; or England and her Colonies* (London, 1886), David Kennedy in *Kennedy's Colonial Travel* (Edinburgh, 1876), John Milner and Oswald Brierly in *The Cruise of H.M.S. Galatea . . .* (London, 1869), Gilbert Parker in *Round the Compass in Australia* (published by E. W. Cole in Melbourne, no date but apparently early 1890s), Anthony Trollope in *Australia and New Zealand* (Melbourne, 1873), W. B. Wildey in his *Australasia and the Oceanic Region* (Melbourne, 1876) and Ralph Stokes, a Johannesburg mining journalist, in his outstanding *Mines and Minerals of the British Empire* (London, 1908). Other travellers' books are listed under the headings of particular mining fields or states.

The following contemporary technical books, all published in New York, were generally useful: W. R. Ingall's *Lead Smelting and Refining* (1906), E. D. Peters' *The Principles of Copper Smelting* (1907) and *The Practice of Copper Smelting* (1911), T. A. Rickard's *Stamp Milling of Gold Ores* (1897), W. H. Weed's *The Copper Mines of the World* (1908), and H. Barger and S. H. Schurr, *The Mining Industries, 1899-1939: a Study of Output, Employment and Productivity* (1944). There are interesting reviews of technical changes in the annual book *The Mineral Industry* (New York, 1892-).

Three English technical books of the nineteenth century were useful: Sir Roderick Murchison's *Siluria* (1854), D. T. Ansted's *The Gold Seeker's Manual* (1849), and A. G. Lock's *Gold: Its Occurrence and Extraction* (1882). Amongst many Australian books on the technical side of mining I owe a debt to S. H. Prior's *Handbook of Australian Mines*

(1890), Donald Clark's lucid *Australian Mining and Metallurgy* (1904), and to two Institute of Mining and Metallurgy books, *An Outline of Mining and Metallurgical Practice in Australia* (1924) and *Geology of Australian Ore Deposits* edited by A. B. Edwards (1953).

A useful background to the growth of mining in other countries came from A. K. H. Jenkin's *The Cornish Miner* (2nd edition, London, 1948), W. P. Morrell's *The Gold Rushes* (London, 1940) and from the following American books: *Seventy Five Years of Progress in the Mineral Industry 1871-1946*, edited by A. B. Parsons (1947), *California Gold* by R. W. Paul (1947), *Mining Camps: a Study in American Frontier Government* by C. H. Shinn (1885), and two works by T. A. Rickard—the two volumed *Man and Metals* (1932) and the good if sketchy *A History of American Mining* (1932).

Anyone wishing to study the fortunes of particular companies and fields will value such reference works as Walter Skinner's *Mining Year Book* (London, 1887-), R. L. Nash's *Australasian Joint Stock Companies' Year Book* (published intermittently from 1898 to 1913-14), and the Tait Publishing Company's *Mining Handbook of Australia* (Melbourne, 1936, 1939). Edward Greville's *Year Book of Australia* prints annual reports of Australian stock exchanges in the 1890s, Gordon and Gotch's annual *Australian Handbook* describes most Australian mining towns from 1870, and the short-lived *Who's Who in Mining and Metallurgy* (London, 1908, 1910) has many Australian entries. On trade unions the Mitchell Library in Sydney has reports of annual conferences of the Amalgamated Miners' Association of Victoria and Tasmania from 1899 to 1912. The reminiscences of W. H. Corbould, once a prominent mining engineer in four states, are also in the Mitchell Library.

Dr John McCarty wrote an outstanding study on 'British Investment in Overseas Mining, 1880-1914', as a doctoral thesis at Cambridge University, and I am grateful to him for allowing me to read his work.

Government Publications: *Commonwealth Census* of 1911, 1921 and 1933; *Year Book of the Commonwealth of Australia* (1908-); the annual *Australian Mineral Industry Review* (1948-) published by the Bureau of Mineral Resources; and the Report of the Tariff Board on a Copper Bounty, being paper no. 193 in the Commonwealth *Parliamentary Papers* for 1927.

Several official U.S.A. publications are relevant to Australian mining history: A. C. Veatch's *Mining Laws of Australia and New Zealand* (Geological Survey Bulletin 505, Washington, 1911); and the U.S. Bureau of Mines' *Economic Papers* 1, 2, 6, and 13, which summarize every country's annual production of copper, zinc, gold, and tin in the nineteenth century and after (Washington, 1928-32).

Journals and Weeklies: For comment on Australian mining during the long revival from about 1886 to 1908 I often used the *Australian Mining Standard*, the Sydney *Bulletin*, the Melbourne *Australasian*, and Randolph Bedford's periodical the *Clarion*. Overseas journals consulted for events of the same period were the *American Engineering and Mining Journal* and the English periodicals, *The Economist*, *British Australasian*, and *Mining Journal, Commercial and Railway Gazette*. For the later period I used mostly the *Mining Magazine* (London, 1909-) and the magazine known today as the *Chemical Engineering and Mining Review* (Melbourne, 1908-).

2. VICTORIA

The most important group of papers on the Victorian goldfields were reports and enquiries sponsored by the Victorian parliament and printed in its *Votes and Proceedings*. Usually the evidence was printed with the report and for a historian is often the more useful of the two, because cross examination of witnesses reveals many things not stated in other sources.

1852-3, vol. 2, Report from Select Committee on the Management of the Gold Fields
1853-4, vol. 3, Select Committee on the Gold Fields; Select Committee on the claims for the Discovery of Gold in Victoria
1854-5, vol. 2, Royal Commission on the Condition of the Gold Fields of Victoria (The evidence appended to this report is essential for understanding mining at Eureka in 1854.)
1855-6, vol. 2, Select Committee on Gold Mining on Private Property; Select Committee on Gold; and Quartz Claim of Messrs Syme & Co.
1856-7, vol. 4, Commission of Enquiry on 'the Mining Resources of the Colony'
1859-60, vol. 1, Geological Survey (Correspondence); Sludge at Epsom
1862-3, vol. 2, Select Committee on Gold Prospectors
1862-3, vol. 3, Royal Mining Commission on the Conditions and Prospects of the Gold Fields of Victoria
1864-5, vol. 3, Select Committee on Gold Rewards
1888, vol. 3, Report of the Ventilation of Mines Board
1890, vol. 2, Report of the Mining Safety Cages Board
1890, vol. 2, Report of the Mining Managers' Certificate Board
1890, vol. 1, Select Committee upon the Claims of Henry Frencham as Discoverer of the Bendigo Goldfield
1891, vols. 3, 5, Royal Commission on Gold Mining (three reports and evidence totalling in all over 1,000 pages)
1893, vol. 1, Report from the Select Committee upon Tributing in Gold Mines (no evidence printed)
1901, vol. 3, Royal Commission on Technical Education.

In addition to these Victorian parliamentary papers is a small report by Hargraves on *Victoria Gold Fields* in N.S.W. Legislative Council papers of 1853, vol. 1. And in the papers of the English parliament is the 'Correspondence and further Papers Relative to the Recent Discovery of Gold in Australia', being ten series of despatches presented to parliament between February 1852 and August 1857.

State Archives: I read Sir Henry Barkly's despatches 1859-62 at the Executive Council Chambers in Melbourne and freely used some of the excellent material in the Chief Secretary's archives at the State Library of Victoria, including the files on Chapman's gold in 1849, much of the correspondence on gold in 1851 and 1852, and the extensive papers on the 1857-8 Clunes Riots (57/ 3402-3 and 58/ 8286, 8908, 9063, 9155, and 9355).

Turning to regular government publications, the Victorian *Hansard* has some informative debates on mining including the debates on the Mining Companies Law Amendment Bill (1871), the Regulation of Mines Statute (1873), and successful Mining on Private Property Bill (1884). The *Victoria Government Gazette* from 1871 lists the occupations of shareholders in the new 'no liability' mining companies. The *Victorian Year Book* (1875-) and the *Censuses of the Colony of Victoria* are also valuable.

Mines Department: Victoria had one of the most vigorous mines departments in the English-speaking world during part of the nineteenth century, and it published an immense amount of statistical and geological information. I used some of its maps and plans, in particular a superb geological map of Ballarat in 1861; its annual reports; the annual reports of the Chief Inspector of Mines (1874-); the quarterly printed reports of mining surveyors and registrars (1864-91); the volumes entitled *Mineral Statistics of Victoria;* issues of the *Mining and Geological Journal* (1937-); the series of *Geological Progress Reports* commencing 1874; the *Records of Geological Survey* commencing 1901 and including Professor J. W. Gregory's invaluable 'A Contribution to the Bibliography of the Economic Geology of Victoria to the end of 1903', being part 3 of vol. 2 of the *Records.*

The series of *Memoirs of the Geological Survey of Victoria* have some fine reports which discuss past as well as present: E. J. Dunn's list of nuggets found in Victoria, A. M. Howitt's reports on Maryborough and Wedderburn, S. B. Hunter's notable report on the deep leads of Victoria, O. A. L. Whitelaw's two reports on the Woods Point field and district, J. W. Gregory's report on Ballarat East, and W. Baragwanath's reports on Castlemaine, Berringa, Aberfeldy, and his 257-page memoir on the Ballarat goldfield (1923) which, in historians' eyes, is probably the most useful report written by an Australian geologist.

The *Bulletins of the Geological Survey,* commencing in 1903, are smaller and their value is more in their discussion of the physical environment of Victorian goldfields than in their human history. I used W. Bradford's bulletins on Clunes, Egerton-Gordon, Maldon, and Stawell; J. W. Gregory's report on the Berry lead at Spring Hill; E. J. Dunn's reports on the Woolshed Valley and his biographical sketch with D. J. Mahony of the 'Founders of the Geological Survey of Victoria'. Outside these series of memoirs and bulletins are many individual and special reports, including G. Thureau's 'Synopsis of a Report on Mining in California and Nevada, U.S.A.' (1879), W. Bradford's report on Creswick (1902), and Dr W. A. Summons's report on phthisis (1906).

Books: I listed in the National section some books written by sharp observers of the Victorian goldfields. Other eye-witness books are James Armour's *The Diggings, the Bush, and Melbourne* (Glasgow, 1864), W. J. Barry's *Up and Down; or, Fifty Years' Colonial Experience in Australia . . .* (London, 1879); Edward Dyson's short stories and his *Rhymes from the Mines* (Sydney, 1898); C. F. Nicholls's *The Rise and Progress of Quartz*

Mining at Clunes (1869); Sir Ernest Scott's edition of *Lord Robert Cecil's Gold Fields Diary* (2nd edition, 1945); and Reverend Robert Young's *The Southern World: Journal of a Deputation . . .* (London, 1854). Another group of books are in part at least contemporary sources on the goldfields: *The Early Story of the Wesleyan Methodist Church in Victoria* by W. L. Blamire and J. B. Smith (1886); *Early Creswick: the First Century* by J. A. Graham (1942); *Australia* by P. Just (Dundee, 1859); *Bendigo and Vicinity* by W. B. Kimberley (1895); *Bendigo Gold-field Registry* by J. N. Macartney (1871, 1872); *History of Bendigo* by G. Mackay (1891); *Men of the Time in Australia. Victorian Series* (1878); *Report of the Proceedings of the Fourth Session of the International Statistical Congress* (London, 1861) which is a valuable guide to Victorian and also New South Wales gold statistics; *Miners' Phthisis: Report of an Investigation at Bendigo . . .* by Walter Summons (1907); *The Gold Fields and Mineral Districts of Victoria*, a voluminous and essential work by R. Brough Smyth (1869); *History of Ballarat* by W. B. Withers (2nd edition, 1887); and *Report on the Bendigo Goldfield 'Central Area'* by E. C. Dyason (1916). Amongst later secondary sources on Victoria, I consulted Vince Kelly's *Achieving a Vision: the life story of P. W. Tewksbury* (Sydney, 1941), Yvonne Palmer's *Track of the Years*, which is the history of St Arnaud (1955), W. G. Sharpley's *History of the Borough of Stawell* (1930), A. J. Williams's *A Concise History of Maldon and the Tarrangower Diggings* (1953). Unless otherwise stated all these books were published in Victoria.

Theses: Two useful theses at the University of Melbourne are June Philipp's *Trade Union Organization in New South Wales and Victoria 1870-1890* (completed in 1953) and Frank Strahan's *Growth and Extent of Company Mining on the Victorian Goldfields in the 1850s* (1955).

Learned journals: Probably the most penetrating article written on Bendigo was T. A. Rickard's 'The Bendigo Gold-field', published in two parts in the *Transactions of the American Institute of Mining Engineers*, vols. 20 and 21, 1892 and 1893. In *Historical Studies, Australia and New Zealand* is the valuable 'Eureka Centenary Supplement' (December 1954) including A. G. Serle's article on Eureka's causes; and the issue of November 1954 has Bruce Kent's 'Agitations on the Victorian Goldfields, 1851-4'. In the *Economic Record*, June 1947, is J. A. C. Mackie's 'Aspects of the Gold Rushes'. In the *Victorian Historical Magazine* is J. G. Harrison's 'Some Aspects of Business Life in Bendigo in 1853' (May 1959) and L. R. Cranfield's 'The First Discovery of Gold in Victoria' (November 1960).

Newspapers: The *Age*, the *Argus*, the *Australasian*, and the *Australian Mining Standard* were consulted; for example the *Argus* was useful for the Bendigo boom of October 1871 and the Loghlen and Learmonth lawsuit, August-October 1874 and March-April 1876. *Dicker's Mining Record* (Melbourne) was the best source on mining from 1861 to 1870.

3. NEW SOUTH WALES

In New South Wales as in Victoria the *Parliamentary Papers* are an essential source on mining. I used the following:

1851, vol. 2, Papers Relative to Geological Surveys
1852, vol. 2, Select Committee on the Management of the Gold Fields
1853, vol. 2, Discovery of Gold in Australia (Sir R. Murchison's Claim); Select Committee on the Gold Fields' Management Bill
1861, vol. 2, Select Committee on the Claims of the Reverend W. B. Clarke
1871-2, vol. 2, Royal Commission of Inquiry on the Working of the Present Gold Fields Act; much of the evidence taken before this Commission is illuminating on Victorian goldfields law.
1890, vol. 4, Select Committee on Claims of William Tom, James Tom, and J. H. A. Lister, as the First Discoverers of Gold in Australia.

Another valuable guide to New South Wales goldfields in the 1850s are the despatches of the governors, printed in the English *Parliamentary Papers* as 'Correspondence and further Papers Relative to the Recent Discovery of Gold in Australia', from February 1852 to August 1857, in ten papers.

Mines Department reports of value were the printed annual reports (1875-) with their wealth of statistics, the bulletins on *Silver, Lead, Zinc* (1923) and *Gold* (1924) by E. J. Kenny, and the series of large reports entitled 'Mineral Resources': J. E. Carne's *The Copper Mining Industry and the Distribution of Copper Ores in New South Wales* (signed 1899), Carne's *The Tin Mining Industry and the Distribution of Tin Ores in New South Wales* (1911), and E. C. Andrews' reports on the Kiandra lead (1901), on the Forbes-Parkes goldfields (1911), and his two volumes on the Cobar copper and goldfield (1911, 1913).

State Archives: Minutes of the N.S.W. Executive Council 1849-51, Colonial Secretary's files of April and May 1851 including inward correspondence and 1851 letterbook 'Judicial Departments', vol. 21, and the third volume of the Deas Thomson Papers (in Mitchell Library) with their letters from J. Norton, J. Macarthur, J. R. Hardy, Sir William Denison, and Thomson himself—all were enlightening on official policy during the first gold rush.

Books which gave evidence on the New South Wales goldfields were W. B. Clarke's *Researches on the Southern Gold Fields of New South Wales* (Sydney, 1860), Simpson Davison's *The Discovery and Geognosy of Gold Deposits in Australia* (London, 1860), J. E. Erskine's *A Short Account of the Late Discoveries of Gold in Australia . . .* (London, 1852), W. R. Glasson's *The Romance of Ophir* (Orange, 1935), E. H. Hargraves' *Australia and its Goldfields* (London, 1855), *The Prose Works of Henry Lawson* (Sydney, 1948) with their evocative descriptions of Home Rule and Gulgong and other goldfields on which he lived, D. G. Moye's *Historic Kiandra* (Cooma, 1959), E. W. Rudder's *Incidents Connected with the Discovery of Gold in New South Wales in the Year 1851* (Sydney, 1861), and C. Swancott's *The Brisbane Water Story* (Woy Woy, 1955) part 4 of which describes Hargraves' early career.

On the New England goldfields there is an excellent thesis in the University of Melbourne library, Dr D. F. Mackay's 'The Rocky River Goldfield 1851-1867' (1953), and a series of New England University monographs by J. P. Belshaw and partners on gold mining in Armidale,

Walcha, and various northern districts. Unfortunately I lacked the space in the book to make more than passing note of these fields.

Learned Journals: the documented argument and more detail for my interpretation of gold in 1851 and earlier may be found in two articles I wrote for *Historical Studies, Australia and New Zealand:* 'Gold and Governors' in May 1961, and 'The Gold Rushes: the year of decision' in May 1962. In the *Journal and Proceedings of the Royal Australian Historical Society* are W. L. Havard's article on Strzelecki (vol. 26, part 1), James Jervis's article on W. B. Clarke (vol. 30, part 6) and E. W. Dunlop's comprehensive work 'The Golden Fifties' (vol. 37, parts 1-3). Also useful were articles on the *Transactions of the N.S.W. Chamber of Mines,* 1900-1, and an article in the *Journal of the Royal Society of New South Wales* on McBrien's gold of 1823 (vol. 43, 1909, pp. 123ff., by J. H. Maiden and R. H. Cambrage).

Newspapers: *Sydney Morning Herald* for 1849-51, the *Examiner* of London for news of gold's impact on England in 1851, the *Alpine Pioneer and Kiandra Advertiser* of 1860 for news on the Snowy rush, and the Sydney *Daily Telegraph* for the share market in 1881 and later years.

4. BROKEN HILL DISTRICT

Although Broken Hill has been a dynamo in Australian history and world mining history, and although many books have been written about its history, it is hard to point to any book of the last forty years that has embodied much research. So many myths have survived or flourished that —until someone writes a careful history of the field—the safest policy is to use mostly contemporary records.

Amongst the most useful of these records are the royal commissions in the New South Wales *Parliamentary Papers:* the royal commission on the working of mines and quarries in the Albert Mining District (1897), the royal commissions into fatal accidents at Broken Hill South mine (1901) and the Broken Hill Central mine (1903), and the royal commission on the mining industry at Broken Hill (1914).

The N.S.W. Mines Department's annual reports touch on the district in the late 1870s and become a silver mine of information from the mid-1880s. Two geologists, J. B. Jaquet and E. C. Andrews, wrote *Memoirs of the Geological Survey* on the geology of the district in 1894 and 1922.

Some of my evidence on Broken Hill came from the following books or pamphlets: *The Barrier Silver and Tin Fields in 1888* by a special correspondent of the *South Australian Register* (Adelaide, 1888), a roneoed pamphlet 'Charles Rasp: Founder of Broken Hill' by Allan Coulls (Broken Hill, 1952), *From Silver to Steel* by Roy Bridges (Melbourne, 1920), the essential *Concentrating Ores by Flotation* compiled by Theodore Hoover (London, 1912), *Concentration by Flotation* edited by T. A. Rickard (New York, 1921), *Principles of Flotation* by K. L. Sutherland and I. W. Wark (Melbourne, 1955), *A Review of the Broken Hill Lead-Silver-Zinc Industry* by O. H. Woodward (Melbourne, 1952), *In The Early Days* by W. R. Thomas (Broken Hill, 1889), *The History of the Comstock Lode 1850-1920* by G. H. Smith (Reno, 1943), *Zinc Corporation Limited, The First Fifty Years* (Melbourne, 1956), and *A Vision of Steel: the Life of G. D. Delprat* by Lady Paquita Mawson (Melbourne, 1958)—useful for its extracts from Delprat's diaries.

Manuscripts consulted included minute books and industrial agreements of the Barrier Ranges Mining Managers' Association from 1886 (held at M.M.A., Broken Hill), historical notes by R. Kearns, L. Spells, R. Mee and J. R. Finlay, and copies of letters written by Bowes Kelly to G. C. Kelly in 1884-6 and held at Broken Hill Municipal Library.

Learned Journals: Amongst the many articles on Broken Hill in the *Proceedings of the Australasian Institute of Mining and Metallurgy* is an indispensable one on the growth of the flotation process, written by members of the Broken Hill branch and published in December 1930 as 'The Development of Processes for the Treatment of Crude Ore, Accumulated Dumps of Tailings and Slime at Broken Hill, N.S.W.'. The *Mining Magazine* (London, April-May 1919) printed V. F. Stanley Low's paper on froth flotation. R. W. Paul's paper in the *Mississippi Valley Historical Review* (June 1960) on 'Colorado as a Pioneer of Science in the Mining West' is stimulating as a backdrop to the American influence on Broken Hill, though the paper does not mention Broken Hill.

Newspapers: the material on the Barrier Ranges in 1877 came from the *Burra News*, September and October 1877, copy in the State Library of South Australia. Other early information on the fields came from slow scanning of files of the *South Australian Advertiser, South Australian Register*, the Melbourne *Age*, and the local *Silver Age* (1884-95).

5. QUEENSLAND

In the Queensland *Parliamentary Papers* is the vital royal commission on 'The Laws Relating to Mining for Gold and Other Minerals, etc.' (1897) which examined 139 witnesses from Queensland mining on all phases of the industry. Other useful royal commissions enquired into health conditions in Queensland mines (1911-12, paper CA 12), the proposed rail way to Mount Isa Mineral Field (1925, paper A 2), and the mining industry of Queensland (1930, paper A 1).

Amongst the publications of the Queensland Mines Department I used R. L. Jack's report *On the Mount Morgan Gold Deposits* (1884 publication no. 17), W. H. Rand's report on *Croydon Goldfield* (1896, no. 118), B. J. Skertchly's report on *Tin Mines of Watsonville* and other mines at Herberton and Irvinebank etc. (1896, no. 119), J. M. Maclaren's report on *Queensland Mining and Milling Practice* (1901, no. 156), B. Dunstan's monumental *Queensland Mineral Index and Guide* (1913, no. 241), and J. H. Reid's *The Charters Towers Goldfield* (1917, no. 256). The annual reports of the under-secretary to the Department of Mines (1878-) and the monthly issues of the *Queensland Government Mining Journal* (1900-) were also useful for statistics or reports on various mines and fields.

Books: The volume of reports on the sixth meeting of the Australasian Association for the Advancement of Science (Brisbane, 1895) has two articles by E. A. Weinberg on the 'Refractory Gold Ores of Queensland and by W. Fryar on 'The Development and Progress of Mining and Geology in Queensland'. The tropical north is discussed in R. L. Jack's *Northmost Australia* (London, 1921) and F. H. Bauer's 'Historical Geographic Survey of Part of Northern Australia' (C.S.I.R.O. roneo, Canberra, 1959), part 1 of which is a penetrating account of the eastern Gulf area from 1863 to 1930. Readers who wish to know more about mining in north-west Queensland may find some of it in my book *Mines in the Spinifex: the Story of Mount Isa Mines* (Sydney, 1960, 1962) which also has a fuller list of some of the sources I used in writing the chapter on Mount Isa. Alan Barnard's biography of T. S. Mort, *Visions and Profits* (Melbourne, 1961) has a short section on Peak Downs copper mine, and Henry Longhurst's *Adventure in Oil* (London, 1959) describes briefly D'Arcy's oil search in Persia.

Pamphlets and brochures consulted on Mount Morgan's history were J. M. Cameron's *Report of the Gold Mines of Mount Morgan, Rockhampton, Queensland* (Rockhampton, 1887), W. H. Dick's two pamphlets, *The Famous Mount Morgan Gold Mine* (Rockhampton, 1887) and *A Mountain of Gold* (Brisbane, 1889), an anonymous work, *Mount Morgan: the Field and Machinery*, by a special correspondent of *Engineering* (reprinted Rockhampton, 1892), A. A. Boyd's *A History of Mount Morgan* (offprint of *Aust. Institute of Min. and Met. Proceedings* 115, 1939), B. G. Patterson's *The Story of the Discovery of Mount Morgan* (Brisbane, 1948), and the company's own *The Story of the Mount Morgan Mine 1882-1957*. All these were consulted at the company's offices at Mount Morgan. In addition I used pamphlets on other fields including J. H. Binnie's *My Life on a Tropic Goldfield* (Melbourne, 1944), Norman Dungavell's *Charters Towers, 1872 to July 1950*), and 'H.S.' on the *History of the Peak Downs Copper Mine, Copperfield, Peak Downs, Queensland . . .* (Brisbane, 1888).

Manuscripts: O. H. Woodward of Hobart kindly sent me notes on John Moffat, his uncle with whom he worked, and Mount Morgan Limited kindly allowed me to peruse company reports and B. G. Patterson's typescript on 'Mount Morgan 1853 to 1927'.

Newspapers: I used various files of newspapers including the *Queenslander*, the *Courier*, and the *Morning Bulletin*, Rockhampton (which from 18 December 1934 has a fascinating series of reminiscences, not always accurate, by Captain G. Richard on Mount Morgan). The *Mining Journal* (London) was informative on the Charters Towers boom, and the *Australian* (Sydney, 1879, vol. 2, p. 54ff.) has an article on Gympie.

6. NORTHERN TERRITORY

Some of the most valuable evidence of mining in the Territory is in the South Australian *Parliamentary Papers*: Report on Pursuit of Daly River Murderers (1885, paper 170), J. E. Tenison Woods's Report on Geology and Mineralogy (1886, no. 122), J. V. Parkes's reports on Northern Territory mines (1891, no. 178 and 1892, no. 32), report and evidence of royal commission on the Northern Territory (1895, no. 19), the government geologist's Report on Explorations (1895, no. 82), A. A. Davidson's Journal of Explorations in Central Australia by the Central Australian Exploration Syndicate Limited (1905, no. 27), the report on Exploration by the Government Geologist in 1905 (1906), and the mining warden's half-yearly reports on the Northern Territory.

Other government publications I used on Territory mining were the *Annual Progress Report of the Geological Survey of Western Australia* for 1936 (containing H. A. Ellis's report on a search for Lasseter's Reef), P. A. Ivanac's report on Tennant Creek in the Bureau of Mineral Resources Bulletin 22 (Canberra, 1954), and many of the geological reports, particularly those by P. S. Hossfeld and V. M. Cottle, in the *Aerial, Geological and Geophysical Survey of Northern Australia* (Canberra, 1935-40).

In South Australia's State Archives the correspondence on the Northern Territory, including inward letters from gold seekers, was consulted for 1871 and 1872.

F. E. Baume's book *Tragedy Track* (Sydney, 1933) is the story of the Granites rush and C. O. L. Cook wrote in *Walkabout* of May 1947 on 'Granite Gold'. Other books that touch on Territory mining are Alfred Searcy's *In Australian Tropics* (London, 1908), W. B. Wildey's excellent *Australasia and the Oceanic Region* (Melbourne, 1876), Ernestine Hill's *The Great Australian Loneliness* (Austn. edn., Melbourne, 1940), and Frank Clune's chapter on Lasseter's end in *The Red Heart* (Melbourne, 1944).

The University of Adelaide has two theses on the Territory mines: Margaret Rendell's 'The Chinese in South Australia and the Northern Territory in the Nineteenth Century' (1952), and Maureen Kennett's thesis on the goldfields in the 1870s (1962).

The *South Australian Chronicle* of 1874 and 1875 has good reports by its Palmerston (Darwin) correspondent, and the *South Australian Advertiser* was one of my sources on the share boom of 1873. The Melbourne *Argus* has a series of articles on the Territory mines in May and June 1886.

The history of the Tennant Creek field is so recent that I was able to interview many of the early men on the field and see some of their notes and written memoirs. In particular C. E. Blackett sent me notes on the early history of the rich Noble's Nob mine.

7. WESTERN AUSTRALIA

The *Parliamentary Papers* of Western Australia were used frequently, and the following papers and reports all provided evidence for the Western Australian chapters: E. T. Hardman's Report on the Geology of the Kimberley District (1885, no. 34), Report upon the Kimberley Goldfield by the Warden (1887, no. A 19), Report on the Goldfields of the Kimberley District by H. P. Woodward (1891-2, Second Session, no. 18), various reports by government geologist and mining wardens in proceedings of Second Session of 1891-2 and Third Session of 1892-3, Royal Commission on Mining (1898, no. 26), Royal Commission on the Ivanhoe

Venture Lease (1899, no. 12), Royal Commission on the Boulder Deep Levels at Kalgoorlie (1904, no. A 13) and the royal commissions on Great Boulder Perseverance (1905, First Session, no. 3), on the Ventilation and Sanitation of Mines (1905, First Session, no. 6), on Pulmonary Diseases amongst Miners (1910-11, no. 12), and on the Mining Industry—the Kingsley Thomas report (1925, no. 3).

There is a wealth of statistical and descriptive matter in the annual reports of the Mines Department (1894-), annual progress Reports of the Geological Survey (1897-) and the geological survey bulletins. However the only bulletin I much used was H. P. Woodward's *Mining Handbook to the Colony of Western Australia* (Perth, 1895). In the South Australian *Parliamentary Papers* an official half-yearly report on the Northern Territory has graphic detail on the Kimberley rush (1886, no. 54).

The State Archives of Western Australia have a variety of illuminating manuscript material including short notes on dry-blowing by J. E. Tregurtha, on the Murchison discovery by J. Connelly and F. W. Bateson, on the Norseman discoveries by A. Hicks, on the Western Australian career of Herbert Hoover by Sir John Kirwan, on Kalgoorlie by Mrs Alex. Sutherland, and above all the letters of E. C. Deland, 1895-7, and the diary and notes of the prospector F. W. P. Cammilleri. The Mitchell Library has a typescript of 'Reminiscences' by U.K. Sligo (1910) and notes by W. Linklater on 'The Ragged Thirteen' (1941). The University of Western Australia Library has David Mossenson's thesis of 1952 on 'Gold and Politics: the influence of the Eastern Goldfields on the political development of Western Australia 1890-1904'. An indispensable guide to the written and published material on the goldfields is F. K. Crowley's *The Records of Western Australia* (Perth, 1953, vol. 1), which lists far more than I have been able to consult.

From the rows of contemporary books or published memoirs of goldfields residents the following particularly added to my knowledge: R. Allen's *West Australian Metallurgical Practice* (Kalgoorlie, 1906), A. F. Calvert's *West Australian Mining Investors' Handbook*, his *Western Australia: Its History and Progress*, and the *Coolgardie Goldfield: Western Australia* (all published in London, 1894), Albert Gaston's *Coolgardie Gold* (London, *c.* 1937), Cyril E. Goode's *Yarns of the Yilgarn* (Melbourne, 1950), Alexander Macdonald's *In Search of El Dorado* (London, 1905), L. R. Menzies' *A Gold Seeker's Odyssey* (London, 1937) and a cheerful pack of lies as well, M. J. O'Reilly's *Bowyangs and Boomerangs* (Hobart, 1944), W. L. Owen's *Cossack Gold* (Sydney, 1933), Arthur Reid's *Those were the Days* (Perth, 1933) which is studded with goldfields reminiscences, Malcolm Uren's *Glint of Gold* (Melbourne, 1948) and G. F. Young's *Under the Coolibah Tree* (London, 1953).

Secondary historical works consulted were F. K. Crowley's fine *Australia's Western Third* (London, 1960), *The Origins of the Eastern Goldfields Water Scheme* (Perth, 1954) by F. Alexander, F. K. Crowley, and J. D. Legge, an article on the 'Kimberley Pastoral Industry' by G. C. Bolton in *University Studies in History and Economics* (Perth, July 1954), the entry on Whitaker Wright in volume 31 of the *Dictionary of*

National Biography, though it's not accurate on part of his mining career, the scholarly article by G. S. Compton on 'Searching Eastwards for Gold in Western Australia' in the *Journal of the Royal Australian Historical Society* (vol. 43, part 1, 1957), and many short articles in *Early Days*, being the Journal and Proceedings of the Western Australian Historical Society: W. L. Owen on his life as goldfields warden (1932), John Kirwan on Coolgardie gold (1941), Owen on Pilbara and C. M. Harris on the Margaret and Murchison (1945), G. S. Compton on tellurides and C. M. Harris on water (1947), F. I. Bray on the Murchison and A. C. Angelo on the north-west (1948), and D. Mossenson on the separation movement on the eastern goldfields (1953).

The Chamber of Mines in Western Australia issued monthly reports (1902-20) that recorded company activities and dividends, and later the annual reports and presidential addresses are useful. Gold Mines of Kalgoorlie, Great Boulder, Lake View and Star, Central Norseman, North Kalgurli, and Western Mining Corporation supplied me with information or allowed me to peruse their collections of published reports.

Newspapers: The *South Australian Chronicle* recorded on 21 October 1893 probably the first interview Arthur Bayley of Coolgardie ever gave, and from September to 11 November a special reporter wrote a series of articles on Coolgardie and Kalgoorlie. The *South Australian Advertiser* printed on 14 April 1883 an interview with Saunders and Johns on their Kimberley trip. The *Western Mail* (Perth) printed Cammilleri's reminiscences from about 15 July to 2 September 1937. The Melbourne *Leader* in 1898 published a supplement on the Western Australian goldfields, edited by D. A. Vindin. The *West Australian*, *Kalgoorlie Miner*, and the Melbourne *Australasian* and the London *Mining Journal* yielded much material through verbatim press interviews, though unfortunately through lack of space I only used a fraction of it.

8. SOUTH AUSTRALIA

Parliamentary Papers that yielded evidence on the mines were the Select Committee on Mineral Development (1860, no. 76), the Select Committee on the 'Prospectus of the Great Northern Copper Mining Co.' (1860, no. 83), the Select Committee on the Present Mineral Laws (1862, no. 51), Report and Journal of E. H. Hargraves (1864, no. 96); Petitions, Correspondence, and Documents Relating to the Moonta Mines (1867, no. 77), the Select Committee on Wallaroo and Clare Railway (1867, no. 141), and Royal Commission on Mines (1890, no. 32). *Hansard* of 1867 (vol. 12) has a debate on the Burra railway bill.

South Australia was surprisingly late in creating a strong geological survey and mines department, and there are not many official publications of value to a historian of nineteenth-century mining. Amongst the exceptions are H. Y. L. Brown's *A Record of the Mines of South Australia* (1887, 1908), R. L. Jack's bulletin on *The Geology of the Moonta and Wallaroo Mining District* (1917), and S. B. Dickinson's bulletin on *Kapunda Mines* (1944).

Books: Francis Dutton's *South Australia and its Mines* (London, 1846), J. B. Austin's *The Mines of South Australia* (Adelaide, 1863), *Bailliere's South Australian Gazetteer* compiled by R. P. Whitworth (Adelaide, 1866), *Journal of Commodore Goodenough, R.N. . . .* who visited Wallaroo (London, 1876), Douglas Pike's *Paradise of Dissent: South Australia 1829-1857* (London, 1957), and S. B. Dickinson's three articles on South Australian mines in *Geology of Australian Ore Deposits* (Melbourne, 1953).

Manuscripts and theses: In the State Archives are a fine collection of letters, directors' minutes and reports of the South Australian Mining Association from 1845, a short typescript on 'The Glen Osmond Mines' (1955) by K. T. Borrow, and Oswald Pryor's fascinating typescript 'Little Cornwall: the story of Moonta (South Australia) and its Cornish miners' (*c.* 1953) which in 1962 was published as a book by Rigby of Adelaide. Three University of Adelaide theses are Henry Brown's 'The Copper Industry of South Australia' (1937), J. W. Higgins's 'The South Australian Mining Asociation' (1956), and I. J. Bettison's 'Kapunda' (1960).

The annual *Statistical Register of South Australia* has figures for copper output and London metal prices. For those who may do research on South Australian mines three little known statistical traps are awaiting them. What South Australian miners called ore we would call concentrate. In many mines including Moonta in the 1860s a ton of ore was 21 cwt, Cornish style. And finally the price of South Australian copper was usually far above the average price on the London market; South Australian copper was purer.

Newspapers: The *South Australian Advertiser* had six articles on Wallaroo and Moonta from 28 January to 18 February 1882, the *Chronicle* had an article reminiscing on Burra on 26 March 1904, the *South Australian Register* had a long obituary on Captain Hughes on 5 January 1887, the *Adelaide Observer* had long articles on Moonta etc. in April-May 1874 during the long strike, and the *Burra News* 1877-8 describes the death of its mine.

9. TASMANIA

Much of the information I recount on Tasmania was gathered during the writing of my book *The Peaks of Lyell* (Melbourne University Press, 1954, 1958), and at the end of that book is a ten-page bibliography listing the sources I used. Rather than repeat a list of most of those sources, I mention only those items that concern fields outside Mount Lyell and the west coast fields which that book described.

The annual reports of the Mines Department have much useful evidence on statistics and fields from the 1880s. Of the bulletins of the Geological Survey of Tasmania the following were consulted: W. H. Twelvetrees and L. K. Ward on Zeehan (1909-10), Loftus Hills on Read-Rosebery (1914, 1915, 1919), L. L. Waterhouse on South Heemskirk Tinfield (1915), A. McIntosh Reid on Mount Bischoff (1923), P. B. Nye on the Sub-Basaltic Tin Deposits of the Ringarooma Valley (1924), W. H. Twelvetrees and Q. J. Henderson on the Blue Tier Tin Field (1928), K. J. Finucane on the Mathinna and Tower Hill Goldfields (1935).

Articles in the *Papers and Proceedings of the Tasmanian Historical Research Association* are also relevant: John Reynolds's 'The Establishment of the Electrolytic Zinc Company in Tasmania' (September 1957), K. M. Dallas's 'Water Power in Tasmanian History' (September 1960), and my articles 'Population Movements in Tasmania, 1870-1901' (June 1954), and 'The Rise and Decline of the West Coast' (February 1956).

10. POST-WAR MINERAL BOOM

Chapter 29, rewritten to describe the present minerals boom, is based largely on annual reports of the State mining departments, on the published reports and brochures of companies prominent in the boom, and on the annual review, *Australian Mineral Industry*, published by the Bureau of Mineral Resources, Canberra. Another indispensable source is Sir Harold Raggatt's book, *Mountains of Ore* (Melbourne, 1968).

On the iron ore embargo, the *Parliamentary Debates* of Western Australia (30 August and 6 September 1938) and of the Commonwealth's Senate (20 October 1938) and House of Representatives (2 and 9 November 1938) are useful. The 'Report by Dr. W. G. Woolnough on the Technical Aspects of the Iron Ore Reserves in Australia' is in Commonwealth *Parliamentary Papers*, 1940, vol. 6. A fuller discussion of the iron ore embargo is in my article 'The Cargo Cult in Mineral Policy', in *Economic Record*, December 1968.

Australia's first hole in search of oil, usually assigned to 1892, was being drilled in the summer of 1881-2, according to the *Geological Survey of Victoria: Report of Progress*, vol. 7 (Melbourne, 1884), p. 98. The oil search in Victoria is discussed by N. Boutakoff in *Mining and Geological Journal* (Melbourne, September 1951), pp. 49-57, and a brief account of the oil search throughout Australia is in the Petroleum Information Bureau's *This Age of Oil* (Melbourne, 1960), ch. 10. W. G. Woolnough's report on his inspection of American oilfields and on oil prospects in Australia is in Commonwealth *Parliamentary Papers*, 1929-30-31, vol. 2. The Tariff Board's 'Report on Crude Oil', 23 July 1965, is a valuable résumé of oil searching on the eve of the Bass Strait bonanzas.

INDEX